# Fatigue Prediction for Random Loads

*Fatigue Prediction for Random Loads* serves as a comprehensive treatise for methods for fatigue estimation and fatigue life prediction in randomly excited structural systems using the rainflow cycle counting method.

Fatigue is an important mode of material degradation in structural components subjected to vibrations and a reliable estimation of their fatigue life span, and it is a key consideration in the design and development of such systems. This book uses advanced concepts of probability theory, random variables and random processes to develop spectral-based methods and formulae for predicting expected fatigue damage and expected fatigue life. The developments presented here bypass the need for a computationally expensive rainflow cycle counting that is usually adopted in time domain approaches.

This book is aimed towards researchers and industry practitioners working in the intersectional areas of mechanics and applied mathematics and is expected to be particularly useful for applications on problems in the fields of wind engineering, offshore engineering, ship research and routing, aerospace engineering, automotive engineering and machine dynamics.

# Fatigue Prediction for Random Loads
## Using the Rainflow Method

Igor Rychlik, Sayan Gupta and Georg Lindgren

CRC Press
Taylor & Francis Group
Boca Raton London New York

CRC Press is an imprint of the
Taylor & Francis Group, an **informa** business

Designed cover image: Photo credit for truck: Gustav Lindh, Scania. Photo credit for ship: Oceanbird

First edition published 2025
by CRC Press
2385 NW Executive Center Drive, Suite 320, Boca Raton FL 33431

and by CRC Press
4 Park Square, Milton Park, Abingdon, Oxon, OX14 4RN

*CRC Press is an imprint of Taylor & Francis Group, LLC*



ISBN: 978-1-032-89584-6 (hbk)
ISBN: 978-1-041-01096-8 (pbk)
ISBN: 978-1-003-01994-7 (ebk)

DOI: 10.1201/9781003019947

Typeset in Nimbus Roman
by KnowledgeWorks Global Ltd.

# Contents

# List of Figures

# List of Tables

# Authors

**Igor Rychlik** is Professor Emeritus at Chalmers University of Technology, Sweden. He earned a PhD in mathematical statistics at Lund University, Sweden, 1986. During 1988–1990 he was post-doctoral student at the Department of Statistics, Colorado State University, and in 1995 he was a visiting professor at the University of Queensland, Brisbane, Australia. He became a Professor of Mathematical Statistics at Lund in 1999 and from 2007 at Chalmers University of Technology, Göteborg. There he also worked as a scientific advisor to the Fraunhofer-Chalmers Centre for industrial mathematics. Since 2017 he has been a Professor Emeritus. His main interest is in engineering applications of random processes and fields and in the modelling of complex environmental loads, wind, sea waves and vehicle operating conditions. He is a coauthor of *Guide to Load Analysis for Durability in Vehicle Engineering* and *Probability and Risk Analysis: An Introduction for Engineers.*

**Sayan Gupta** is a Professor at the Indian Institute of Technology Madras in the Department of Applied Mechanics. He graduated with a degree in civil engineering from Jadavpur University Kolkata in 1997, earned a PhD at the Indian Institute of Science Bangalore in 2005 and was a post-doctoral student at the Technical University of Delft from 2005 to 2007. He has held visiting positions at Waterloo, Lund, and Chalmers. Since 2007, he has been working at IIT Madras, where he leads the Uncertainty Lab. He is the principal investigator for the Center for Complex Systems and Dynamics, IIT Madras, and he is the coordinator for the degree program on Complex Systems and Dynamics. His primary research interests include non-linear dynamics and stochastic mechanics.

**Georg Lindgren** is Professor Emeritus in the Engineering Faculty at Lund University, Sweden. He earned a PhD in mathematical statistics at Lund in 1972 with a thesis inspired by an intriguing question regarding random load cycle ranges. After a period as an Associate Professor at Lund and Umeå, he became a Professor of Mathematical Statistics at Lund in 1986, and since 2006 he has been a Professor Emeritus. His main interest is random processes and their use in communication technology, marine and mechanical engineering and medicine. He is a coauthor of a 1983 research volume on the extremes and related properties of random sequences and processes and two research/advanced student texts on stationary stochastic processes. He is a member of the Royal Swedish Academy of Engineering Sciences (IVA).

# Preface

This book has its roots in the 1950s, when random processes made their entrance in engineering and the development of new industrial solutions. Telecommunication was in the lead, with the pioneer Steve O. Rice as member of Bell Labs Mathematical and statistical research center in USA. The ocean engineering community had Michael S. Longuet-Higgins working with random wave models for shipping safety. Julius S. Bendat educated mechanical engineers and NASA how to use random signals for reliability and fatigue.

In Sweden, the development was underpinned by new research and teaching resources in mathematical statistics at the Royal Institute of Technology (KTH) in Stockholm, supporting core technology for expanding industry, in particular telephone communication and air-plane, car and shipping industry. For science, economy and insurance, Harald Cramér at Stockholm University was an international authority on random processes and acted as a model for the buildup of statistics in engineering.

While telecommunication used random processes to create new technology, the mechanical industry worked to increase reliability and needed statistical extremes. Waloddi Weibull, the "Weibull distribution father", was already professor in machine elements at KTH when he published his now widely used statistical distribution on the strength of material in 1951. His well-cited book on *Fatigue Testing and Analysis* [211] appeared in 1961.

It is notable that the professorship was originally financed by the arms manufacturer Bofors. The National Aeronautical Research Institute and the aeroplane manufacturer SAAB were other bodies with serious interest in statistics and reliability.

Swedish shipyards and the technical university Chalmers in Göteborg got a statistical injection early in the 1950s when Rutger Bennet was employed to develop an organisation for statistical analysis of ship movement in sea waves.

This was the Swedish scene when the truck and bus manufacturer Scania-Vabis in Södertälje 1947 employed Sverker Sjöström to work on quality development. Sjöström was expert in strength of materials and had experience both from the Aeronautical Research Institute and Bofors, and he was committed to an advanced research project on random fatigue. He was well aware of the international work of random load effect on fatigue but could also use the advanced models by Steve Rice on level crossings and random extremes.

Sjöström defended his doctoral thesis "On random load analysis" at KTH in 1961. The official opponent at the defence was Bertram Broberg, a student and former assistant to Waloddi Weibull. He spotted a statistical gap in the thesis and managed to fill it in – almost. The details are described in Chapter 6.

Bertram Broberg was appointed as professor in solid mechanics at the new Lund Institute of Technology in 1961 and became its first pro-rector. There, he handed over the gap to the professor in mathematical statistics, Gunnar Blom, who had a background as statistical consultant to Waloddi Weibull, i.a. on statistical extremes.

This is the moment when the work on the present book started. Its third author, Georg Lindgren, resolved the mystery around the gap, earned a PhD degree in mathematical statistics in 1972 and continued to work on statistical extreme value theory and wave properties of Gaussian processes. An inquiry from the British National Institute of Oceanography 1973 raised the question whether the sequence of wave crests and wave troughs in a Gaussian sea could be regarded as a Markov chain. Preliminary studies showed that "Yes, it is a reasonable approximation".

This early example of Markov chain of turning points lies dormant for about 10 years. In the early 1980s, Igor Rychlik, this book's first author, had just become a PhD student in mathematical statistics in Lund. Teachers in solid mechanics in Lund wanted a statisticians view on how to interpret the SN-curve and – most importantly – they asked for the statistical distribution of the rainflow cycle range! Georg Lindgren picked up the MCTP-idea. Igor used it to invent the local definition of the rainflow count and presented the exact distribution based on the transition matrix in his PhD thesis in 1986.

What made the Markov-based solution to the rainflow problem unique was the parallel development of computer software for exact computation of wave characteristics in Gaussian loads. The transition probabilities between local extremes could be computed from the covariance or spectrum structure.

About 20 years ago, Sayan Gupta, who was at that time working as a postdoc at the Department of Civil Engineering of Technical University of Delft, joined our team on rainflow analysis in non-Gaussian loads that are complex functions of Gaussian loads, and later was also working on the distribution of maxima and crossings functions of transformed Laplace loads.

The intention of this book is to provide a comprehensive treatise of the mathematical developments based on random process theories in addressing the problem of random fatigue and life assessment of structural systems subjected to random loadings. The intended audience for this book includes graduate students and researchers pursuing studies on structural reliability due to random vibrations, as well as practitioners from the aerospace, automobile and naval and wind-turbine industries.

Igor Rychlik
Sayan Gupta
Georg Lindgren
*October 2024*

# Introduction

Fatigue is the initiation and propagation of cracks in a material due to cyclic loading. Once a fatigue crack has initiated, it grows a small amount with each loading cycle. It is the experienced loading variability, and the material's resistance to cracking that correlates with the fatigue life of the components in a construction.

To predict fatigue failure is the ultimate task, and one challenge is to tame the uncertainty in the random loading. The physics of crack growth is complicated, and the concise S-N curve is more schematic, but both require statistical information about the random load. To transform that information to a statement about fatigue damage and fatigue life is also partly of statistical concern – what characteristics of a random load work best in the prediction of the fatigue process?

The existence of this book rests on the need for answers to two questions of statistical character. First: What properties of random load models are needed – or asked for – in the prediction of fatigue failure? Second: Which random load models are simple and realistic enough to be useful in the sense that one can extract, empirically and theoretically, the needed properties for use in fatigue prediction?

**What properties are useful, and what are used?** Both strength and loading distributions could be described by a number of factors. For example, S-N curves can give information on material strength, while the crack growths can be regressed on some explanatory measures of load severity, depending on load variability. A simple example is the so-called *total variation* (or rather positive variation) of a load. The total variation tells us how much the load $x(t)$ moves up and down (or only upwards in case of positive variability). We denote it by $V_T$ and compute it as $V_T = \int_0^T |\dot{x}(t)|\, \mathrm{d}t$.

Obviously, in order to evaluate $V_T$, it is sufficient to extract a sequence of turning points $T_P(x)$ from the load $x$. It is a sequence of minima and maxima $m_i, M_i$. Those can be paired, $(m_i, M_i)$, to a *cycle count*, naturally called the min-Max count. The sequence of turning points is the primary summary information about load variability, essential to predict time to fatigue failure. From the turning points, one can calculate the *crossing spectrum*, $N(u)$, which is the number of times during the observation interval that the load crosses level $u$.

Although simple, these measures are used for a first indication of fatigue failure. An essential requirement on a random load model is that one can compute the distribution of the min-Max range in that model and calculate the distribution of the min-Max damage,

$$D_T^{mM} = \alpha \sum_{i=1}^{N} (M_i - m_i)^{\beta},$$

where $\alpha$ is a small constant, while $\beta$ takes values greater than or equal to one.

The "rainflow method", which is the central theme in this book, uses the turning points but includes a more complex relation to the damage rate. It modifies the min-Max cycles by pairing the maxima $M_i$ with local minima that form the bottom of closed hysteresis loops in the stress-strain plane. Consequently, it is not the adjacent extremes that are paired but the one that defines a closed hysteresis loop, forming the rainflow cycle $(M_i, m_i^{rfc})$.

Chapter 4 is devoted to cycle counts and gives a detailed description of the rainflow cycle count and its relation to level crossings, and the more general concept, interval crossings.

The ultimate demand on a useful random load model is that it allows for theoretical analysis of the transformation from min-Max to rainflow cycles.

**What models?** Random load modelling for fatigue prediction is a topic with many facets. Wave loads used in naval engineering are traditionally described by their energy spectra, adapted for deep water, shallow water, stormy routes, etc. But how to handle wave loads along a trade route with strong seasonal variation?

Fatigue loads on a motor vehicle in heavy service on a network of roads have to include a very variable road quality and driver behaviour. The random load model must also take account of available realistic service data, brought home from the trucking company that bought the truck.

A common requirement in these examples is that the random models must be useful in the design phase and reliability testing of a product and that they can deliver correct estimates of the variable used in the damage calculations.

The language for random loads is that of random processes. In this book, you will find the Gaussian process models in Chapters 6 and 7. Here, we deal with the important statistical distribution of the height of successive minima and maxima in a Gaussian load process, an important property since it gives the distribution of the min-Max cycle ranges $M_i - m_i$.

If you have a model that you trust, i.e. an energy spectrum for a Gaussian process, you may want the exact distribution of the min-Max range – numerical tools to get it are described in Chapter 6. In other situations, you need an approximation, often you use it to get a strict upper bound for the damage. Many such approximations and bounds are described in Chapter 7.

How to use the min-Max distribution to get the distribution of the rainflow range $M_i - m_i^{rfc}$ is the main result in Chapter 8, on Markov models. The inverse problem, to find a Markov model that has a specific rainflow range distribution, is also treated in this chapter. The Markov model is then efficiently used for fatigue testing.

What shall one do when the Gaussian model is "too normal" and the Markov model is too simple? Chapter 9 offers the Laplace model, an extension of the Gauss model with heavier tails and with the possibility of skewed load distribution. It is easy to simulate, and one can get similar damage bounds as for the Gaussian models based on level crossings.

Chapters 10 and 11 are devoted to applications where the flexibility of the presented methods is tested. How can the methods, constructed for statistically stationary loads, be applied in an environment that changes with season of the year, or when the normal, "Gaussian", fatigue process is interrupted by occasional big transients, e.g., potholes in a smooth road surface? Then, pieces of a Laplace model can be mixed in with a Gaussian load model into a more realistic hybrid model.

How confident can one be in damage estimates? This is one of the topics in Chapter 11.

This introduction ends with a hint to Chapters 1–3 with brief introductions to fatigue, reliability and the underlying probability theory. And at the very end, why are the rainflow cycles physically relevant? Wikipedia has a visual answer: The rainflow cycles correspond to energy storing hysteresis loops.

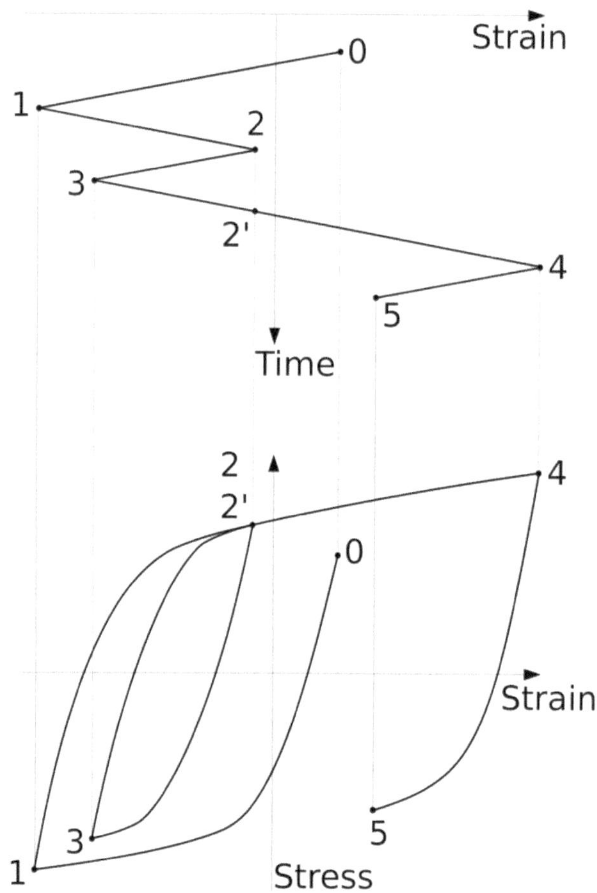


**FIGURE 0.1** Why rainflow cycles are physically relevant.

# List of Notations

## ROMAN LETTERS

| | |
|---|---|
| $A_k$ | random amplitude |
| $\mathsf{Cov}[X,Y]$ | $\mathsf{E}[XY]-\mathsf{E}[X]\mathsf{E}[Y]$, covariance of $X$ and $Y$ |
| CoV | coefficient of variation, $\sigma(X)/\lvert\mathsf{E}[X]\rvert$ |
| $c(\beta,\alpha_2)$ | correction factor $d^{pp}=c(\beta,\alpha_2)d^{nb}$ |
| $D(T)$ | $\sum f(m_{k(i)},M_i)$, total damage during time $T$ |
| $\dot{D}(t)$ | damage increase rate at time $t$ |
| $D^{cc}(T)$ | total damage by crossing-cycle count |
| $D^{HM}(T)$ | total damage by Holm-de Maré approximation |
| $D^{mM}(T)$ | total damage by min-Max count |
| $D^{nb}(T)$ | total damage by narrow band approximation |
| $D^{pp}(T)$ | total damage by positive peak count |
| $D^{rfc}(T)$ | total damage by rainflow count |
| $\mathsf{D}[X],\sqrt{\mathsf{Var}[X]},\sigma_X$ | standard deviation of $X$ |
| $d=\mu\times\mathsf{E}[\delta]$ | $\mathsf{E}[D(t)]/t$, damage intensity = cycles per time unit $\times$ expected damage per cycle = expected damage increase per time unit |
| $d^{rfc},d^{mM}$, etc | damage intensities by counts |
| $d^{nb},d^{nbMCTP}$, etc | narrow-band approximate damage intensities |
| $\widetilde{d}$ | damage intensity standardised load with $h_s=f_z=1$ |
| $d^{+}$ | general damage intensity bound |
| $d^{HM}$ | Holm-de Maré damage intensity by [84] |
| $d^{LL}(\beta)$ | single moment damage intensity by [134] |
| $d^{SO}(\beta),d^{SOm}(\beta)$ | damage intensity by [185] with modification |
| $d^{JM}(\beta)$ | damage intensity by [87] |
| $d^{BT}(\beta)$ | damage intensity by [204] |
| $d^{WL}(\beta)$ | damage intensity by [214] |
| $d^{Dirlik}(\beta)$ | rainflow range intensity by [53] |
| $d^{ZB}(\beta)$ | rainflow range intensity by [219] |
| $d^{HM}(t)$, etc | time dependent damage intensity |
| $\mathsf{E}[X]=m_X$ | expectation of random variable $X$ |
| $\mathsf{E}[X\mid Y],\mathsf{E}[X\mid Y=y]$ | conditional expectation of $X$ given $Y=y$ |
| $E_V$ | expectation of conditional variance |
| $E(\tau)$ | $\sqrt{x^2(\tau)+x_H^2(\tau)}$, envelope process |
| $F^{mM}$ | Markov matrix with min-Max cycle counts |
| $F^{rfc}$ | rainflow matrix with rainflow cycle counts |
| $F_{ij}$ | $\pi_i P_{ij}$, stationary distribution of $Z_n,Z_{n+1}$ in Markov chain |

| | |
|---|---|
| $F_X(x)$ | distribution function for random variable $X$ (cdf) |
| $F_{X,Y}(x,y)$ | distribution function for bivariate variable $(X,Y)$ |
| $f_X(x) = \mathrm{d}F_X(x)/\,\mathrm{d}x$ | probability density function (pdf) |
| $f_{X,Y}(x,y)$ | $\partial^2 F_{X,Y}(x,y)/\partial x \partial y$, bivariate probability density |
| $f_{X\mid Y}(x \mid y)$ | $f_{X,Y}(x,y)/f_Y(y)$, conditional density of $X$ given $Y = y$ |
| $f(\omega), f_x(\omega)$ | power/spectral density functions for stationary process |
| $f(u,v)$ | $\alpha(v-u)^\beta$, common damage function |
| $f_{max}$ | maximi-frequency |
| $f_p$ | peak frequency |
| $f_z$, $f_z(t)$ | zero-crossing (time dependent) frequency |
| $G(z)$ | exhausture function in Holm-de Maré method |
| $G(Z_n)$ | transformation from Markov chain $Z_n$ to Markov load $X_n = G(Z_n)$ |
| GMA | Gaussian moving average |
| $g(t)$ | kernel in Laplace moving average LMA |
| $g_k$ | $\int g^k(t)\,\mathrm{d}t$ for LMA kernel |
| $H$ | stress range |
| HCF | high cycle fatigue |
| $H_i^{rfc}$ | $M_i - m_i^{rfc}$ rainflow range |
| $H^{rfc}(\tau)$ | $\min(H^-(\tau), H^+(\tau))$, rainflow cycle process |
| $H_i^-, H_i^+$ | rainflow semi ranges, $M_i - m_i^-$, $M_i - m_i^+$ |
| $H_0^-, H_0^+$ | rainflow memories for the load and for the reversed load |
| $H^-(\tau), H^+(\tau)$ | local semi ranges, $x(\tau) - m_\tau^-$, $x(\tau) - m_\tau^+$ |
| $H(\omega)$ | frequency response of linear filter |
| $H_s$, $h_s$ | $4\sigma_X$, significant load-range, in marine science called significant wave-height |
| $h_s(t)$ | time dependent significant wave height |
| $h(\omega)$ | impulse response of linear filter |
| $K_i$, $K_i^*$ | gamma-variables in Laplace movian average |
| $k$ | alternative to $\beta$ as damage exponent |
| $k_i$ | permutation that takes $(m_i, M_i)$ to $(m_{k_i}, M_i)$ |
| $k(y,x)$ | kernel in Fredholm integral equation |
| $k^+(u,v)$ | $\min_{u \le z \le v} \mu^+(z)$, minimal crossing spectrum |
| LCF | low cycle fatigue |
| LMA | Laplace moving average |
| MC, MCTP | Markov chain, Markov chain of turning points |
| $M^+$ | the height of a randomly chosen positive maximum in a stationary process |
| $M_i$ | local maximum number $i$ |
| $(m_{k(i)}, M_i)$ | min-Max pair in cycle count |
| $(m_i^{rfc}, M_i)$ | rainflow min-Max pair |
| $(m_i^{tr}, M_i^{cr})$ | wave trough and wave crest, lowest and highest values between mean level up-crossings |

| | |
|---|---|
| $m_i < M_i$ | local minimum number $i$ |
| $m_\tau^-, m_\tau^+$ | minimum in left and right rainflow memory |
| $m(t)$ | expectation $\mathsf{E}[X(t)]$ of random process at time $t$ |
| $m_u, m_{Z\|Y=u}$ | $m_Y + \sigma_Y/\sigma_Z \rho_{ZY}(u - m_Z)$, conditional mean for bivariate Gaussian $Y, Z$ |
| $N(u,v)$ | number of cycles $(m_{k(i)}, M_i)$ with $M_i > v$, $m_{k(i)} < u$, cumulative cycle count |
| $N, N_f$ | number of cycles to failure |
| $N_T^{rfc}, N_T^{cc}$, etc | cumulative rainflow cycle, crossing-cycle count, etc. in time $T$ |
| $N_{ij}^{rfc}$ | element of $N_T^{rfc}$ |
| $N^h(s)$ | number of ranges $h_i$ exceeding $s$ |
| $N^+(u)$, $N_T^+(u)$ | (up)crossing spectrum, number of $u$-upcrossings, number of $u$-upcrossings in time $T$ |
| $N_T^{max}$ | number of local maxima $\tau_i \in [0,T]$ |
| $N_T^{max}(u)$ | number of local maxima $\tau_i \in [0,T]$ with $x(\tau_i) > u$ |
| $N_T(A \mid u)$ | number of crossings where $A$ is satisfied |
| $\mathsf{P}(A)$ | probability that event $A$ happens |
| $\mathsf{P}(A \mid B)$ | $\mathsf{P}(A \cap B)/\mathsf{P}(B)$, conditional probability of $A$ given $B$ |
| $\mathsf{P}_T(M > u)$ | empirical probability that a random maximum $M > u$ |
| $\mathsf{P}_f$ | probability of failure |
| $P_{ij}$ | transition probability between states $i$ and $j$ |
| $P^{mM}, P^{Mm}$ | transition matrices in Markov chain of turning points |
| $q$ | skewness parameter in asymmetric LMA |
| $q_{[u,v]}(y \mid x)$ | transition probability between maxima with restriction on minimum |
| $R$ | standard Rayleigh-variable with density function $z\exp(-z^2/2), z \geq 0$ |
| $r_X(\tau)$ | $\mathsf{Cov}[X(t), X(t+\tau)]$, covariance function for stationary process $X(t)$ |
| $r_{ij}(s,t)$ | $\mathsf{Cov}[X_i(s), X_j(t)]$, cross-covariance function |
| $S$ | stress amplitude |
| SMC | switching Markov chain |
| SMCTP | switching Markov chain of turning points |
| $S(\omega), S_X(\omega)$ | spectral density functions for stationary process |
| $S^J(\omega)$ | JONSWAP spectrum |
| $S^{PM}(\omega)$ | Pierson-Moskowitz spectrum |
| $S^{Tor}(\omega)$ | Torsethaugem bimodal spectrum |
| $S^{MIRA}(\xi)$ | road spectrum by [112] |
| $T_f$ | time to failure, first passage time |
| $T_p, t_p$ | peak period |
| $T_P$ | $(m_1, M_1, \ldots, m_n, M_n)$, turning points |
| $T_z$ | $2\pi(\omega_0/\omega_z)^{1/z}$, generalised average period |
| $t_{max}, T_4$ | maximi-period |

| | |
|---|---|
| $t_z$, $T_2$ | zero-crossing period |
| $\mathsf{Var}[X] = \mathsf{E}[(X - m_X)^2]$ | variance of random variable $X$ |
| $\mathsf{Var}[X \mid Y = y]$ | $\sigma^2_{X\mid Y=y}$, conditional variance of $X$ given $Y = y$ |
| $V_E$ | variance of conditional expectation |
| $V_f(I)$ | total variation of function $f$ in interval $I$ |
| $v(t) = \sigma^2(t)$ | variance function $\mathsf{Var}[X(t)]$ |
| $W(t)$ | standard Wiener process with $\mathsf{Var}[W(t)] = t$ |
| $X, Y, Z, \ldots$ | random variables |
| $x, y, z, \ldots$ | observations of random variables |
| $X(t), x(t)$ etc. | random process, realisation of random process |
| $\overline{X}$ | arithmetic mean |
| $x_H(\tau)$ | Hilbert transform of $x(\tau)$ |
| $X_{max}(t)$ | Slepian modell for Gaussian process near maximum |
| $Z_n$ | Markov chain as basis for markov load |
| $Z_u^{reg}(s)$ | regression part of Slepian model |
| $Z_u(s)$ | Slepian process model for Gaussian process after $u$-level upcrossing |

## GREEK LETTERS

| | |
|---|---|
| $\alpha$ | scaling parameter in S-N curve |
| $\alpha_1$ | $\frac{\omega_1}{\sqrt{\omega_0\omega_2}}$, groupiness measure |
| $\alpha_2$ | $\frac{\omega_2}{\sqrt{\omega_0\omega_4}}$, spectral width measure |
| $\alpha_m$ | $\frac{\omega_m}{\sqrt{\omega_0\omega_{2m}}}$, alternative spectral width |
| $\beta$ | damage exponent in damage function $\alpha(v-u)^\beta$ |
| $\beta_C$ | the Cornell safety index |
| $\beta_{HL}$ | the Hasofer-Lind safety index |
| $\Gamma(p)$ | gamma function, $\int_0^\infty x^{p-1}e^{-x}\,\mathrm{d}x$ |
| $\gamma = \kappa_3$ | skewness of random variable and asymmetric LMA |
| $\Delta(\tau)$ | non-stationary Gaussian residual process for $x(\tau)$ after regression on $x(\tau), \dot{x}(\tau)$ |
| $\zeta$ | relative damping |
| $\kappa_3 = \gamma$ | skewness of random variable |
| $\kappa_4 = \kappa$ | kurtosis of random variable |
| $\kappa_e$ | excess kurtosis $(\kappa - 3)$ |
| $\mu^a(u)$ | SORM-approximation of $\mu^+(u)$ |
| $\mu_{max}$ | intensity of local maxima in a stationary process |
| $\mu^+$ | upcrossing spectrum |
| $\mu_u$, $\mu_u^+$ | intensity of level $u$ crossings and upcrossings |
| $\mu_u(A), \mu^+(A \mid u)$ | intensity of level $u$ crossings and upcrossings where a condition $A$ is satisfied |

| | |
|---|---|
| $\mu(u,v)$ | $\mathsf{E}[N_T(u,v)]/T$, cycle intensity |
| $\mu(U,V)$ | intensity of oscillations between sets |
| $\widehat{\mu}_{ij}^{rfc}$ | $N_t^{rfc}(u_j,u_i)/t$ estimated cumulative rainflow cycle intensity |
| $\widetilde{\mu}(u)$ | crossing spectrum for trough-crest cycles |
| $\nu_{ij}$ | oscillation intensity in Markov chain |
| $\xi$ | spatial frequency |
| $\rho_{X,Y}$ | $\mathsf{Cov}[X,Y]/(\sigma_X\sigma_Y)$ correlation coefficient between $X$ and $Y$ |
| $\Sigma_{\mathbf{X}}$ | covariance matrix for $m$-dim variable $\mathbf{X}$ |
| $\Sigma_{\mathbf{XY}}$ | cross-covariance matrix between $\mathbf{X}$ and $\mathbf{Y}$ |
| $\sigma_X$ | standard deviation for variable $X$ |
| $\sigma_{X\cdot Y}^2$ | conditional variance of $X$ given $Y$ |
| $\pi$ | $(\pi_1,\ldots,\pi_S)$, stationary distribution |
| $\phi$, $\Phi$ | standard normal density and distribution function |
| $\phi_k$ | random phase shift in linear wave models |
| $\chi_A$ | indicator function for event $A$ |
| $\Psi(x)$ | $\phi(x)-x(1-\Phi(x))$ |
| $\omega_2$ | $-r''(0)=V(x(t))$, second spectral moment |
| $\omega_k$ | $\int_0^\infty \omega^k S(\omega)\,\mathrm{d}\omega$, spectral moment of order $k$ |
| $\omega_k$ | frequency in linear wave models |

## SPECIAL SYMBOLS

| | |
|---|---|
| $\mathbf{1}_A$ | indicator function for event $A$ |
| $\mathscr{A}$ | domain for random process, $X(t), t\in\mathscr{A}$ |
| $\widetilde{x}$ | statistic $x$ computed for standardised load with $h_s=1$ and $f_z=1$ |
| $\mathscr{L}$ | linear time-invariant filter |
| $\overset{\mathscr{L}}{=}$ | "have the same distribution as" |
| # | number of elements |
| $\mathscr{O}$ | big ordo |
| $\emptyset$ | empty set |
| $\mathscr{S}$ | sample space of possible outcomes of random experiment or random variable |

Note that we use $\dot{f}$ and $f'$ for the derivative of function $f$ and $\mathbf{1}_A$ and $\chi_A$ for an indicator function in chapters with engineering and mathematical focus, respectively.

# 1 Fatigue Damage

## 1.1 INTRODUCTION

The focus of design in engineering systems is not only to meet the expected system requirements satisfying traditional codal standards but also to meet the stringent safety requirements imposed by environmental and regulatory bodies. This involves quantifying the risk associated with structural systems from various failure possibilities and demonstrating that the associated risks from failure are commensurate with the risks acceptable to society [149]. This is especially true for structural systems whose failures can lead to catastrophic consequences to society. Examples of such structural systems include important engineering infrastructure, such as bridges, dams, oil refineries, nuclear or thermal power plants, pipelines, transmission towers, wind turbines, as well as transportation infrastructure, such as aircraft components, railway lines, vehicular components, etc. The consequences of failure even in non-critical infrastructure, such as machine components in factories, lead to breakdown and factory outages, which lead to financial losses. It is therefore crucial that the design of structural components meets stringent controls that meet target reliability levels.

A first step in designing structural systems towards specified target reliability levels requires identifying the possible failure modes. For structural systems subjected to random dynamic loadings, the failure modes can be classified essentially into two broad categories. The first category involves failures when the structure response exceeds safe threshold limits during its service life. These failures are characterised in terms of the time at which the structure response exceeds the specified safe thresholds for the first time, and are referred to as first passage failures; see Figure 1.1. For loadings/response that are random time histories as shown in Figure 1.1, the first passage time, $T_f$, is a random quantity. First passage failures are usually classified as failures in serviceability criteria and need not imply structural failures. For example, a serviceability criterion in a metallic machine component subjected to time varying loading could be that the von Mises stress lies below the yield stress for all time. While an instantaneous exceedance of the von Mises stress beyond the yield stress does not lead to structural failure, the material behaviour transgressing beyond the elastic regime (as defined in terms of the yield stress) induces permanent deformations into the system, which, in turn, have the potential to affect the subsequent behaviour of the material.

The second mode of failure in vibrating systems occurs when the accumulated damage in the structure reaches a threshold limit due to the gradual degradation in the structure. For example, a material subjected to vibrating loads induces fatigue damage in the material even for small stress reversals due to small amplitude oscillations. This degradation in the material accumulates over time, and a failure occurs when this accumulated damage exceeds a critical threshold value; see

DOI: 10.1201/9781003019947-1

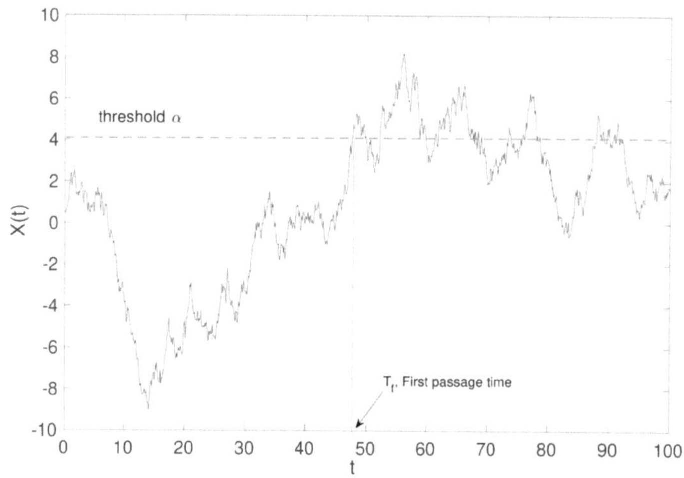


**FIGURE 1.1** Schematic of a first passage type of failure in a randomly vibrating system.

Figure 1.2. The figure shows the accumulated damage growth in time for three different data sets, corresponding to random fatigue loadings. As a result, the damage growth curves are monotonic non-decreasing random time histories, and the times at which the accumulated damage exceeds a specified threshold indicated by the horizontal dash-dotted line are indicated by the times that the corresponding vertical lines intersect the time axis. These curves are monotonically non-decreasing as

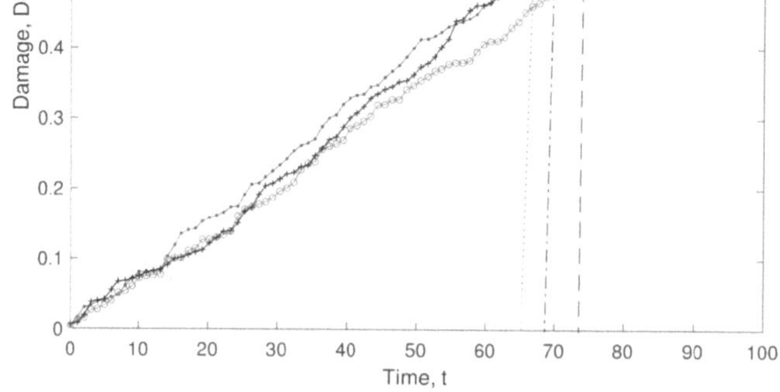


**FIGURE 1.2** Schematic of accumulation of damage in material.

auto-repair is not possible in reality and hence the damage at each time increment either increases or remains constant. Structural systems with accumulated damage $D(t)$ close to the threshold may undergo catastrophic structural failures, even under small loadings. It is therefore obvious that this second mode of failure can be more critical. Damage due to fatigue and creep effects in a material typically leads to this type of failure.

For structural systems subjected to random dynamic loads, the damage increments at each time instant are random, and hence the time histories of the damage growth $D(t)$ constitute realisations of a random process in time. This implies that the times when the response time history crosses the safe threshold (first passage failure) or the accumulated damage crosses the critical threshold are random quantities that define its lifetime, and which cannot be predicted easily through traditional deterministic approaches. For targeted reliability-based designs and to devise efficient maintenance schedules, it is crucial to accurately characterise the structural lifetime against both these failure modes. To take into account the significant uncertainties and randomness in both modelling and characterising the loadings and material behaviour, it is imperative to adopt strategies based on probabilistic modelling along with the use of random process theories to estimate the structure lifetimes in a probabilistic sense. This monograph is focused on discussing the techniques of the state-of-art in this subject, with respect to characterising fatigue damage in metals.

## 1.2 THE PHYSICS OF FATIGUE

Fatigue is the progressive and localised damage in a material when it is subjected to repeated cyclic stress reversals and constitutes an important ageing mechanism in vibrating structural systems. Stress reversals in a structural system due to vibrations lead to micro-structural changes in the material. As is well known, the micro-structure in the material media exhibits crystalline structures comprising grains, with each grain having particular crystalline orientations. Localised defects primarily occur at the grain boundaries. The orientation of the crystalline structure across the grain boundaries is usually different. As a result, each of these grains – the region enclosed by the grain boundaries – exhibits differential behaviour to the load imparted on the structure. The resultant differential dislocations in the grains result in grain boundary sliding effects, leading to the formation of voids and cavities in the grain boundaries. On further application of the loadings over time, these voids and cavities move along the grain boundaries and coalesce forming micro-cracks, which later grow in size forming macro-cracks. When the growth of these cracks becomes unstable, they propagate quickly, leading to unstable crack growth and fracture.

The growth of fatigue damage can be investigated at different spatial scales of resolution. The micro mechanical growth of the damage due to fatigue effects is a slow process and occurs over a significant period of the lifetime of the structure. This slow growth of damage at the micro-structure scales or the meso-scales is typically referred to as damage accumulation phase of the material. In this phase, any changes in the material occur at the granular scales, and no apparent damage is observed at the component level (macro-structural scales). Growth of fatigue damage in the second stage is typically a longer process and constitutes the primary fatigue life of the component. Fatigue damage in stages I and II is usually characterised in terms of the rate of growth in fatigue crack size, $a$, with respect to the cycles of loading. In the third stage, the crack growth becomes unstable and leads to fracture. Figure 1.3 shows a schematic diagram of the different stages of fatigue crack growth.

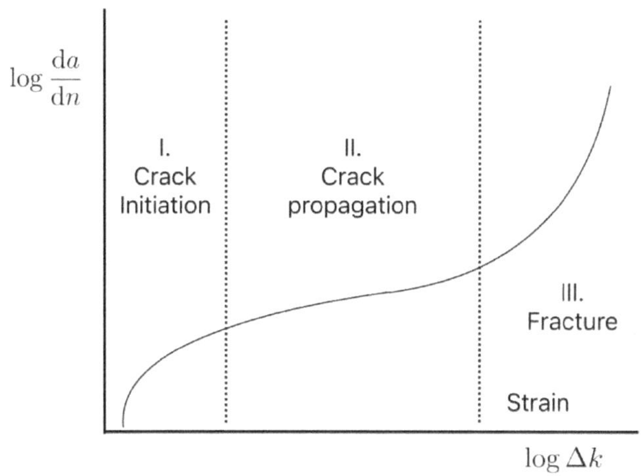


**FIGURE 1.3** Schematic of the stages in damage accumulation.

Fatigue failure occurs by the initiation and propagation of surface cracks and is usually expressed in terms of the number of cycles required for crack initiation and crack propagation. The effect of temperature dependence on fatigue life is best described in terms of stress-strain curves. Stress-strain curves are typically obtained from tensile tests where the loading is monotonic. On the other hand, for cyclic loadings, the corresponding stress-strain curves are significantly different and often show higher strength capacities. Though usually a cyclic stress-strain curve is described in terms of steady state stress and strain ranges, the variation in accompanying strain or stress ranges under constant stress or strain loading can be significant at different temperatures. For example, a material under constant strain loading but at different temperatures shows that an increase in temperature reduces its cyclic life.

Under constant cyclic strain loading, typically a material initially exhibits rapid material hardening due to the formation of stable dislocation structures. This is followed by the steady state conditions when saturation stress levels are reached. The final decrease in stress levels just before rupture can be attributed to the presence of macro-cracks in the material.

At elevated temperatures, the initial hardening due to the formation of stable dislocations is much lower, and this contributes to the lower strength characteristics of the material. In low temperature high cycle fatigue, nearly the entire fatigue life is spent in crack initiation. In these cases, the fatigue life depends greatly on the surface roughness as the surface grooves act as sites for fatigue crack initiation owing to the high stress concentrations at these locations. However, in high temperature fatigue, the process resembles the characteristics of low cycle fatigue (LCF) because fatigue damage is accompanied by plastic deformation. Additionally, because of high temperature, creep deformations are also significant. In high temperature fatigue cycling with creep effect, grain boundaries are nucleated, and the nucleated cavities grow to exhibit high temperature brittle fracture behaviour. Thus, the high temperature fatigue mechanism of these materials is expected to be different from that of LCF mechanism. This is because even if a fatigue crack is formed at the surface and may propagate into the material, if creep cavitation prevails with a high value of the cavity nucleation parameter, the fatigue life may not be influenced as much as by the fatigue cracks as by the cavities. This underlines the importance of understanding and modelling creep effects in materials subjected to high temperatures.

The intention of this monograph is not in characterising the growth of fatigue damage from such material point of view; there are several monographs that address these topics at length [57, 199]. Instead, the focus of this monograph is on establishing techniques for characterising the fatigue damage at macro-structural scales. Thus, the developments in this monograph use measures and mechanistic principles that are used for characterising fatigue damage only at the macro-structural scales.

## 1.3 S-N CURVES

The inherent randomness in the growth of fatigue cracks has led to studies in the characterisation of the fatigue life of a structure. Traditionally, fatigue life has been defined in terms of the number of constant amplitude cycles, at a given stress level, that a material can withstand before ultimate fracture. Due to the random nature of the growth of fatigue cracks, considerable scatter is observed in defining fatigue life in this manner. Historically, a best fit curve is fitted to define the cycle life of a material when subjected to cyclic loading of a fixed stress level. These curves were first defined by Wöhler [216] and are known as S-N curves; for a historical perspective, see [201, p. 167]. Here, the magnitude of the cyclical stress ($S$) is plotted against the logarithmic scale of cycles to failure ($N$).

The S-N curves are constructed by subjecting sample specimens prepared as per ASTM standards to cycling loadings having stress amplitude $S$ and noting the number of cycles $N$ for rupture. As the stress amplitudes increase, the number of cycles required for failure are lower and vice-versa. A number of tests are carried out for different stress amplitude levels, and the best fit curve is obtained from regression analysis. Typically, the S-N curves from regression analysis appear as a straight line when plotted in the logarithmic scales. However, the S-N curves appear to have two distinct regimes, characterised by straight lines with distinct changes in the slope; see Figure 1.4.

Based on the number of cycles to failure, the fatigue can be defined as either:

(i) high cycle fatigue (HCF), where the number of cycles to failure ranges from $10^3 - 10^8$ cycles at a constant stress level, or
(ii) LCF, where typically the number of cycles required for failure is less than $10^3$. LCF is typically associated with materials that exhibit widespread plasticity.

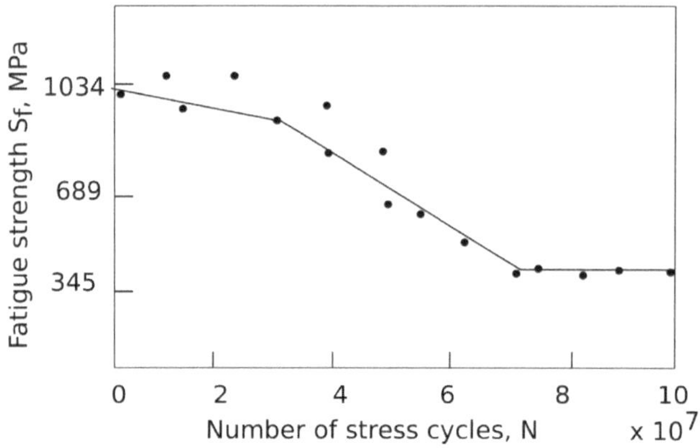


**FIGURE 1.4** Schematic diagram of S-N curves.

With reference to Figure 1.4, the distinct regimes can be characterised by the different slopes of the S-N curves.

Since the S-N curves in a particular regime are fitted as a straight line in the logarithmic axes, it follows that the relationship between $S$ and $N$ can be mathematically expressed as

$$N = \alpha S^{-\beta}, \tag{1.1}$$

where $\alpha$ and $\beta$ are constants estimated from the S-N diagrams specific to the material for which the S-N diagrams are obtained from experimental observations. It must be emphasised here once again that the S-N curves are obtained from a regression analysis for data obtained from experimental observations. Considerable scatter is observed in the data, and in fact repeating the experiment with an identical specimen subjected to the same stress loading can show failure at a different number of cycles. This lack of repeatability is not a scientific drawback; it must be kept in mind that the mechanisms of fatigue damage lie on the crystalline structure at the micro-structural scales of the material and the failure of the material after $N$ cycles, when subjected to stress levels $S$ is the macro-scale manifestation of these effects. Thus, the significant variation in the number of cycles to failure for identical specimens subjected to the same stress loadings is not unexpected. This also highlights the fact that the use of S-N curves is a grossly simplified technique adopted by the engineering community for estimating the fatigue life.

More involved attempts on characterising the scatter in the S-N curves have been to define probability bounds $P$ for the scatter at each point leading to the development of what are known as the S-N-P curves; see Figure 1.5.

It must be emphasised here that while preparing the S-N curves, one needs data for the same material for different stress levels. On the other hand, to characterise the scatter corresponding to the S-N curves through the S-N-P curves requires several observations for each of the axis points in the S-N curves. Thus, preparing the S-N-P curves for a given material is time-consuming and expensive. Analytically,

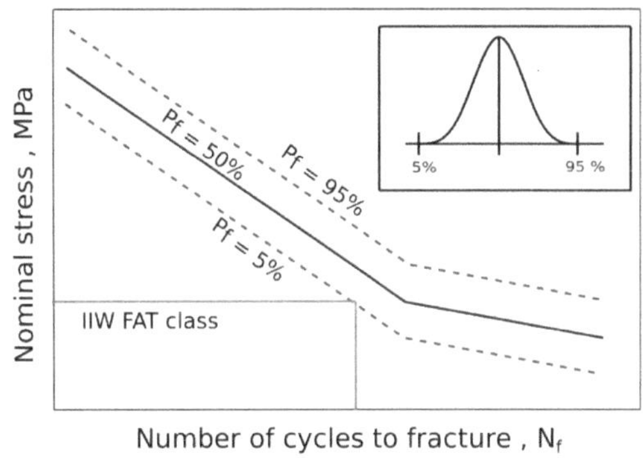


**FIGURE 1.5** Schematic diagram of S-N-P curves; the dotted lines represent the 5% and 95% bounds and are representative of the variability in the observations.

the scatter that is expected in the physical observations can be modelled by treating the material constants $\alpha$ and $\beta$ in Equation (1.1) as random variables, whose probabilistic characteristics have to be estimated through statistical fitness tests on data obtained from experimental measurements.

## 1.4 FATIGUE DAMAGE

Most applications in real life result in structural components being subjected to loadings of variable stress amplitudes at different stages of its operation. This implies that the direct application of the S-N curves is not useful for characterising the fatigue life of the material. This has led to the concept of defining a non-dimensional parameter $D$ as a measure of partial fatigue damage in the material, given by

$$D = \frac{n}{N}, \tag{1.2}$$

where for a loading of stress amplitude $S$, $N$ denotes the number of cycles to failure and $n < N$ represents the number of cycles of loading. It is obvious that $D$ in Eq. (1.2) takes values between 0 and 1. $D$ is referred to as damage and is not a measurable metric but is a non-dimensional quantity used as a parameter to quantify the extent of fatigue damage in a specimen. For example, if fatigue damage is measured in terms of size of fatigue cracks, $D$ can be viewed to be the ratio of the fatigue crack length at a given time in the life of the specimen to the critical crack length beyond which the crack growth becomes unstable leading to fracture.

When the cyclic loadings consist of a sequence of harmonic cycles of varying stress amplitudes, the fatigue damage is measured in terms of $D$ by applying Palmgren-Miner's rule [147, 154], given by

$$D = \sum_j \frac{n_j}{N_j}. \tag{1.3}$$

Here, the index $j$ is over the various stress amplitude levels, $n_j$ represents the number of cycles with stress amplitude $S_j$ the material has been subjected to and $N_j$ denotes the number of cycles for failure corresponding to $S_j$. Thus, $D$ in Eq. (1.3) is the summation of the contributions of the partial fatigue damage for each stress amplitude level; see Figure 1.6 for a schematic diagram that explains the Palmgren-Miner's rule. A fatigue failure is deemed to occur when $D \geq 1$.

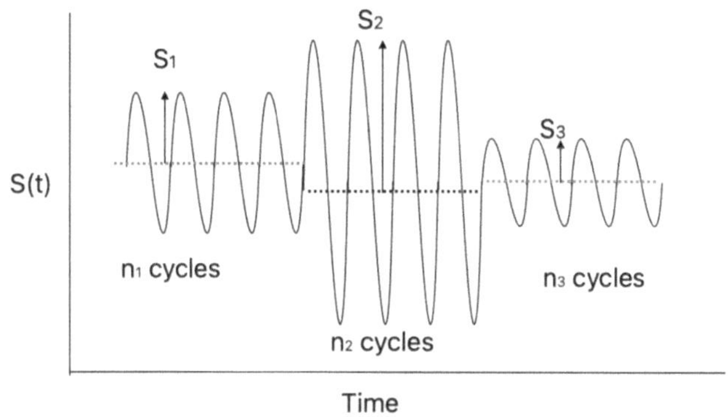


**FIGURE 1.6** Schematic diagram explaining the Palmgren-Miner's rule for partial fatigue damage due to variable amplitude loading.

The use of Palmgren-Miner's rule enables the use of the S-N curves for estimating fatigue damage in structural components, subjected to variable loadings and is useful in scheduling maintenance or replacement of components that have fatigue damage close to specified levels. As can be seen from Eq. (1.3), the total damage is obtained by superimposing the damage from different sequences of stress cycles through simple linear addition. The linear assumption of damage accumulation implies that the sequence of loading of different stress cycles has no effect on the fatigue life. However, experimental studies [215, 116] have shown the linear damage accumulation rules to be unreliable.

Several alternative damage accumulation rules that are non-linear in nature have been proposed in the literature. Marco and Starkey [144] proposed a non-linear version of Eq. (1.3), where the incremental damage $\Delta_j$ due to $n_j$ cycles corresponding to stress $S_j$ is given by

$$\Delta D_j = \left(\frac{n_i}{N_j}\right)^{x_{S_j}}, \tag{1.4}$$

where $N_j$ cycles are required for failure for stress levels $S_j$ and the exponent $x_{S_j}$ takes values between 0 and 1. This damage accumulation implies that the incremental damage increases proportionally with the stress amplitude $S_j$. Several variations of Eq. (1.4) have been proposed in subsequent studies [187, 42, 82], the details of which are summarised in [192]. An alternative form of a non-linear damage accumulation rule, proposed by Manson and Halford [140], comprises a piecewise linear damage accumulation rule, termed as the double linear damage rule. They considered a bilinear relationship for the damage growth for modelling the two phases of fatigue crack propagation – the crack initiation phase and the stable crack propagation phase – with different rates of damage growth and characterised by the difference in the slopes [117].

Despite the many damage accumulation rules proposed in the literature, the variability of the predictions with experimental results remains significant. The primary reason for these variations from experimental results is that the damage accumulation rules are essentially simple ad hoc attempts at modelling the growth of fatigue damage at macro-structural scales, which do not take into account the underlying mechanics at the micro-structure scales that govern their growth. Therefore, the damage accumulation rule as proposed by Palmgren and developed by Miner, which is simplest in form, continues to remain the most widely used. The developments proposed in this monograph will also be limited with respect to the use of the Palmgren-Miner rule.

The above approach suffers from the disadvantage that the fatigue damage $D$ cannot be related to any physically measurable quantity. An alternative to characterising fatigue damage is based on fracture mechanics theories and crack growth analysis. Here, usually the length of the fatigue crack, $a$, is assumed to be indicative of the extent of fatigue damage. Typically, the fatigue crack growth rate, under deterministic loading, is observed to be governed by the Paris-Erdogan equation, given by

$$\frac{\mathrm{d}a}{\mathrm{d}N} = \widetilde{C}(\Delta k)^m. \tag{1.5}$$

Here, $\widetilde{C}$ and $m$ are material dependent constants, $\Delta k = K_{\max} - K_{\min}$ is the stress intensity factor range with $K_{\max}$ and $K_{\min}$ being the corresponding maximum and the minimum for the stress intensity factor within a cycle, respectively, $N$ denotes the number of cycles and $\mathrm{d}a/\mathrm{d}N$ is the rate of change of fatigue crack length per unit cycle. Equation (1.5) is an idealisation when the fatigue crack propagation is restricted to straight isolated cracks, and the loading is such that crack closures do not occur. The processes of nucleation and crack initiation are not considered either. For embedded and surface cracks, the stress intensity factor, $K$, is typically of the form [63]

$$K = S\sqrt{\pi a}B(a/c), \tag{1.6}$$

where, $S$ represents the far field stress due to the applied loading, $a$ is the crack length and is assumed to be equal to the semi-major diameter of elliptical cracks, $c$ is the corresponding semi-minor diameter and $B(\cdot)$ is a factor that accounts for the shape of the specimen and the crack geometry and is a function of the $a/c$ ratio. From Eq. (1.6), it follows that a cyclic variation in $S$ also induces a proportional variation in $K$, and for a constant amplitude deterministic loading,

$$\Delta k = \sqrt{\pi a}B(a/c)\left[S_{\max} - S_{\min}\right]. \tag{1.7}$$

The crack growth process as defined in Eq. (1.5) is essentially a discrete temporal event characterised by complete cycles. Rewriting Eq. (1.5) in terms of time $t$ enables representing the fatigue crack growth process as a continuous function of time and is more amenable for mathematical treatment. Thus, Eq. (1.5) can be expressed as

$$\frac{da}{dt} = \frac{da}{dN}\frac{dN}{dt} = C(a/c)\sqrt{\pi}a^{\frac{m}{2}}(\Delta S)^m \frac{dN}{dt}, \tag{1.8}$$

where, $C(a/c) = \widetilde{C}B(a/c)$ and $dN/dt$ is the rate of occurrence of cycles per unit time.

For variable amplitude loadings, estimating fatigue crack growths is significantly more difficult. This is primarily because of the interaction effects due to the irregularity of the loadings. One of the important interaction effects is the retardation in fatigue crack growth due to tensile overloading. Crack retardation remains in effect for some period after the overloading and in general, this duration is proportional to the magnitude of the overloading [4]. For sufficiently large overloads, sometimes, even crack arrest has been observed to occur when the crack growth stops for some time. Though the effect of large compressive overloading is generally small, accelerated crack growths have been observed for large compressive loadings that initiate yielding [196]. It is worth pointing out that the crack growth relation in Eq. (1.8) is non-linear which implies that the loading sequence is important for accurate crack growth estimations.

## 1.5 FATIGUE UNDER RANDOM LOADS

Loadings in real life situations are seldom of a constant amplitude type; rather, the loading is usually irregular and often, in many applications, constitutes a random process, a sample time history realisation of which would typically look like as shown in Figure 1.7. A fatigue crack growth-based approach to predicting the residual lifetime of a component necessitates making a prediction about the fatigue crack growth against future expected loadings that are realisations of a random process. For systems subjected to loads that are well defined cycles exhibiting periodic behaviour, the predictions are not too difficult as one simply needs to apply the crack-growth relation in Eq. (10.6). However, in situations where the future loading is expected to

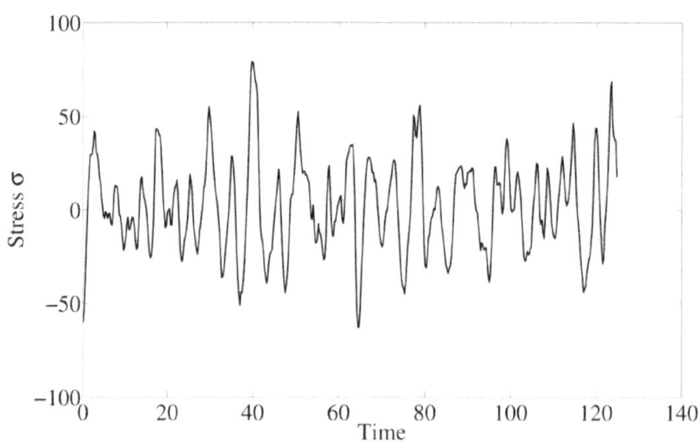


**FIGURE 1.7** Schematic diagram of random loading.

exhibit significant variability and aperiodic behaviour, the problem is more difficult. This involves probabilistic modelling of the loadings and the fatigue crack growth process. Randomised versions of deterministic models for fatigue crack growth when subjected to deterministic load ranges have been studied extensively in the literature; see [54, 119, 120, 152, 192, 194]. Here, the randomisation of the deterministic models has usually been carried out by modelling certain model parameters as random variables or by multiplying the deterministic equations by a non-negative random function of time. The random time function is assumed to model the uncertain nature of secondary effects such as temperature, corrosion and other effects. The distributional properties of the random variables/functions have been assumed based on the data obtained from experimental results [192]. In most of the studies, the resultant incremental crack growth process has been modelled as having Markovian properties. Approximations for the probability density functions (pdf) of the time to failure have been obtained as solutions of the associated Fokker-Planck-Kolmogorov (FPK) equation. Extension of these studies to include the effects of random loadings has also been considered in [23, 36, 206, 223], and analytical approximations for the pdf of the failure times have been obtained by solving the corresponding FPK equation.

In situations where the primary uncertainty is in the loading, one can adopt a Monte Carlo simulation-based approach to estimate the expected fatigue crack growths and assess the residual life. This involves modelling the load as a random process with specified pdf and power spectral density function (psd). Models for the pdf and psd for the random process can be constructed from the random loadings observed in the past using statistical techniques. The Monte Carlo procedure involves generating an ensemble of time histories for the loading that satisfy the pdf and psd. Corresponding to each realisation time history, the fatigue crack growth needs to be calculated. The expected fatigue crack growth can be obtained by taking the ensemble average. As is obvious, Monte Carlo simulations involve repeated deterministic analysis using the model equation. Thus, any inaccuracies in the adopted deterministic model are carried over into the Monte Carlo estimate. Since most models used in the literature show significant variations with the experimental observations, the fatigue life predictions are also expected to be at variance with experimental observations. However, this would raise questions on the acceptability of the deterministic model and not on the method per se.

The irregular nature of the random time history implies that direct application of the S-N curves is not possible for fatigue damage estimation as it is not easy to identify the harmonic cycles or the stress range of such cycles. In such situations, damage characterisation involves the following key steps:

(a) The random loading time history is recast as a summation of unit cyclic loadings of different stress amplitudes, following an appropriate cycle counting technique.
(b) The fractional damage corresponding to each of these unit cyclic loadings is calculated from the appropriate S-N curves for the material.
(c) Finally, the cumulative fatigue damage is estimated by combining the fractional damages using a suitable damage accumulation rule.

It is quite obvious that the key step in this procedure is decomposition of the random time history into an equivalent number of cyclic loadings corresponding to various stress amplitudes. These are referred to in the literature as cycle counting algorithms, several of which have been discussed in the literature.

Of these, the most popularly used are the peak counting algorithm, the range counting method, the level crossing technique and the rainflow cycle counting algorithm [145]. Some of these algorithms are used in the time domain, while others are used in the frequency domain. A review of these algorithms in the manner in which they are generally used is available in [10, 136, 205].

Of the several cycle counting algorithms that have been proposed, it has been observed that the rainflow cycle counting algorithm leads to predictions which are at least variance with those obtained from experimental results. In the rainflow cycle counting scheme, the cycle sequences are paired into complete cycles by randomising the sequence. This raises questions on the applicability of this method to predict fatigue crack growth. However, under certain assumptions, the rainflow fatigue damage predictions when carried out in time domain lead to reasonable results for fatigue crack growth. An analytical equivalent of the complicated time-domain rainflow cycle counting algorithm has been developed in the spectral form [170, 172]. This spectral-based analytical approach relies on the use of theories of random processes and statistics to obtain quantitative approximations for the fatigue damage in a probabilistic sense and provides elegant and computationally cheap solutions for obtaining the same. The focus of the present monograph is on presenting the developments in this approach, using the mathematical theories of random processes. The method can be applied, provided the crossing statistics of the random stress time histories can be estimated.

# 2 Mathematical Prerequisite

As has been mentioned in Chapter 1, random fatigue and expected lifetimes can be characterised using the language of statistics, like average and spreading. Probability theory gives us language and tools to handle our partial knowledge about uncertain events, in daily life as well as in strictly controlled physical experiments. This chapter gives a concise summary of the necessary elements of probability that we use to handle random fatigue of materials under influence of random loads.

Often in engineering and physical sciences, outcomes of random experiments are numbers associated with some physical quantities. In a controlled fatigue experiment, a central quantity we want to investigate is the number of load cycles a specimen can withstand before it breaks. We decide to use the capital letter $N$ to denote the number of cycles, accepting that $N$ will most surely take different values each time we repeat the experiment. To understand the seemingly random variation, we can keep an eye on quantities that may influence the result, such as temperature $T$, surface roughness $Ra$, physical dimensions $U, V, W$, material sensitivity $K$, load amplitude $X$, etc.

In the list of possible influencing quantities, we identify some which can be controlled in the experiment, but not perfectly. There are other quantities that are rather difficult to control, while some quantities are not at all possible to control, such as the load amplitude in live survival experiments. Regardless of the type of variability, these parameters are technically defined in the statistical literature as *random variables*, or in the case when these are indexed in time or space, as *random processes.* In this chapter, we present the tools used to describe and define the possible variability of simple random variables and random processes.

## 2.1 RANDOM VARIABLES AND THEIR PROPERTIES

### 2.1.1 WHAT IS A RANDOM VARIABLE AND WHAT IS AN OBSERVATION?

In this text random variables, abbreviated r.v., i.e. symbols for the outcome of a random experiment, are usually identified by upper case letters, e.g. $X, Y, N$. The set $\mathscr{S}$ of possible values of a random variable is called the *sample space*. It can be the set of integers, or the non-negative integers in the case the experiment leads to a counting procedure. It can be the set of all real numbers, or some subset thereof. When the experiment is actually performed, we often use the corresponding lower case letter for the numerical value we obtain, e.g. $x$ for an *observation* of the random variable $X$, and $x_1, x_2, \ldots, x_n$, if the experiment is repeated $n$ times.

DOI: 10.1201/9781003019947-2

## 2.1.2 PROBABILITY DISTRIBUTIONS AND SIMPLE PROPERTIES

### 2.1.2.1 Distribution Functions and Densities

By the *distribution* of a random variable, we mean a statement about the probabilities to obtain the different values that can occur.

If a random variable $X$ is *discrete*, i.e. there is only a finite (or countable) number of possible values, $\mathscr{S} = \{x_1, x_2, \ldots, x_n\}$, like the number of eyes of a dice or the number of rainy days next week, one can simply attach a number $p_k \geq 0$ to each of the possible outcomes $x_k \in \mathscr{S}$, and write

$$p_k = \mathsf{P}(X = x_k)$$

for the probability that the experiment will result in the number $x_k$. One has to make sure that $\sum_k p_k = 1$, if one is 100% sure that the experiment will produce *some* result. Putting $p_k = 0$ for some $k$ is an acceptable choice.

For random variables that can take any value in a real interval, one needs a more general tool to describe its variation. This tool is the *(cumulative) distribution function, cdf*, $F_X(x)$ defined as

$$F_X(x) = \mathsf{P}(X \leq x), \quad x \in \mathbb{R},$$

from which we can deduce, by one of the basic laws of probability,

$$\mathsf{P}(a < X \leq b) = F_X(b) - F_X(a). \tag{2.1}$$

Any function $F(x)$ that is non-decreasing, right continuous and with the limits $\lim_{x \to -\infty} F(x) = 0$, $\lim_{x \to \infty} F(x) = 1$, is (or can be used as) a distribution function.

The importance of the probability distribution function lies in the following fact:

**Theorem 2.1.** *The probability of (almost) any statement about a random variable $X$ can be computed (at least in theory) when the distribution function $F_X(x)$ is known.*

For simplicity, we write $F(x)$ for $F_X(x)$. The discrete random variables have cdf of the type $F(x) = \sum_{k; x_x \leq x} p_k$, and it has jumps of height $p_k$ at each possible $x_k$, being constant in between. If the cdf $F(x)$ is differentiable, the random variable is called *continuous* and the derivative

$$f(x) = \frac{\mathrm{d}F(x)}{\mathrm{d}x} \geq 0 \tag{2.2}$$

is called the *probability density function* (pdf). It has the following important interpretation $\mathsf{P}(X \in [x, x + \Delta x]) \approx f(x)\Delta x$. Obviously, the inverse of (2.2) is

$$F(x) = \int_{-\infty}^{x} f(u)\,\mathrm{d}u, \text{ with } \int_{-\infty}^{\infty} f(u)\,\mathrm{d}u = 1.$$

Just as the probability function $p_k$ spreads out the total probability 1 over the possible discrete values, the pdf spreads out the total probability over a continuum of values. As a consequence, no single $x$ gets positive probability, $\mathsf{P}(X = x) = 0$ if $X$ is continuous.

Any non-negative function that integrates to one can be used as a pdf for a distribution and in fact the pdf is normally used to define the classes of standard statistical distributions.

**Remark 2.1.** *Note that there are random variables that are a mixture of discrete and continuous type, e.g. $X$, the amount of rain falling next week – the number of rainy days is discrete and the amount of rain that falls on rainy days is continuous. If $N =$ number of rainy days $= 0$, then $X = 0$, and if $N > 0$, then $X > 0$ is a continuous variable we can call $X^+$ with a* cdf *$F^+(x)$. The total* cdf *for the amount of rain is shown in Figure 2.1 and is given by*

$$F_X(x) = \begin{cases} 0 & \text{if } x < 0, \\ \mathsf{P}(N=0) & \text{if } x = 0, \\ \mathsf{P}(N=0) + \mathsf{P}(N>0) \times F^+(x) & \text{if } x > 0. \end{cases}$$

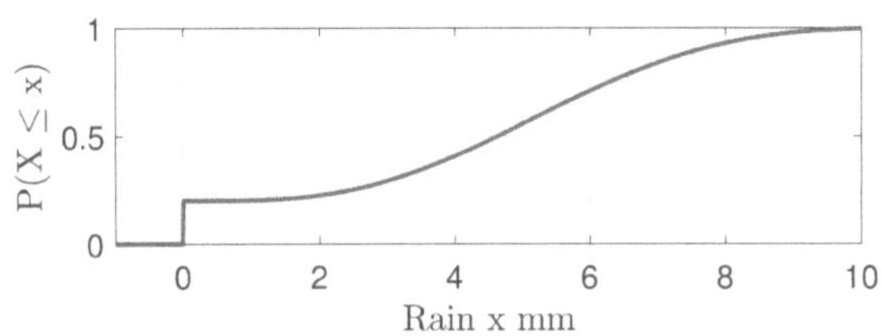


**FIGURE 2.1** Cumulative distribution function for variable of mixed type.

#### 2.1.2.2 Simple Moments

The *expected value* of a random variable $X$ is equal to the centre of gravity of the distribution, often denoted $m_X$ and simply called the *mean* of the distribution,

$$\begin{aligned} \mathsf{E}[X] &= \sum_k x_k \mathsf{P}(X = x_k) \quad \text{if the distribution is discrete,} \\ \mathsf{E}[X] &= \int_{-\infty}^{\infty} x f_X(x)\,\mathrm{d}x \quad \text{if the density } f_X(x) = \frac{\mathrm{d}}{\mathrm{d}x} F_X(x) \text{ exists.} \end{aligned} \tag{2.3}$$

Note that there are distributions for which the expectation is not well defined. For example, if $\int |x| f_X(x)\,\mathrm{d}x = \infty$, we let $\mathsf{E}[X]$ be undefined.

**Useful properties:** When $X$ is a non-negative variable,

$$\mathsf{E}[X] = \int_0^{\infty} \mathsf{P}(X > x)\,\mathrm{d}x = \int_0^{\infty} \mathsf{P}(X \geq x)\,\mathrm{d}x. \tag{2.4}$$

If $Y = g(X)$ is a differentiable function of $X$, with derivative $g'(x)$, then

$$\mathsf{E}[Y] = \mathsf{E}[g(X)] = g(0)\mathsf{P}(X=0) + \int_0^\infty g'(x)\mathsf{P}(X>x)\,\mathrm{d}x. \tag{2.5}$$

The *variance* of a random variable $X$ is defined as the moment of rotational inertia of the distribution around the expectation:

$$\mathsf{Var}[X] = \mathsf{E}[(X-m_X)^2] = \mathsf{E}[X^2] - m_X^2.$$

The square root of the variance $\mathsf{D}[X] = \sqrt{\mathsf{Var}[X]}$ is called the *standard deviation* of the distribution, or of $X$. It has the same dimension as $X$. In statistics, the standard deviation is often denoted $\sigma = \sigma_X = \mathsf{D}[X]$. The *coefficient of variation*, also called the *relative standard deviation*, is defined as $\mathsf{CoV}[X] = \mathsf{D}[X]/\mathsf{E}[|X|]$.

Of the higher moments, we will have use for the *skewness* and *kurtosis* defined by the *central moments*

$$\mu_k = \mathsf{E}[(X-m_X)^k)], \text{ as}$$
$$\kappa_3 = \mu_3/\sigma^3, \quad \kappa_4 = \mu_4/\sigma^4, \text{ respectively.} \tag{2.6}$$

A symmetric distribution has skewness $\kappa_3 = 0$, a positive value indicates a "heavy" upper tail. The kurtosis can be used as an indication of heavy tails in general. The normal distribution has $\kappa_4 = 3$, and in this book, we will use $\kappa_e = \kappa_4 - 3$ as an *excess kurtosis* measure, meaning tails are heavier than normal.

#### 2.1.2.3 Some Important Distributions

The discrete *Poisson* distribution and the continuous *exponential* distribution form an important pair of distributions in many reliability studies.

**Poisson distribution:** A random variable $N$ taking non-negative integer values has a Poisson distribution with parameter $\mu$ if it has probability function

$$p_k = \mathsf{P}(N=k) = e^{-\mu}\frac{\mu^k}{k!}, \quad k = 0,1,2,\ldots$$

A Poisson variable has mean equal to its variance, $\mathsf{E}[N] = \mathsf{Var}[N] = \mu$.

**Exponential distribution:** A random variable $T$ taking positive real values has an exponential distribution with parameter $\tau > 0$ if it has pdf and cdf, respectively,

$$f_T(t) = \frac{1}{\tau}e^{-t/\tau}, \quad F_T(t) = 1 - e^{-t/\tau}, \quad t > 0.$$

An exponential random variable $T$ has mean equal to its standard deviation, $\mathsf{E}[T] = \mathsf{D}[T] = \tau$.

The relation between the Poisson and the exponential distribution reads as follows: If "catastrophes" occur randomly at a rate $\mu$ per time unit, then the distribution of the number of catastrophes in the time interval $[0,t]$ has a Poisson distribution with parameter $\mu t$ for any choice of $t > 0$, and the waiting time $T$ from time 0 to the first catastrophe has an exponential distribution with parameter $\tau = 1/\mu$.

**Laplace and Cauchy distributions:** The Laplace distribution, more known as the "double exponential distribution", and its relative, the Cauchy distribution, will be frequently used in the examples in coming chapters and the presentation of the Laplace moving average (LMA) load model. Their pdf are respectively given by,

$$\text{Laplace: } f(x) = \frac{1}{2\tau} e^{-|x|/\tau}, \; x \in \mathbb{R},$$
$$\text{Cauchy: } f(x) = \frac{1}{\sigma\pi} \frac{1}{1+(x/\sigma)^2}, \; x \in \mathbb{R}.$$

The even central moments of the Laplace distribution are identical to those of the exponential distribution. The Cauchy distribution does not have a finite mean, and therefore samples from it do not follow the law of large numbers or the central limit theorem. Historically, it was a touchstone in the development of probability theory.

**Normal distribution:** The distribution of a standard Gaussian or standard *normal* random variable is defined by means of its density function. The density of a standard normal variable has its own symbol $\phi(x)$ and is given by

$$\phi(x) = \frac{1}{\sqrt{2\pi}} \mathrm{e}^{-x^2/2}, \quad x \in \mathbb{R}. \tag{2.7}$$

The distribution function of the variable, $F(x)$, has its own symbol $\Phi(x)$,

$$\Phi(x) = \int_{-\infty}^{x} \frac{1}{\sqrt{2\pi}} e^{-t^2/2} \,\mathrm{d}t, \quad x \in \mathbb{R}. \tag{2.8}$$

A normal variable with mean $m$ and variance $\sigma^2$ has density and distribution functions

$$f_X(x) = \frac{1}{\sigma}\phi\left(\frac{x-m}{\sigma}\right), \quad F_X(x) = \Phi\left(\frac{x-m}{\sigma}\right).$$

To indicate that a r.v. $X$ is normal with mean $m$ and variance $\sigma^2$, one can use the notation $X \in N(m, \sigma^2)$.

Another function, which will be useful later in this book, is $\Psi$, defined as

$$\Psi(x) = \int_x^{\infty} (1 - \Phi(z))\,\mathrm{d}z = \phi(x) - x(1 - \Phi(x)). \tag{2.9}$$

Note that the functions $\Phi$ and $\Psi$ cannot be computed analytically. Very precise approximations of these functions exist, and they are included in most of numerical toolboxes. In the special case when $x = 0$, we have that $\Phi(0) = 1/2$ and $\Psi(0) = 1/\sqrt{2\pi}$. Using the introduced functions, we can write the following formulas: Let $f_{m,\sigma}(x)$ be the normal density with mean $m$ and variance $\sigma^2$, then

$$\int_0^\infty x f_{m,\sigma}(x)\,\mathrm{d}x = \sigma\Psi\left(-\frac{m}{\sigma}\right),$$
$$\int_{-\infty}^\infty |x| f_{m,\sigma}(x)\,\mathrm{d}x = \sigma\left(\Psi\left(-\frac{m}{\sigma}\right) + \Psi\left(\frac{m}{\sigma}\right)\right). \tag{2.10}$$

### 2.1.3 INDEPENDENCE, DEPENDENCE, JOINT DISTRIBUTION

When the outcome of an experiment is numerical, we call it a random variable. Obviously, for one and the same experimental result, one can measure more than one numeric quantity. For example, at a meteorological station, weather is described by temperature $T$, pressure $P$, wind speed $W$, etc., and each variable can be regarded as a random variable. The result of the experiment is a vector $\mathbf{X} = (T, P, W)$ of random variables, defined on the same outcome. We now describe the tools to handle many random variables simultaneously.

For simplicity only, let us consider $\mathbf{X} = (X_1, \ldots, X_n)$ with $n = 2$. The function

$$F_{X_1,X_2}(x_1,x_2) = \mathsf{P}(X_1 \le x_1 \text{ and } X_2 \le x_2),$$

is called the *two-dimensional distribution function* of the pair of random variables. The probability of any statement about the joint behaviour of $X_1, X_2$ can be computed (at least in theory) if the distribution function $F_{X_1,X_2}(x_1,x_2)$ is known, e.g., by means of formula (2.18). The distribution of a vector of $n$ random variables is defined in a similar way.

#### 2.1.3.1 Independence

**Independence**: We say that two events (statements) $A$, $B$ about an outcome of an experiment are independent if

$$\mathsf{P}(A \cap B) = \mathsf{P}(A \text{ and } B) = \mathsf{P}(A)\mathsf{P}(B).$$

For random variables $X_1$ and $X_2$, we say that if any statement about $X_1$ is independent of any statement about $X_2$, then they are independent. The following theorem gives the conditions for independence:

**Theorem 2.2.** *The variables $X_1$, $X_2$ with distributions $F_{X_1}(x), F_{X_2}(x)$, respectively, are independent if for all values $x_1$, $x_2$:*

$$\mathsf{P}(X_1 \le x_1, X_2 \le x_2) = F_{X_1}(x_1) \cdot F_{X_2}(x_2).$$

*Similarly for any $n$, $X_1, X_2, \ldots, X_n$ are independent if*

$$\begin{aligned} F_{X_1,\ldots,X_n}(x_1,\ldots,x_n) &= \mathsf{P}(X_1 \le x_1, \ldots, X_n \le x_n) \\ &= F_{X_1}(x_1) \cdot F_{X_2}(x_2) \cdot \ldots \cdot F_{X_n}(x_n). \end{aligned} \tag{2.11}$$

The following theorem has its roots in the sixteenth century when the Italian mathematician Gerolamo Cardano (1501–1576) observed that the average in an expanding series of experiments had a tendency to stabilise. This *law of large numbers* got its following final form by a theorem by Khinchin [102]. It is, together with the *central limit theorem*, the most well-known theoretical results in probability theory.

**Theorem 2.3** (Law of large numbers)**.** *Let $X_1, \ldots, X_k$ be a sequence of iid (independent identically distributed) variables all having the distribution $F_X(x)$. Denote by $\overline{X}$ the average of all $X_i$, i.e.*

$$\overline{X} = \frac{1}{k}(X_1 + X_2 + \cdots + X_k). \tag{2.12}$$

*(Obviously, $\overline{X}$ is a random variable itself.) If the mean of the distribution $F_X(x)$ of $X$ exists and is finite, $m_X = \mathsf{E}[X] < \infty$, then, as $k$ increases (we are averaging more and more variables), $\overline{X} \approx \mathsf{E}[X]$ with equality when $k$ approaches infinity.*

#### 2.1.3.2 Dependence and Conditional Probability

Two events $A$ and $B$ are defined as independent if $\mathsf{P}(A \cap B) = \mathsf{P}(A)\mathsf{P}(B)$. We shall now quantify what is meant by statistical *dependence*. Let $A$ be the event that tomorrow will be very windy and we have learned that $\mathsf{P}(A) = 0.4$. Also, let $B$ be the event that there will be heavy rain tomorrow. Obviously, there is some dependence between the two events. Our experience tells us that there is 60% chance of rain on a very windy day. That leads us to invent the *conditional probability*,

$$\mathsf{P}(B \mid A) = 0.6,$$

meaning "the probability that $B$ occurs given that $A$ occurs is 60%".

This exercise leads us to the simple conclusion that to have both strong wind and rain, we first of all need wind, and then also rain,

$$\mathsf{P}(A \cap B) = \mathsf{P}(A) \cdot \mathsf{P}(B \mid A) = 0.4 \cdot 0.6 = 0.24. \tag{2.13}$$

To continue the example, suppose we also know that on a not so windy day, the chance of rain is 20%, $\mathsf{P}(B \mid A^*) = 0.2$, so

$$\mathsf{P}(A^* \cap B) = \mathsf{P}(A^*) \cdot \mathsf{P}(B \mid A^*) = (1 - 0.4) \cdot 0.2 = 0.12. \tag{2.14}$$

Here, $A^*$ implies the complement of set $A$.

In this example, we used our common sense to assign conditional probabilities to an event under different circumstances. We make this formal by *defining* the conditional probability on the reduced sample space of possible outcomes from $\mathscr{S}$ to $A \cap \mathscr{S}$ and the probability that $B$ occurs, given that $A$ has occurred, to

$$\mathsf{P}(B \mid A) = \frac{\mathsf{P}(A \cap B)}{\mathsf{P}(A)}. \tag{2.15}$$

We combine (2.13) and (2.14) to a useful theorem, containing the rule of total probability and the Bayes' theorem.

**Theorem 2.4** (Total probability and Bayes' theorem)**.** *The rule of "total probability" is a smart way of computing complicated probabilities, while Bayes' theorem can help to trace the origin of an occurred event.*

$$\textit{Total probability: } \mathsf{P}(B) = \mathsf{P}(A \cap B) + \mathsf{P}(A^* \cap B)$$
$$= \mathsf{P}(A) \cdot \mathsf{P}(B \mid A) + \mathsf{P}(A^*) \cdot \mathsf{P}(B \mid A^*),$$
$$\textit{Bayes' theorem: } \mathsf{P}(A \mid B) = \frac{\mathsf{P}(A)\mathsf{P}(B \mid A)}{\mathsf{P}(A) \cdot \mathsf{P}(B \mid A) + \mathsf{P}(A^*) \cdot \mathsf{P}(B \mid A^*)}.$$

### 2.1.4 MULTI-DIMENSIONAL DISTRIBUTIONS

#### 2.1.4.1 Joint Density

In this section, we shall assume that we have only two random variables, $n = 2$, and in order to simplify the notation, we shall denote them by $X, Y$. The distribution function will be denoted by

$$F_{X,Y}(x,y) = \mathsf{P}(X \le x, Y \le y),$$

which we shall often simplify to $F(x,y)$. The distributions of the variables $X$ and $Y$ will be denoted by $F(x) = \mathsf{P}(X \le x)$ and $F(y) = \mathsf{P}(Y \le y)$, respectively, letting the argument $x,y$ define the respective distribution. From the definition of $F(x,y)$, it follows immediately that

$$F(x) = F(x, +\infty), \quad F(y) = F(+\infty, y).$$

If the distribution $F(x,y)$ is differentiable with respect to $x$ and $y$, the derivative

$$f(x,y) = \frac{\partial^2 F(x,y)}{\partial x \partial y}$$

is called the joint probability density function (pdf), from which one can calculate the distribution function

$$F(x,y) = \int_{-\infty}^{x} \int_{-\infty}^{y} f(\tilde{x}, \tilde{y})\, \mathrm{d}\tilde{x}\, \mathrm{d}\tilde{y}.$$

Any non-negative function $f(x,y)$ that integrates to one

$$\int_{-\infty}^{\infty} \int_{-\infty}^{\infty} f(x,y)\, \mathrm{d}x\, \mathrm{d}y = 1$$

is a density of some pair of random variables $(X,Y)$. The one-dimensional densities of $X$, $Y$, called the *marginal densities*, can be computed from the joint density by integrating out the redundant variable,

$$f(x) = \int_{-\infty}^{\infty} f(x, \tilde{y})\, \mathrm{d}\tilde{y}, \qquad f(y) = \int_{-\infty}^{\infty} f(\tilde{x}, y)\, \mathrm{d}\tilde{x}. \tag{2.16}$$

It is easy to prove that for independent $X,Y$.

$$f(x,y) = f(x)f(y). \tag{2.17}$$

Finally, for any events $A,B$

$$\mathsf{P}(X \in A \text{ and } Y \in B) = \int_A \int_B f(x,y)\,\mathrm{d}y\,\mathrm{d}x. \tag{2.18}$$

Quite often, the last formula has to be computed numerically, even for simple sets $A,B$. For example, this is the case when $X,Y$ are Gaussian.

#### 2.1.4.2 Expectation, Covariance and Correlation

We will often in this work encounter moments or other expectations of functions of two or more random variables. Here, we have collected the simple rules for such moments.

Consider a function $z = g(x,y)$. Define a new random variable $Z$ as a value of $g$ computed at the result of a random experiment $X,Y$, i.e. $Z = g(X,Y)$. For example, $Z = X+Y$ or $Z = X \cdot Y$. We want to find the expected value of $Z$. Obviously, if the distribution of $Z$ is explicitly known, we could use (2.3) to compute the expectation of $Z$. However, it can also be done directly by means of the following formulas:

$$\mathsf{E}[Z] = \mathsf{E}[g(X,Y)] = \int_{-\infty}^{\infty}\int_{-\infty}^{\infty} g(x,y)f(x,y)\,\mathrm{d}x\,\mathrm{d}y \tag{2.19}$$

or

$$\mathsf{E}[Z] = \mathsf{E}[g(X,Y)] = \sum_{j=0}^{\infty}\sum_{k=0}^{\infty} g(j,k)\,p_{jk},$$

where $p_{jk} = \mathsf{P}(X = j, Y = k)$.

It is also easy to check by (2.17) and (2.19) that for independent variables $X,Y$,

$$\mathsf{E}[X \cdot Y] = \mathsf{E}[X] \cdot \mathsf{E}[Y].$$

Actually, it can happen that the last equality holds even for dependent variables. Such variables are called *uncorrelated*. (All independent variables are uncorrelated but not conversely.) Now, if the equation does not hold, the difference between the terms is a measure of dependence between the variables $X$ and $Y$. This measure is called *covariance* and is defined by

$$\mathsf{Cov}[X,Y] = \mathsf{E}[X \cdot Y] - \mathsf{E}[X] \cdot \mathsf{E}[Y].$$

Obviously, $\mathsf{Cov}[Y,Y] = \mathsf{Var}[Y]$.

When one has two random variables, one often represents their variance and covariance in the form of a *covariance matrix*:

$$\mathsf{Cov}[X,Y;X,Y] = \begin{bmatrix} \mathsf{Var}[X] & \mathsf{Cov}[X,Y] \\ \mathsf{Cov}[X,Y] & \mathsf{Var}[Y] \end{bmatrix}. \tag{2.20}$$

Since $\mathsf{Cov}[aX,bY] = ab \cdot \mathsf{Cov}[X,Y]$, the covariance is scale dependent. A scale independent measure is the *correlation*, given by

$$\rho_{XY} = \frac{\mathsf{Cov}[X,Y]}{\mathsf{D}(X)\mathsf{D}(Y)},$$

where $\mathsf{D}(X) = \sqrt{\mathsf{Var}[X]}$. The correlation is always between one and minus one, i.e. $-1 \leq \rho_{XY} \leq 1$.

**Theorem 2.5.** *Let $X,Y$ be two random variables such that $|\rho_{XY}| = 1$. Then, there are constants $a,b$ (both not equal zero) such that $aX + bY = 0$, with probability one.*

However, not all functionally dependent variables $X,Y$ are perfectly correlated ($|\rho_{XY}| = 1$). For example, for $X \in \mathrm{N}(0,1)$, define $Y = X^3$. Obviously, if we know the outcome of the random experiment, then $X = x$ while $Y = x^3$. Now, the correlation between $X,Y$ is given by $\rho_{XY} = 3/\sqrt{15} < 1$.

#### 2.1.4.3 Conditional Distributions and Densities

The conditional probability "$B$ if $A$" was defined by (2.15) as

$$\mathsf{P}(B \mid A) = \mathsf{P}(A \cap B)/\mathsf{P}(A).$$

Consider now variables $X,Y$ having continuous distributions. We wish to find the conditional distribution $F(x \mid Y = y) = \mathsf{P}(X \leq x \mid Y = y)$. However, we face a problem since for a continuous variable $Y$, $\mathsf{P}(Y = y) = 0$ for all $y$. An easy solution to this problem can be found if $X,Y$ have the joint density $f(x,y)$. In such a case, we can define, if $f(y) > 0$,

$$f_{X|Y=y}(x) = f(x \mid y) = \frac{f(x,y)}{f(y)}. \tag{2.21}$$

Since, for a fixed value $y$, $f(x \mid y)$ as a function of $x$ integrates to one, it is a probability density function. Let us denote by $F(x \mid y)$ a distribution having as density $f(x \mid y)$, i.e. for any $x$

$$F(x \mid y) = \int_{-\infty}^{x} f(\tilde{x} \mid y)\,\mathrm{d}\tilde{x}. \tag{2.22}$$

Now, a combination of Equations (2.18), (2.21) and (2.22) leads to the following important result

$$\mathsf{P}(X \leq x) = F(x) = \int_{-\infty}^{\infty} F(x \mid y) f(y)\,\mathrm{d}y. \tag{2.23}$$

The last equation is a special case of the law of total probability given next.

**Theorem 2.6** (Total probability for continuous variables). *Assume that a random experiment renders values of a r.v. $Y$ and that we, in addition, are interested in any statement $B$, say, about the outcomes of the experiment. Then, for each $y$, there exists* *a probability* $\mathsf{P}(B \mid Y = y)$ *(for different y, the probabilities* $\mathsf{P}(B \mid Y = y)$ *may take different values), such that*

$$\mathsf{P}(B) = \int_{-\infty}^{\infty} \mathsf{P}(B \mid Y = y) f_Y(y)\,\mathrm{d}y. \tag{2.24}$$

*If* $X, Y$ *have joint density* $f(x,y)$*, then*

$$\mathsf{P}(B \mid Y = y) = \int_B f(x \mid y)\,\mathrm{d}x,$$

*where* $f(x \mid y)$ *is the conditional probability density defined by (2.21).*

#### 2.1.4.4 Functions of Random Variables

In this section, we list some basic tools for handling special functions of a finite number of random variables.

**Sums and differences**

$$\begin{aligned}
f_{X+Y}(z) &= \int_{-\infty}^{\infty} f_Y(y) f_{X|Y=y}(z-y)\,\mathrm{d}y \\
&= \int_{-\infty}^{\infty} f_Y(y) f_X(z-y)\,\mathrm{d}y, \quad \text{if } X, Y \text{ are independent}, \\
\mathsf{E}[X \pm Y] &= \mathsf{E}[X] \pm \mathsf{E}[Y], \\
\mathsf{Var}[X \pm Y] &= \mathsf{Var}[X] + \mathsf{Var}[Y] \pm 2\,\mathsf{Cov}[X,Y] \\
&= \mathsf{Var}[X] + \mathsf{Var}[Y], \quad \text{if } X, Y \text{ are uncorrelated.}
\end{aligned}$$

**Two multivariate variables,** $\mathbf{X} = (X_1, \ldots, X_m)$, $\mathbf{Y} = (Y_1, \ldots, Y_n)$ with $\mathsf{Cov}[X_i, X_j] = \sigma^x_{ij}$, $\mathsf{Cov}[Y_i, Y_j] = \sigma^y_{ij}$, $\mathsf{Cov}[X_i, Y_j] = \sigma^{XY}_{ij}$. Their covariance matrices are denoted

$$\boldsymbol{\Sigma}_{\mathbf{X}} = \begin{bmatrix} \sigma^X_{11} & \cdots & \sigma^X_{1m} \\ \vdots & \sigma^X_{ij} & \vdots \\ \sigma^X_{m1} & \cdots & \sigma^X_{mm} \end{bmatrix}, \quad \boldsymbol{\Sigma}_{\mathbf{Y}} = \begin{bmatrix} \sigma^Y_{11} & \cdots & \sigma^Y_{1n} \\ \vdots & \sigma^Y_{ij} & \vdots \\ \sigma^Y_{n1} & \cdots & \sigma^Y_{nn} \end{bmatrix},$$

and the cross-covariance matrix is

$$\boldsymbol{\Sigma}_{\mathbf{XY}} = \begin{bmatrix} \sigma^{xy}_{11} & \cdots & \sigma^{xy}_{1n} \\ \vdots & \sigma^{xy}_{ij} & \vdots \\ \sigma^{xy}_{m1} & \cdots & \sigma^{xy}_{mn}. \end{bmatrix}.$$

**Linear combinations** of **X** and **Y**: $U = \sum_i a_i X_i = \mathbf{a}\mathbf{X}^T$, $V = \sum_j b_j Y_j = \mathbf{b}\mathbf{Y}^T$, have

$$\begin{gathered}
\mathsf{Var}[U] = \mathbf{a}\boldsymbol{\Sigma}_{\mathbf{X}}\mathbf{a}^T, \quad \mathsf{Var}[V] = \mathbf{b}\boldsymbol{\Sigma}_{\mathbf{Y}}\mathbf{b}^T, \\
\mathsf{Cov}[U,V] = \mathbf{a}\boldsymbol{\Sigma}_{\mathbf{XY}}\mathbf{b}^T.
\end{gathered} \tag{2.25}$$

**Complex random variables:** $X = X_1 + \mathrm{i}X_2$, $Y = Y_1 + \mathrm{i}Y_2$ with $\mathsf{E}[X_1] = \mathsf{E}[X_2] = \mathsf{E}[Y_1] = \mathsf{E}[Y_2] = 0$,

$$\begin{aligned}\mathsf{E}[X] &= \mathsf{E}[X_1] + \mathrm{i}\mathsf{E}[X_2],\\ \mathsf{Var}[X] &= \mathsf{E}[X \times X^*] = \mathsf{E}[|X|^2] = \mathsf{E}[X_1^2] + \mathsf{E}[X_2^2],\\ \mathsf{Cov}[X,Y] &= \mathsf{E}[X \times Y^*] = \mathsf{E}[X_1Y_1] + \mathsf{E}[X_2Y_2] - \mathrm{i}\mathsf{E}[X_1Y_2] + \mathrm{i}\mathsf{E}[X_2Y_1].\end{aligned} \tag{2.26}$$

**Maxima and minima:** Extremes of a sequence of random variables play an important role in reliability analysis of random systems. If $X_1,\ldots,X_n$ are independent with the same continuous distribution $F(x)$, the distribution of their maximum and minimum is simply

$$\begin{aligned}\mathsf{P}(\max X_k \le x) &= F(x)^n,\\ \mathsf{P}(\min X_k \le x) &= 1 - (1 - F(x))^n.\end{aligned}$$

Unfortunately, as often when it comes to adverse influences, bad moments and accidents do not come isolated, and then one cannot treat the variables $X_1,\ldots,X_n$ as independent.

#### 2.1.4.5 Conditional Expectation and Variance

Conditional expectation and conditional variance are two simple but valuable statistics quantifying dependence. For example, monthly flood damage $X$ in private homes in an area depends to a great extent on the amount $Y$ of rain during the month. A useful measure is the average damage as a function of the amount of rain, in statistical terms, the "conditional expected damage given the amount of rain", viz. the expectation in the conditional distribution,

$$\mathsf{E}[X \mid Y = y] = m_{X|Y}(y) = \int x f_{X|Y=y}(x)\,\mathrm{d}x. \tag{2.27}$$

The conditional variance is defined in the analogous way as

$$\mathsf{Var}[X \mid Y = y] = \sigma^2_{X|Y}(y) = \int (x - m_{X|Y}(y))^2 f_{X|Y=y}(x)\,\mathrm{d}x. \tag{2.28}$$

Figure 2.2 illustrates the principle behind conditional mean and variance. The average damage increases, and the variation widens with increasing rainfall.

The conditional mean and variance are both functions of the conditioning variable $Y$. They are therefore random variables, and it is meaningful to ask about their expectation, variance and even their distribution.

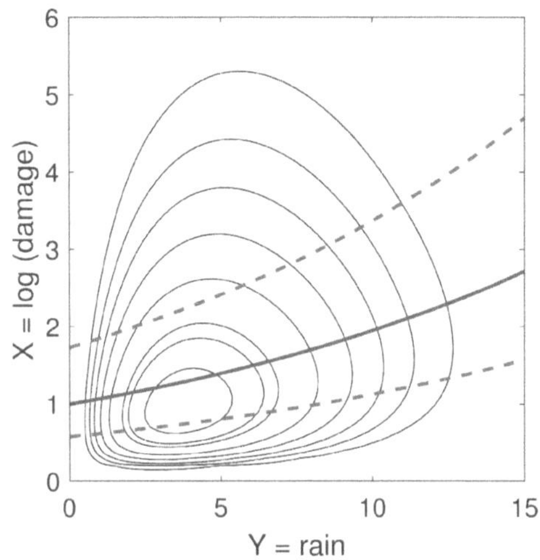


**FIGURE 2.2** Level curves of joint density of rainfall and damage (in the log scale). Solid curve: Conditional mean damage; dashed curves: Conditional mean ± one conditional standard deviation.

**Theorem 2.7** (Total mean and variance). *For a pair $(X,Y)$ of random variables, the total, unconditional, mean and variance of X are given by,*

$$\mathsf{E}[X] = \mathsf{E}[\mathsf{E}[X \mid Y]] = \int_y m_{X|Y}(y) f_Y(y)\,\mathrm{d}y, \tag{2.29}$$

$$\begin{aligned}\mathsf{Var}[X] &= \mathsf{E}[\mathsf{Var}[X \mid Y]] + \mathsf{Var}[\mathsf{E}[X \mid Y]] \\ &= \int_y \sigma^2_{X|Y}(y) f_Y(y)\,\mathrm{d}y + \int_y (m_{X|Y}(y) - m_X)^2 f_Y(y)\,\mathrm{d}y.\end{aligned} \tag{2.30}$$

*The first term in the expression for variance captures the variation left after "using Y to predict X", while the second term captures the variation due to the mean of the prediction.*

### 2.1.5 THE MULTIVARIATE NORMAL DISTRIBUTION

#### 2.1.5.1 Two-Dimensional Gaussian Distribution

We consider two Gaussian variables $X$ and $Y$ with distributions $N(m_X, \sigma_X^2)$, $N(m_Y, \sigma_Y^2)$, respectively. Their probability density functions are

$$f(x) = \frac{1}{\sqrt{2\pi}\sigma_X} e^{-\frac{1}{2\sigma_X^2}(x-m_X)^2}, \qquad f(y) = \frac{1}{\sqrt{2\pi}\sigma_Y} e^{-\frac{1}{2\sigma_Y^2}(y-m_Y)^2}. \tag{2.31}$$

If $X$ and $Y$ are independent, their joint probability density $f(x,y)$ is given by

$$f(x,y) = f(x)f(y) = \frac{1}{2\pi\sigma_X\sigma_Y} e^{-\frac{1}{2}\left(\frac{(x-m_X)^2}{\sigma_X^2} + \frac{(y-m_Y)^2}{\sigma_Y^2}\right)}. \tag{2.32}$$

**Example 2.1** (Rayleigh distribution). *We are ready to introduce the Rayleigh distribution, named after Lord Rayleigh and an important distribution for reliability analysis, in particular for the analysis of load cycles.*

*A Rayleigh distribution with scale parameter $\sigma$ has density and distribution functions*

$$
\begin{aligned}
f_R(z) &= \frac{z}{\sigma^2} e^{-z^2/(2\sigma^2)}, \quad z \geq 0, \\
F_R(z) &= 1 - e^{-z^2/(2\sigma^2)}, \quad z \geq 0,
\end{aligned}
$$

*and its mean and variance are*

$$\mathsf{E}[R] = \sigma\sqrt{\pi/2}, \quad \mathsf{Var}[R] = \sigma^2(4-\pi)/2.$$

*Note that* $\mathsf{E}[R^2] = \mathsf{Var}[R] + \mathsf{E}[R]^2 = \sigma^2(4-\pi)/2 + \pi/2 = 2\sigma^2$. *This result comes from the relation between a Rayleigh variable and two independent normal variables $X,Y$ with distribution $N(0,\sigma^2)$:*

$$\mathsf{P}(R > z) = \mathsf{P}(X^2 + Y^2 > z^2), \tag{2.33}$$

*i.e. the distance from the origin to a random point with coordinates $(X,Y)$ has a Rayleigh distribution with parameter $\sigma$.*

There are, at least, two natural ways to define the two-dimensional Gaussian distribution with dependent components. The straightforward one is to define a bivariate density $f(x,y)$ such that the marginal densities (2.16) are equal to normal marginals (2.31). The more sophisticated one is to use the following:

**Definition 2.1.** *The pair $(X,Y)$ of random variables is said to have a bi-Gaussian distribution if every combination $aX + bY$ with real $a,b$ has a uni-variate Gaussian distribution. The rules for mean and variance for sums of random variables give that $\mathsf{E}[aX + bY]$ has mean $am_X + bm_Y$ and variance*

$$\mathsf{Var}[aX + bY] = a^2\sigma_X^2 + b^2\sigma_Y^2 + 2ab\rho_{XY}\sigma_X\sigma_y.$$

Obviously, if $(X,Y)$ have a bi-variate normal distribution according to this definition, then each variable has the right normal distribution. However, it can also be shown that if $|\rho_{XY}| < 1$, the variables have a joint density

$$f(x,y) = \frac{1}{2\pi\sigma_X\sigma_Y\sqrt{1-\rho_{XY}^2}} e^{-\frac{1}{2}\left(\frac{(x-m_X)^2}{\sigma_X^2} + \frac{(y-m_Y)^2}{\sigma_Y^2} - 2\rho_{XY}\frac{(x-m_X)}{\sigma_X}\frac{(y-m_Y)}{\sigma_Y}\right)}. \tag{2.34}$$

Consequently, five parameters define the two-dimensional Gaussian distribution. These are $m_X$, $m_y$, $\sigma_X^2$, $\sigma_Y^2$, and $\rho_{XY}$. For the fact that $X,Y$ is Gaussian we use the expression

$$X,Y \in N(m_X, m_y, \sigma_X^2, \sigma_Y^2, \rho).$$

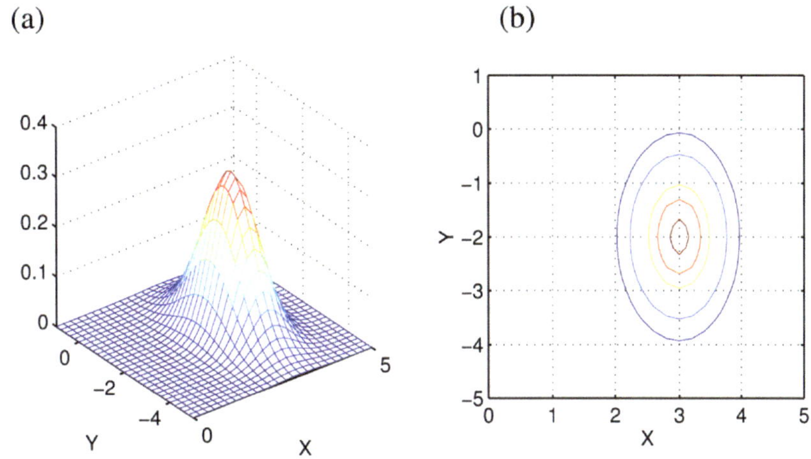


**FIGURE 2.3** Two-dimensional Gaussian distribution: $m_X = 3$, $\sigma_X = 0.5$, $m_Y = -2$, $\sigma_Y = 1$. Independent variables: $\rho = 0$. (a) Density function. (b) Contour lines.

An illustration of the two-dimensional Gaussian distribution will now be given. Let $m_X = 3$, $\sigma_X = 0.5$, $m_Y = -2$, $\sigma_Y = 1$. First, let $\rho = 0$. In Figure 2.3(a), the density function is shown, and in Figure 2.3(b), the corresponding contour lines are shown. Introducing a dependence between the variables by $\rho = 0.2$, the corresponding plots are shown in Figure 2.4. The effect of the dependence is clear from the contour plot of the latter case, where an increase in $X$ is seen to result in an increase in $Y$, indicating a positive correlation.

### 2.1.5.2 Some Properties of Gaussian Vectors

Consider a vector of r.v. $\mathbf{X} = (X_1, X_2, \ldots, X_n)^T$, which consists of $n$ different random variables. We say that $\mathbf{X}$ is a Gaussian vector if, for all real constants $a_1, a_2, \ldots, a_n$,

$$Y = \sum_{i=1}^{n} a_i X_i, \tag{2.35}$$

has a Gaussian distribution.

Let us denote the means by $m_i = \mathsf{E}[X_i]$, the variances by $\sigma_{ii} = \mathsf{Cov}[X_i, X_i]$ and the pairwise covariances by $\sigma_{ij} = \mathsf{Cov}[X_i, X_j]$. Obviously for any $i$, $X_i \in N(m_i, \sigma_{ii})$; just take $a_i = 1$, all other $a_j = 0$.

Let us introduce the notation

$$\mathbf{m} = (m_1, m_2, \ldots, m_n)^T, \qquad \mathbf{\Sigma} = \begin{bmatrix} \sigma_{11} & \ldots & \sigma_{1n} \\ \vdots & \sigma_{ij} & \vdots \\ \sigma_{n1} & \ldots & \sigma_{nn} \end{bmatrix}.$$

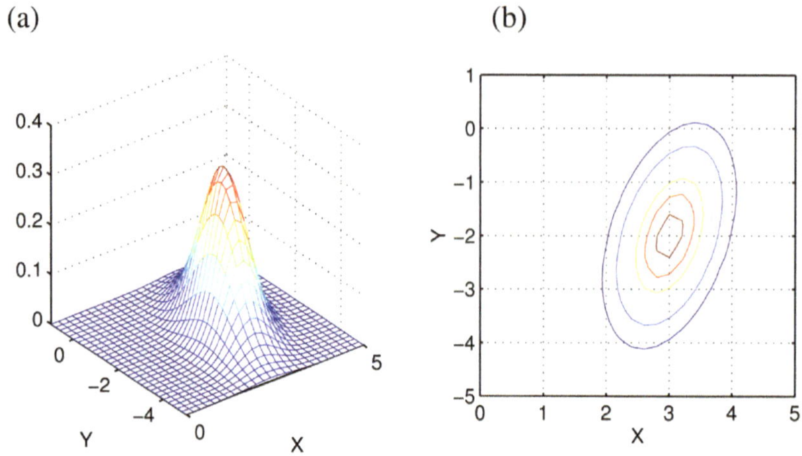


**FIGURE 2.4** Two-dimensional Gaussian distribution: $m_X = 3$, $\sigma_X = 0.5$, $m_Y = -2$, $\sigma_Y = 1$. Dependent variables: $\rho = 0.2$. (a) Density function. (b) Contour lines.

If the determinant $\det(\mathbf{\Sigma}) > 0$, then the distribution $\mathbf{X}$ has a density

$$f(\mathbf{x}) = \frac{1}{(2\pi)^{n/2}} \frac{1}{\det(\mathbf{\Sigma})^{1/2}} e^{-\frac{1}{2}(\mathbf{x}-\mathbf{m})^T \mathbf{\Sigma}^{-1}(\mathbf{x}-\mathbf{m})}, \tag{2.36}$$

at $\mathbf{x} = (x_1, x_2, \ldots, x_n)^T$. The statements "$\mathbf{X}$ is a Gaussian vector" or "$f$ is a Gaussian density" will be written as $\mathbf{X} \in N(\mathbf{m}, \mathbf{\Sigma})$, $f(\mathbf{x}) \in N(\mathbf{m}, \mathbf{\Sigma})$, respectively.

From (2.35), it follows that for any real-valued $(k,n)$-matrix $\mathbf{A}$, the vector $\mathbf{Y} = \mathbf{AX}$ is a $k$-dimensional Gaussian vector too. If $\mathbf{X} \in N(\mathbf{m}, \mathbf{\Sigma})$ then $\mathbf{Y} \in N(\mathbf{Am}, \mathbf{A\Sigma A}^T)$.

We present one more very useful result about the conditional Gaussian densities. Let $\mathbf{X}$ be structured into two vectors $\mathbf{Z} = (X_1, \ldots, X_k)^T$ and $\mathbf{Y} = (X_{k+1}, \ldots, X_n)^T$. With obvious notation, we write

$$\mathbf{m} = \begin{bmatrix} \mathbf{m_Z} \\ \mathbf{m_Y} \end{bmatrix}, \qquad \Sigma = \begin{bmatrix} \Sigma_{\mathbf{ZZ}} & \Sigma_{\mathbf{ZY}} \\ \Sigma_{\mathbf{YZ}} & \Sigma_{\mathbf{YY}} \end{bmatrix}.$$

Now, the conditional density $f_{\mathbf{Z}|\mathbf{Y}}(\mathbf{z}|\mathbf{y})$ of $\mathbf{Z}$, given $\mathbf{Y} = \mathbf{y}$, is Gaussian with the mean

$$\mathbf{m_{Z|Y}} = \mathbf{m_Z} + \mathbf{\Sigma_{ZY}\Sigma_{YY}^{-1}}(\mathbf{Y} - \mathbf{m_Y}), \tag{2.37}$$

and covariance matrix

$$\mathbf{\Sigma_{Z|Y}} = \mathbf{\Sigma_{ZZ}} - \mathbf{\Sigma_{ZY}\Sigma_{YY}^{-1}\Sigma_{YZ}}, \tag{2.38}$$

i.e. $f_{\mathbf{Z}|\mathbf{Y}}(\mathbf{z}|\mathbf{y}) \in N(\mathbf{m_{Z|Y}}, \mathbf{\Sigma_{Z|Y}})$.

## 2.2 RANDOM PROCESSES

### 2.2.1 BASIC DEFINITIONS

#### 2.2.1.1 A Random Process as a Family of Random Variables

A random variable identified by a symbol $X$ stands for a numerical value that can be measured at an experiment. A random process $X(\cdot) = \{X(t), t \in \mathscr{A}\}$ is similarly a concept that is used when one measures a whole function in the experiment. The *parameter* $t$ varies in the *domain* $\mathscr{A}$, and it can be one-dimensional. Usually, the independent parameter $t$ denotes time, and $X(t)$ is defined as a random process, with $\mathscr{A}$ being the time frame, e.g. $[0, T]$ over which one observes the process. The independent parameter could also represent a spatial coordinate, say $u$, varying over the spatial extent $\mathscr{A} = [0, L]$. The mathematical properties of a random process in time and in one-dimensional space are identical. The spatial parameter can be multi-dimensional $\mathbf{u}$, in which case $X(\mathbf{u})$ is referred to as a *random field*, varying over a spatial domain, e.g. $\mathscr{A} = \mathbb{R}^n$. When $\mathbf{u} = (u_1, u_2)$ is a coordinate in the plane, the field obviously represents a surface over $\mathscr{A}$. The parameter can combine time and space, $(t, u_1, u_2)$, to allow the surface to vary with time.

A random process with $\mathscr{A} = \mathbb{N}$ or $\mathbb{Z}$ is called a random *time series*. A realisation of a random process $\{X(t), t \in \mathscr{A}\}$ is denoted as $x(t), t \in \mathscr{A}$. Here, $x(t)$ is a deterministic function. Instead of "realisation", we often use "sample function".

We shall also use vector-valued processes, denoted by $\{\mathbf{X}(t), t \in \mathscr{A}\}$, i.e. we have a collection of random processes observed in a single experiment. The individual processes are allowed to develop more or less independently with time $t$. Common examples that we will meet are $(X(t), X'(t))$, i.e. a process and its time-derivative, and the input-output processes in a linear filter $(X(t), \mathscr{L}(X)(t))$.

#### 2.2.1.2 The Distribution of a Random Process

A random process $X(t)$ collectively represents an ensemble of sample functions $x_i(t)$. For a given $t$, one gets an ensemble of values $x_i$, which can be viewed as the realisations of a random variable, which has a distribution

$$F_t(x) = F_{X(t)}(x) = \mathsf{P}(X(t) \leq x), \tag{2.39}$$

that specifies what variability one can expect at time $t$ after the start of the experiment. Similarly, a pair $(X(t_1), X(t_2))$ of process values at two different time points $t_1, t_2$ has a bi-variate distribution $F_{t_1,t_2}(x_1, x_2)$, and so on for triples, $n$-tuples, etc., for finite $n$. All these distributions, with $t_1, \ldots, t_n \in \mathscr{A}$, form the *family of finite-dimensional distributions* for the process $X(\cdot)$.

In general, it is very difficult to specify the joint distribution of all finite collections of variables that constitute the process, e.g. $X(t_1), X(t_2), \ldots, X(t_n)$. In particular, one needs the *marginal distributions*, i.e. the distribution of each $X(t)$. Most random processes that are discussed in this book are specified by some simple criteria. Gaussian processes, Markov processes, auto-regressive time series, Poisson processes, etc. are all defined by their finite dimensional distributions.

**Definition 2.2** (Gaussian process). *A real random process is called Gaussian if every real finite linear combination of process values*

$$Y = a_1X(t_1) + a_2X(t_2) + \cdots + a_nX(t_n) \tag{2.40}$$

*has a Gaussian distribution; this is just a copy of the definition of the Gaussian distribution* (2.35).

The definition implies that in a Gaussian process, every $X(t)$ is Gaussian, every pair $(X(t_1), X(t_2)$ is bi-variate Gaussian, etc. When we come to derivatives and integrals, it can be shown that they are Gaussian too, when they exist.

## 2.2.2 PROPERTIES OF RANDOM PROCESSES

### 2.2.2.1 Moment Functions

Since every value of a stochastic process is a random variable $X(t)$, one can study its *moments*, like expectation, variance, and these generally depend on the specific time point $t$ at which we observe the process:

- mean function: $m(t) = \mathsf{E}[X(t)]$,
- (auto-)covariance function: $r(s,t) = \mathsf{Cov}[X(s), X(t)]$,
- (auto-)correlation function: $\rho(s,t) = r(s,t)/\sqrt{v(s)v(t)}$,
- variance function: $v(t) = \sigma^2(t) = \mathsf{Var}[X(t)] = r(t,t)$.

Similarly as there are random variables for which the expected value (and variance) do not exist, e.g. $Z = X/Y$, where $X, Y$ are independent $N(0,1)$ variables, not all processes have finite moments. However, here, we shall only discuss processes with finite variance. The moments only partially specify the properties of the random process, e.g. in the case of random variables one may have two variables having the same expectation and variance but different distributions. Despite this fact, many important properties of a random process depend on its moments, in particular on its covariance function. In the following we shall mostly study processes with continuous time with $\mathscr{A} = \mathbb{R}$ or $\mathbb{R}^+$. Then, it seems intuitive that the correlation function should give information on the smoothness of the process, e.g., its continuity or differentiability.

To avoid notational complications, we often restrict our interest to a class of processes that do not change their statistical properties with time, called *stationary processes*. For more details about stationary processes, we refer to the classical work by Cramér and Leadbetter (1967) [43] and to the recent textbooks [130], with the basics on stationary processes and their use in dynamic systems in engineering and [125] with a more mathematical treatment.

**Example 2.2.** *The distribution of a Gaussian process is completely defined by its mean function and covariance function.*

#### 2.2.2.2 Stationary Processes ...

A process $X(\cdot)$ is called *stationary* if, for any $\tau$, the vectors $X(t_1), X(t_2), \ldots, X(t_n)$, and $X(t_1+\tau), X(t_2+\tau), \ldots, X(t_n+\tau)$ have the same distribution.[1] In plain language, this means that we observe the same statistical variability wherever we decide to start observing the process, as long as we don't let the process influence where we start.

To see the necessity of the restriction, just think of a situation when $\tau > 0$ is the time one has to wait to the first time after $t_1$ when the process crosses the level 0. Then $X(t_1+\tau) = 0$ but $X(t_1)$ probably is not.

Obviously, for stationary processes, the mean and variance functions are constant $m(t) = m$, $v(t) = \sigma^2$, and there is a function $r_X(\tau)$, also called the *covariance function*, such that, for all $t$,

$$r_X(\tau) = r(t, t+\tau) = \mathsf{Cov}[X(t), X(t+\tau)]. \tag{2.41}$$

Obviously $r_X(0) = \sigma^2$, $r_X(-\tau) = r_X(\tau)$, and $r_X(0) \geq |r_X(t)|$ for $t \neq 0$.

**Example 2.3.** *Linear combinations of sample values appear frequently in engineering applications. We give the expression for the variance of such a combination*

$$U = a_1 X(t_1) + a_2 X(t_2) + \ldots + a_n X(t_n)$$

*when the process is stationary.*

*Since* $\mathsf{Cov}[X(t_j), X(t_k)] = r_X(t_k - t_j)$*, the vector* $\mathbf{X} = (X(t_1), \ldots, X(t_n))$ *has the covariance matrix*

$$\Sigma = \begin{bmatrix} r_X(0) & r_X(t_2-t_1) & \ldots & r_X(t_n-t_1) \\ r_X(t_1-t_2) & r_X(0) & \ldots & r_X(t_n-t_2) \\ \vdots & \vdots & \ddots & \vdots \\ r_X(t_1-t_n) & r_X(t_2-t_n) & \ldots & r_X(0) \end{bmatrix}.$$

*From (2.25), we learn that*

$$\mathsf{Var}[U] = \mathbf{a}\Sigma\mathbf{a}^T = \sum_{jk} a_j a_k r_X(t_j - t_k), \tag{2.42}$$

*so the right-hand-side double sum is necessarily non-negative, since* $\mathsf{Var}[U] \geq 0$.

In the same way, as one has introduced a vector of random variables, one often needs to consider a vector of random processes $X_1(t), \ldots, X_n(t)$. We then introduce an extra moment function to describe the relation between the components,

- cross-covariance function: $r_{ij}(s,t) = \mathsf{Cov}[X_i(s), X_j(t)]$,

For example, for a smooth process $X$, let $(X_1(t) = X(t), X_2(t) = \dot{X}(t))$ form a vector of dependent processes. Here, we shall be mostly interested in the class of Gaussian vector valued processes. Those are defined by specifying the mean vector and the matrix of cross-covariance functions.

$$\mathbf{m}(t) = (m_1(t), m_2(t), \ldots, m_n(t))^T, \quad \Sigma(t,s) = \begin{pmatrix} r_{11}(t,s) & \ldots & r_{1n}(t,s) \\ \vdots & r_{ij}(t,s) & \vdots \\ r_{n1}(t,s) & \ldots & r_{nn}(t,s) \end{pmatrix},$$

where $r_{ij}(t,s) = \mathsf{Cov}[X_i(t), X_j(s)]$. For stationary processes, $m_i(t) = m$ and $r_{ij}(t,s) = r_{ij}(s-t)$.

#### 2.2.2.3 ... and Their Energy Spectra

It is often difficult to visually comprehend a covariance function and interpret its relation to the realisations of the process. On the other hand, the *energy spectrum* of the process, also called the *power spectrum*, has many advantages in engineering applications.

**Definition 2.3** (Spectral distribution). *Let $r_X(\tau)$ be the covariance function of a real stationary process in continuous time with finite variance. If $r_X(\tau)$ is continuous at $\tau = 0$, then there exists a* spectral distribution function $F(\omega)$ *with the properties*

$$r_X(\tau) = \int_{-\infty}^{\infty} e^{i\tau\omega}\, \mathrm{d}F(\omega), \tag{2.43}$$

$$F(\infty) - F(-\infty) = r_X(0) = \mathsf{Var}[X(t)] < \infty, \tag{2.44}$$

$$\mathrm{d}F(\omega) = \mathrm{d}F(-\omega) \geq 0. \tag{2.45}$$

The spectral distribution function shares many properties with a statistical distribution function for a symmetric distribution: It is non-decreasing with symmetric increases (2.45) and finite. It is not normalised however, as a statistical distribution.

For most of the load processes we will encounter in this monograph, the spectral distribution is *absolutely continuous* with a *spectral density*

$$\widetilde{S}(\omega) = \frac{d}{d\omega} F(\omega),$$

similar to a probability density. It is non-negative, symmetric and has finite integral, equal to $\mathsf{Var}[X(t)] < \infty$. In that case,

$$\begin{aligned} r_X(\tau) &= \int_{-\infty}^{\infty} e^{i\tau\omega}\, \widetilde{S}(\omega)\, \mathrm{d}\omega, \\ r_X(0) &= \int_{-\infty}^{\infty} \widetilde{S}(\omega)\, \mathrm{d}\omega. \end{aligned} \tag{2.46}$$

Equation (2.46) explains the name "energy spectrum": The total energy = squared variation, of the process emanates from harmonic complex components $e^{i\tau\omega}$ with frequencies $\omega$, each contributing in proportion to the spectral density $\widetilde{S}(\omega)$. By (2.45), it is clear the $\widetilde{S}(\omega)$ is symmetric, $\widetilde{S}(-\omega) = \widetilde{S}(\omega)$, and in this work, we will use the one-sided spectrum $S(\omega) = 2\widetilde{S}(\omega)$ for $\omega > 0$. Then, (2.46) is replaced by

$$\begin{aligned} r_X(\tau) &= \int_0^{\infty} \cos(\tau\omega)\, S(\omega)\, \mathrm{d}\omega, \\ r_X(0) &= \int_0^{\infty} S(\omega)\, \mathrm{d}\omega. \end{aligned} \tag{2.47}$$

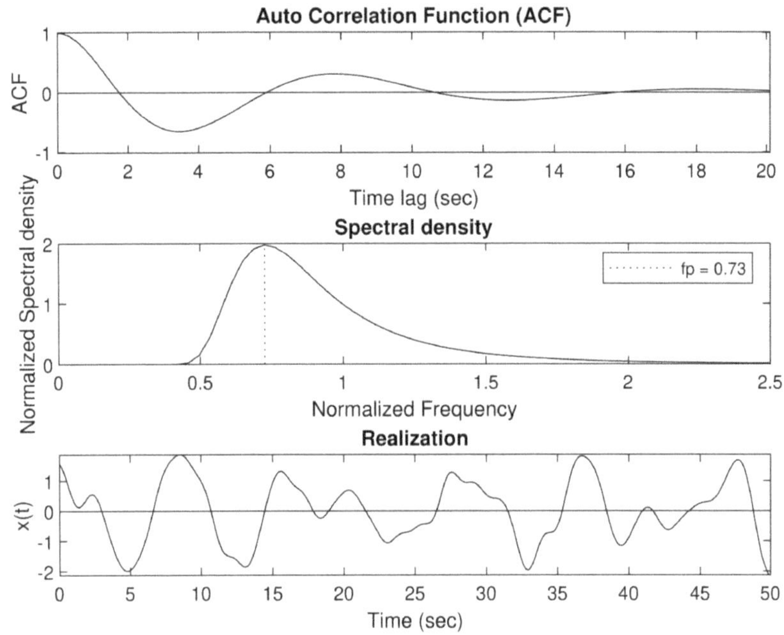


**FIGURE 2.5** Correlation function, spectral density and realisation for Gaussian sea way with Pierson-Moskowitz spectrum.

If the spectrum has a unique maximum at frequency $f_p = \omega_p/2\pi > 0$, it is called the *peak frequency*, and the inverse $t_p = 2\pi/f_p$ is the *peak period*.

Figures 2.5 and 2.6 illustrate the relation between auto correlation function (ACF), spectral density and process behaviour. The Pierson-Moskowitz spectrum in Figure 2.5 is concentrated around a central frequency, the correlation function oscillates with period related to the peak frequency and the time history is also rather oscillating. The diffusion spectrum in Figure 2.6 starts at low frequencies and decreases slowly, the correlation time is rather short and no regularities can be found in the realisation, even if it seems quite smooth compared with Figure 2.5.

If $\int_{-\infty}^{\infty} |r(\tau)|\, \mathrm{d}\tau < \infty$, the spectral density exists and is given by the *Fourier inversion formula*

$$\widetilde{S}(\omega) = \frac{1}{2\pi}\int_{-\infty}^{\infty} e^{-i\omega\tau} r(\tau)\, \mathrm{d}\tau. \tag{2.48}$$

The *spectral moments* bring simple information about the smoothness of the process. They are defined as the un-normalised moments of the spectral distribution and are given by

$$\omega_k = 2\int_0^{\infty} \omega^k\, \mathrm{d}F(\omega) = \int_0^{\infty} \omega^k S(\omega)\, \mathrm{d}\omega, \tag{2.49}$$

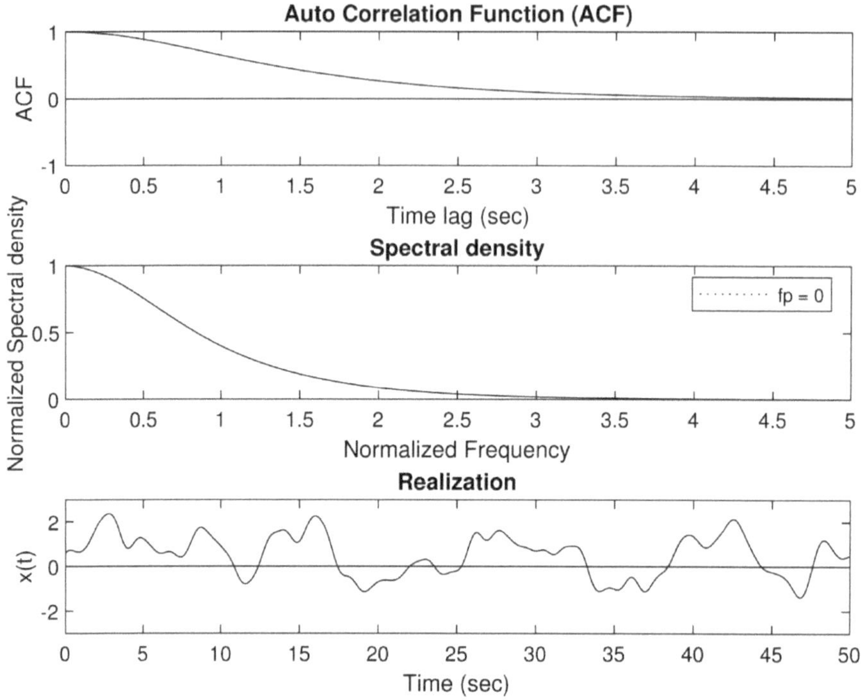


**FIGURE 2.6** Correlation function, spectral density and realisation for Gaussian stationary diffusion of order 2.

where the last expression holds for the one-sided spectral density, when it exists, with $\omega_k$ representing the $k$-th moment.

### 2.2.3 FUNCTIONS OF RANDOM PROCESSES

In this section, we describe how to handle differentiation and integration of random processes and how to use covariance function and spectrum to solve questions about existence and statistical properties. We will also briefly describe how the statistical properties change when a random process is passed through a linear filter.

#### 2.2.3.1 Derivatives: Existence and Properties

When we say that a process is continuous, we mean that the sample functions are continuous functions, and when we say that the process is differentiable we mean that all sample functions are differentiable. In rainflow analysis, local maxima and minima are identified as zero crossings of the load process derivative with negative and positive second derivative.

We restrict the analysis to stationary processes and assume that the covariance function $r_X(\tau)$ and the spectral density $S(\omega)$ exist. It is then possible to formulate simple conditions that guarantee that the sample functions are continuous, differentiable, twice differentiable and so on.[2]

**Example 2.4.** *Any stationary process with*

$$r_X(\tau) = r_X(0) - C|\tau|^a + o(|\tau|^a) \tag{2.50}$$

*for some constant* $C > 0$ *as* $\tau \to 0$ *has continuous sample paths if* $1 < a \leq 2$.

*If the process is stationary and Gaussian, it has continuous sample paths as long as* (2.50) *holds, with* $0 < a \leq 2$.

The continuity condition in the example has an intuitive interpretation: In a continuous random process, values close in time must also be close in space, and (2.50) says that they are indeed very correlated.

In order that a stationary process has continuously differentiable sample functions, it is necessary that its covariance function has a smooth higher order Taylor expansion. To make a long story short, we extend Example 2.4 and include also the spectrum.

**Example 2.5.** *A stationary process* $X(t), t \in \mathbb{R}$, *is continuously differentiable if its covariance function is twice differentiable everywhere, and*

$$-r_X''(\tau) = -r_X''(0) - C|\tau|^a + o(|\tau|^a) \tag{2.51}$$

*for some constant* $C > 0$ *and* $1 < a \leq 2$. *If* $X(t)$ *is Gaussian, it suffices that* $0 < a \leq 2$.

The derivative process $X'(t)$ is also stationary and its covariance function and spectral density are

$$r_{X'}(t) = -r_X''(t), \quad S_{X'}(\omega) = \omega^2 S_X(\omega). \tag{2.52}$$

Higher order derivatives are treated in an analogous way. Further, in terms of spectral moments (2.49),

$$\mathsf{Var}[X(t)] = r_X(0) = \omega_0, \quad \mathsf{Var}[X'(t)] = -r_X''(0) = \omega_2, \quad \mathsf{Var}[X''(t)] = r_X^{iv}(0) = \omega_4.$$

**Example 2.6** (Gaussian process and its derivative)**.** *If* $X(t)$ *is a differentiable process with mean* $m(t)$ *and covariance function* $r_X(s,t)$ *the derivative has mean* $m'(t)$ *and covariance function*

$$r_{\dot{X}}(s,t) = \frac{\partial^2}{\partial s \partial t} r_X(s,t).$$

*If* $X(t)$ *is stationary, the derivative is stationary with mean* 0 *and the covariance function simplifies to* $r_{X'}(t) = -r_X''(t)$. *The cross-covariance between the process and its derivative is*

$$r_{X,X'}(s,t) = \mathsf{Cov}[X(s), X'(t)] = \frac{\partial}{\partial t} r_X(s,t), \tag{2.53}$$

*which for a stationary process simplifies to* $r_{X,X'}(t) = \mathsf{Cov}[X(0), X'(t)] = r_X'(t)$.

*Note that* $r_X'(0) = 0$ *for every stationary process, so* $X(t), X'(t)$ *are always uncorrelated. A Gaussian process is jointly Gaussian with its derivative process, and we draw the conclusion that the value and derivative at time t are independent.*

### 2.2.3.2 Integration

Random fatigue failure is the result of damage accumulation caused by a varying stochastic load. To quantify the random effect of the accumulation, we may want to compute mean, variance and other properties of the integral of a random process.

Let $X(t)$ be a random process with mean function $m(t) = \mathsf{E}[X(t)]$ and covariance function $r_X(s,t) = \mathsf{Cov}[X(s),X(t)]$ and assume that its sample functions are piece wise continuous with only a finite number of jump discontinuities in the interval $[a,b]$. Then, the integral $\int_a^b X(t)\,\mathrm{d}t$ is a random variable with mean and variance

$$\mathsf{E}\left[\int_a^b X(t)\,\mathrm{d}t\right] = \int_a^b m(t)\,\mathrm{d}t,$$
$$\mathsf{Var}\left[\int_a^b X(t)\,\mathrm{d}t\right] = \int_{t=a}^b \int_{s=a}^b r(s,t)\,\mathrm{d}s\,\mathrm{d}t.$$

If the process is stationary, then, with $T = b-a$,

$$\mathsf{E}\left[\int_a^b X(t)\,\mathrm{d}t\right] = m(b-a),$$
$$\mathsf{Var}\left[\int_a^b X(t)\,\mathrm{d}t\right] = \int_{t=a}^b \int_{s=a}^b r(t-s)\,\mathrm{d}s\,\mathrm{d}t$$
$$= 2T\int_{s=0}^T (1-s/T)r(s)\,\mathrm{d}s. \tag{2.54}$$

The variance formulas are continuous versions of (2.25) and (2.42). Note that the integral of a Gaussian process is a normal random variable.

### 2.2.3.3 Random Processes in a Linear Filter

Safety analysis of structural mechanical systems often includes a statistical analysis of the effect of feeding the system with a random process as input. The system acts as a *filter* between the environmental load and the local strain history at critical details in the structure. In this section, we describe the simplest such system, a *linear time invariant system* when the input signal is a stationary random process.

A linear time invariant filter is a transformation $\mathscr{L}$ that takes an input signal $X(\cdot)$ into an output signal $Y = \mathscr{L}(X)$, such that for all constants $c_1, c_2$ and time delay $\tau$,

$$\mathscr{L}(c_1X_1 + c_2X_2) = c_1\mathscr{L}(X_1) + c_2\mathscr{L}(X_2),$$
$$\mathscr{L}(X(\cdot+\tau)) = \mathscr{L}(X)(\cdot+\tau).$$

**Definition 2.4** (Frequency response, impulse response). *The frequency response of a linear time invariant filter $\mathscr{L}$ is a real or complex valued function $H(\omega)$ that describes how the filter acts on the complex harmonic $e^{i\omega t}$,*

$$\mathscr{L}(e^{i\omega t}) = H(\omega)e^{i\omega t}.$$

*The impulse response is the Fourier inverse of* $H(\omega)$,

$$h(u) = \frac{1}{2\pi}\int_{-\infty}^{\infty} H(\omega)e^{-i\omega u}\, \mathrm{d}\omega,$$

*which expresses the input-output relation as the convolution*

$$Y(t) = \int_{-\infty}^{\infty} h(t-u)X(u)\, \mathrm{d}u.$$

If $h(u) = 0$ for $u < 0$, the output $Y(t)$ does not depend in a causal way on the future input $X(u)$, $u > t$, and the filter is called "causal". However, there may very well be a statistical dependence, fundamental for all statistical forecasting.

Below are some examples of linear time invariant filters with their frequency response:

- Differentiation $Y(t) = X'(t)$ with frequency response $H(\omega) = i\omega$;
- Time shift $Y(t) = X(t-\tau_0)$ with $H(\omega) = e^{-i\omega\tau_0}$;
- Exponential smoothing $Y(t) = \alpha\int_{-\infty}^{t} e^{-\alpha(t-s)}X(s)\,\mathrm{d}s$ with $H(\omega) = \frac{\alpha}{\alpha+i\omega}$.

When a stationary process $X(t)$ passes through a linear time invariant filter, its energy spectrum changes according to the frequency response function. The output process $Y(t)$ is stationary and the spectral density is

$$S_Y(\omega) = |H(\omega)|^2 S_X(\omega). \tag{2.55}$$

Writing $H(\omega) = |H(\omega)|e^{i\rho(\omega)}$, we get the *amplitude response* $|H(\omega)|$ and the *phase response* $\rho(\omega)$.

Relation (2.55) implies that the phase response has no implication for the energy spectrum of the output process. The phase response is an important quantity that describes possible time-delay between input and output processes.

### 2.2.4 STATIONARY GAUSSIAN PROCESSES

From Example 2.2, we draw the conclusion that the distribution of a stationary Gaussian process is determined by its mean value $m$ and its covariance function $r(\tau)$. Without going into details about the limiting arguments, we also claim that derivatives and integrals of Gaussian processes all have normal distributions. The Gaussian property is also retained when a Gaussian process is filtered through a linear filter.

In this section, we introduce some special topics, much used in the coming chapters. The first is a recipe for how to make a Monte Carlo simulation of a stationary process from its spectral density. The second is a brief introduction to level crossing and extremal properties, some for general use and some restricted to Gaussian processes.

#### 2.2.4.1 Simulation of a Stationary Gaussian Process

Gaussian load processes appear frequently in examples and illustrations in this book. Monte Carlo simulation of samples from such processes can be done either from its covariance function or from its spectrum, the choice depends on type of application and available software.

Direct simulation through circulant embedding of the covariance matrix $\Sigma$ in Example 2.3 is fast and exact and simply generates a multivariate normal vector with the correct covariance structure. The method requires software that can cope with very large and near-singular matrices.

When the focus is on the energy spectrum of the process, a Fourier simulation by summation of cosine functions with random amplitudes and the random phase angle can be preferred. These functions have the form

$$\begin{aligned} X(t) &= a\sin\omega t + b\cos\omega t \\ &= \sqrt{a^2+b^2}\left(\frac{a}{\sqrt{a^2+b^2}}\sin\omega t + \frac{b}{\sqrt{a^2+b^2}}\cos\omega t\right) = A\cos(\omega t - \phi), \quad (2.56) \end{aligned}$$

where $A = \sqrt{a^2+b^2}$ is the amplitude and $\phi$ is a phase angle $\sin\phi = a/A$, $\cos\phi = b/A$. If $a, b$ are independent random and normal variables with mean zero and equal variance $\sigma^2$, $X(t)$ becomes a very simple stationary Gaussian process. According to (2.33), the amplitude $A$ has a Rayleigh distribution with parameter $\sigma$ and further, the phase will be uniformly distributed over $[0, 2\pi)$ and independent of $A$. The process (2.56) has mean zero and variance $\mathsf{Var}[X(t)] = \mathsf{E}[A^2/2] = \sigma^2$. It has all its energy from the single frequency $\omega$.

To get a good approximation of a process with spectral density $S(\omega), \omega > 0$, one can add many independent $X_k(t)$ with frequencies $\omega_k = k\Delta, k = 1, \ldots, N$, and $\sigma_k^2 = S(\omega_k)\Delta$,

$$X^{appr}(t) = \sum_{k=1}^{N} A_k \cos(\omega_k t - \phi_k). \quad (2.57)$$

The variance of $X(t)$ is

$$\mathsf{Var}[X^{appr}(t)] = \sum_{k=1}^{N} \mathsf{Var}[A_k^2]/2 = \sum_{k=1}^{N} S(k\Delta)\Delta \approx \int_0^{N\Delta} S(\omega)\, \mathrm{d}\omega.$$

The approximating process $X^{appr}(t)$ has a discrete spectrum with the continuous energy spectrum truncated and lumped to the discrete frequencies $\omega_k$. A good approximation requires $\Delta$ small and $N\Delta$ large to capture all energy.

**Example 2.7** (Illustration of phase and amplitude response)**.** *We illustrate the effect of a linear filter with frequency response $H(\omega) = |H(\omega)|e^{ip(\omega)}$ on the process*

$X^{appr}(t) = \Re \sum_1^N A_k e^{i(\omega_k t - \phi_k)}$,

$$\mathscr{L}\left(\sum_1^N A_k e^{i(\omega_k t - \phi_k)}\right) = \sum_1^N |H(\omega_k)| A_k e^{i(\omega_k t - \phi_k + \rho(\omega_k))}$$

$$= \sum_1^N |H(\omega_k)| A_k \cos(\omega_k t - \phi_k + \rho(\omega_k)) + i \sum_1^N |H(\omega_k)| A_k \sin(\omega_k t - \phi_k + \rho(\omega_k)).$$

*The terms in the output, the real part, have their amplitudes changed by a factor* $|H(\omega_k)|$ *with a phase shift* $\rho(\omega_k)$ *compared to the input.*

#### 2.2.4.2 Level Crossings and Extremes: Rice's Formula

A fundamental variable in structural reliability studies is "time to first crossing", i.e. what is the distribution of the waiting time before the stress exceeds the strength for the first time? For fatigue damage analysis, the main topic in this book, the sequence of local maxima and minima in a load process is the dominating statistical factor.

We first define the important crossing and extreme value variables and then present some statistical characteristics. We assume that the studied process $X(t)$ is stationary with continuous sample paths, when needed also differentiable.

**Definition 2.5** (Level crossings and local extremes).

*(a) Counts:*

$$\begin{aligned} N_T(u) &= \textit{number of times the stationary and differentiable process } X(t) \\ &\qquad \textit{crosses the level } u \textit{ in the interval } [0,T], \\ N_T^+(u) &= \textit{number of times } X(t) \textit{ crosses the level } u \textit{ in the upward direction}, \\ N_T^{max} &= \textit{number of local maxima of } X(t) \textit{ in } [0,T]. \end{aligned}$$

*(b) Intensities:*

$$\begin{aligned} \mu(u) &= \textit{crossing intensity, expected number of crossings of level } u \\ &\qquad \textit{per time unit, } \mathsf{E}[N_1(u)], \\ \mu^+(u) &= \textit{upcrossing intensity, upcrossing spectrum, expected number of} \\ &\qquad u\textit{-upcrossings per time unit, } \mathsf{E}[N_1^+(u)], \\ k^+(u,v) &= \min_{u \le z \le v} \mu^+(z) = \textit{minimal upcrossing spectrum}, \\ \mu_{max} &= \textit{expected number of local maxima per time unit, } \mathsf{E}[N_1^{max}]. \end{aligned}$$

In the fatigue analysis in later chapters, the number of observed level upcrossings will be the most simple damage variable. It is called the *(up)crossing spectrum* and denoted

$$N_T^+(u) \quad \text{with expectation} \quad \mu_T^+(u) = \mathsf{E}[N_T^+(u)]. \tag{2.58}$$

The average number of upcrossings of the mean level $m$ per time unit is the *apparent frequency*,

$$f_z = \mu_1^+(m). \tag{2.59}$$

**Theorem 2.8** (Rice's formula and a simple consequence). *Early versions of the following formulas for the expected number of crossings and extremes were shown by Steve O. Rice in [168]. They appear frequently in safety analysis based on extremes and crossings:*

$$\mu(u) = \mathsf{E}[N_1(u)] = f_{X(0)}(u)\,\mathsf{E}[|X'(0)| \mid X(0) = u], \tag{2.60}$$

$$\mu^+(u) = \mathsf{E}[N_1^+(u)] = f_{X(0)}(u)\,\mathsf{E}[X'(0)^+ \mid X(0) = u], \tag{2.61}$$

$$\mu_{max} = \mathsf{E}[N_1^{max}] = f_{X'(0)}(0)\,\mathsf{E}[X''(0)^- \mid X'(0) = 0], \tag{2.62}$$

*where* $a^+ = \max(0,a)$, $a^- = -\min(0,a)$.

*The distribution of* $T = \min\{t; X(t) \geq u\}$ *can be bounded by the upcrossing intensity. Since the event* $\{T \leq t\}$ *is equivalent to* $\{X(0) \geq u \text{ or } N_t^+(u) > 0\}$, *we have the simple conservative bound for the risk of level exceedance before time t,*

$$\begin{aligned} \mathsf{P}(T \leq t) &= \mathsf{P}(X(0) \geq u \text{ or } N_t^+(u) > 0) \leq \mathsf{P}(X(0) > u) + \mathsf{P}(N_t^+(u) > 0) \\ &\leq \mathsf{P}(X(0) > u) + \mathsf{E}[N_t^+(u)]. \end{aligned}$$

**Example 2.8** (Rice's formula for Gaussian processes). *For a differentiable stationary Gaussian process, the process and its derivative at time t are independent, so* $\mathsf{E}[X'(0)^+ \mid X(0) = u] = \mathsf{E}[X'(0)^+] = \sigma_{X'}/\sqrt{2\pi}$. *With* $\sigma_X^2 = \omega_0 = \mathsf{Var}[X(t)]$, $\sigma_{X'}^2 = \omega_2 = \mathsf{Var}[X'(t)]$, $\mathsf{E}[X(t)] = m$, *the expected number of u-level upcrossings and maxima per time unit are, respectively,*

$$\mu^+(u) = \frac{\sigma_{X'}}{2\pi\sigma_X} e^{-(u-m)^2/(2\sigma_x^2)}, \quad \mu_{max} = \frac{\sigma_{X''}}{2\pi\sigma_{X'}}. \tag{2.63}$$

The average number of level crossings and upcrossings per time unit in a stationary process are called *crossing intensities*. We will meet the term *intensity* in many shapes in the coming chapters, not only related to level crossings. The most important of these is the *damage intensity*. It is the result of a combination of a *cycle intensity* $\mu$, defined as the average number of cycles per time unit, and the damage increase caused by each cycle. To each cycle, there is then defined a *cycle damage*, generally denoted $\delta$ and determined by the cycle properties. The average damage per time unit is the *damage intensity*,

$$d = \mu \times \mathsf{E}[\delta]. \tag{2.64}$$

Each cycle definition has its specific cycle intensity depending on the load, while the damage intensity also involves material properties. In Chapter 4, we analyse the *rainflow damage intensity*, denoted $d^{rfc}$, in detail, and also the *min-max damage intensity*, $d^{mM}$.

**Example 2.9** (Sample function irregularity). *One motivation for the use of the rainflow cycle counting technique compared to min-max cycle counting is that it reduces the fatigue influence of possible small oscillations superimposed on large dangerous cycles. To measure the amount of that phenomenon, one can use the ratio* $\alpha_2 = \mu^+(0)/\mu_{max}$, *i.e. the inverse of the average number of local maxima per mean level upcrossing.*

*If* $\alpha_2$ *is almost one, the process is very regular with only one maximum between mean level upcrossings. Such a process will be called* narrow-banded. *A value near zero indicates an excess of min-max cycles. For a twice differentiable Gaussian process,* $\alpha_2 = \frac{\omega_2}{\sqrt{\omega_0 \omega_4}}$.

**Remark 2.2** (Measures of spectral width). *The parameter* $\alpha_2 = \frac{\omega_2}{\sqrt{\omega_0 \omega_4}}$ *is an example of a spectral width measure, related to the regularity of fatigue loads.*

*Another measure is* $\alpha_1 = \frac{\omega_1}{\sqrt{\omega_0 \omega_2}}$, *which is related to the length of wave groups. A general width measure that is used in the analysis of rainflow damage is*

$$\alpha_m = \frac{\omega_m}{\sqrt{\omega_0 \omega_{2m}}}, \tag{2.65}$$

*defined also for non-integer m.*

**Remark 2.3** (A note on notation). *In this book, we will have use for alternative notations of counts and intensities/expectations. For example, we use* $\mu_T^+(u)$ *for the expected number of level u upcrossings in a interval of length T in case the process is stationary and our interest is on the u-dependence. For a non-stationary process with crossing intensity varying with time t, we write* $\mu_u^+(t)$ *and integrate over t to get the expectation. Exceptions from these principles occur and are locally defined.*

## 2.3 RICE'S FORMULA, HISTORY AND PROOFS

### 2.3.1 BACKGROUND

A.J. Rainal, colleague of Steve O. Rice at AT&T Bell Laboratories, writes in his 1988 paper *Origin of Rice's formula* [164]:

> The origin of Rice's formula for the average rate of level crossings of a rather general random process is traced to his 1936 notes entitled "Singing Transmission Lines." This celebrated formula and its generalisations are now basic for the analysis of level crossings and first and second passage times of random processes.

The engineering problem in the Singing Lines notes [165] was the statistical distribution of the maximum of a random signal in an interval following an already high value. In the published paper [166], Rice presented his formula for the average rate of maxima, i.e. for the expected number of zero downcrossings by the signal derivative. Rice informal argument for the formula is in principle the same as presented in Section 2.3.2 in strict form.

Rice's impact on the telecommunication industry comes from his monumental work *Mathematical analysis of random noise* in two parts 1944 and 1945, [167, 168]. Within a decade, the idea that environmental loads and disturbances should be modelled as random processes, characterised by correlation and frequency spectrum, and described by statistical distributions, had been accepted in many other engineering sciences and technologies.

Ocean engineering and Naval architecture were among the first with the, to begin with, controversial paper *On the movement of ships in confused seas* by St Denis and Pierson 1953 [195]. Automatic control became stochastic control, structural vibration incorporated random vibration, aero and auto industry learned about Rice's formula. The role of Rice's work at Bell Laboratories in this revolution can hardly be overestimated.

Around the time of Rice's work on maximum of a random process, interest in the number of zeros of random polynomials started to grow in probability and theoretical physics. Marc Kac 1942 [97] designed a simple zero-counting algorithm which also can give the average number of zeros. The Kac method became standard in physics but was replaced in statistics by the Rice method with the advent of the book on stationary processes by Cramér and Leadbetter 1967 [43].

We present three different proofs for the Kac-Rice formula for the average number of level crossings.

### 2.3.2 RICE'S PROOF AND INTERPRETATION

The proof is from [43].

**Theorem 2.9.** *For a stationary process $X(t)$ with continuous sample paths, suppose that $X(0)$ and $\zeta_n = 2^n(X(1/2^n) - X(0))$ have a joint density $g_n(u,z)$, continuous in $u$ for all $z$ and all sufficiently large $n$. Also suppose $g_n(u,z) \to p(u,z)$ uniformly in $u$ for fixed $z$ as $n \to \infty$ and that $g_n(u,z) \le h(z)$ with $\int_0^\infty z h(z)\,\mathrm{d}z < \infty$. Then, the expected number of $u$-upcrossings by $X(t)$ in the unit interval is*

$$\mu^+(u) = \mathsf{E}[N^+_{[0,1]}(u)] = \int_0^\infty z p(u,z)\,\mathrm{d}z. \tag{2.66}$$

*It is of course tempting to think of $p(u,z)$ as the joint density of $X(0), X'(0)$, but no argument for this is involved in the following proof.*

*Proof.* Write for short $N^+(u) = N^+_{[0,1]}(u)$. We first device a counting technique for the upcrossings by dividing the interval $[0,1]$ into dyadic subintervals, $[(k-1)/2^n, k/2^n]$, $k = 1, \ldots, 2^n$, and checking the values at the endpoints. Let $N_n^+$ denote the number of points $k/2^n$ such that $X((k-1)/2^n) < u < X(k/2^n)$. Since $X(t)$ has continuous sample paths, there is at least one $u$-upcrossing in every interval such that $X((k-1)/2^n) < u < X(k/2^n)$, and hence $N_n^+ \le N^+(u)$.

Furthermore, since $X(t)$ has a continuous distribution, we may assume that $X(k/2^n) \ne u$ for all $n$ and $k = 1, \ldots, 2^n$. When $n$ increases to $n+1$, the number of sub-intervals doubles, and each interval contributing one upcrossing to $N_n^+$ will contribute at least one upcrossing to $N_{n+1}^+$ in at least one of the two new sub-intervals. Hence, $N_n^+$ is increasing, and it is easy to see that, regardless of if $N^+(u) = \infty$ or $N^+(u) < \infty$, that $N_n^+ \uparrow N^+(u)$ as $n \to \infty$. Monotone convergence implies that $\lim_{n\to\infty} \mathsf{E}[N_n^+] = \mathsf{E}[N^+(u)]$.

Now, define

$$J_n(u) = 2^n\, \mathsf{P}(X(0) < u < X(1/2^n)),$$

so, by stationarity,

$$\mathsf{E}[N^+(u)] = \lim_{n\to\infty} \mathsf{E}[N_n^+] = \lim_{n\to\infty} J_n(u).$$

By writing the event $\{X(0) < u < X(1/2^n)\}$ as

$$\{X(0) < u < X(0) + \zeta_n/2^n\} = \{X(0) < u\} \cap \{\zeta_n > 2^n(u - X(0))\},$$

we have

$$\begin{aligned} J_n(u) &= 2^n \mathsf{P}(X(0) < u, \zeta_n > 2^n(u - X(0))) \\ &= 2^n \int_{x=-\infty}^{u} \int_{y=2^n(u-x)}^{\infty} g_n(x,y)\, \mathrm{d}y\, \mathrm{d}x. \end{aligned}$$

By a change of variables, $x = u - zv/2^n$, $y = z$, $(v = 2^n(u-x)/y)$, this is equal to

$$\int_{z=0}^{\infty} z \int_{v=0}^{1} g_n(u - zv/2^n, z)\, \mathrm{d}v\, \mathrm{d}z,$$

where $g_n(u - zv/2^n, z)$ tends pointwise to $p(u,z)$ as $n \to \infty$, by the assumptions of the theorem. Since $g_n$ is dominated by the integrable $h(z)$, it follows that the double integral tends to $\int_0^\infty z p(u,z)\, \mathrm{d}z$ as $n \to \infty$. □

**Remark 2.4** (Horizontal window conditioning). *The proof of Rice's formula as illustrated in Figure 2.7 shows the relation to the Kac and Slepian horizontal window conditioning: One counts the number of times the process passes through a small horizontal window* $[t, t + \Delta t]$.

**Remark 2.5** (Non-stationary process). *Rice's formula can be extended to non-stationary processes, in which case the crossing intensity is time dependent. Then, the density function* $f_{X(t),X'(t)}(u,z)$ *in the integral depends on* $t$, *e.g.*

$$\mathsf{E}[N_{[a,b]}(u)] = \int_{t=a}^{b} \int_{z=-\infty}^{\infty} |z| f_{X(t),X'(t)}(u,z)\, \mathrm{d}z\, \mathrm{d}t. \tag{2.67}$$

*The integral* $\mu_t(u) = \int_{-\infty}^{\infty} |z| f_{X(t),X'(t)}(u,z)\, \mathrm{d}z$ *is the local crossings intensity at time* $t$, *with obvious modifications for up- and downcrossings.*

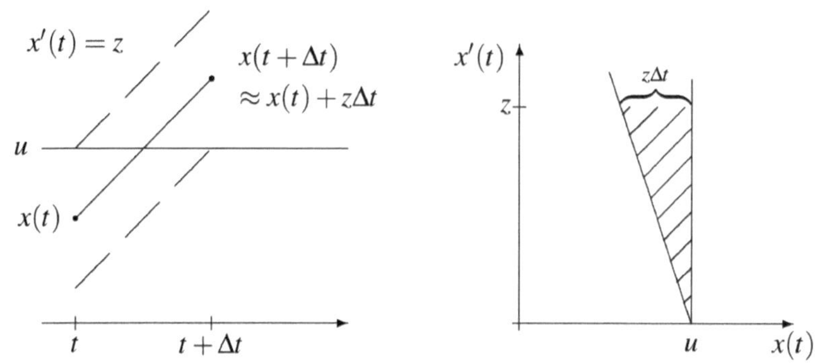


**FIGURE 2.7** Upcrossing occurs between $t$ and $t+\Delta t$ when $x(t)$ is between $u - x'(t)\Delta t$ and $u$. The probability is the integral of the joint density $f_{X(t),X'(t)}(u,z)$ over the dashed area.

### 2.3.3 KAC'S METHOD AND COUNTING CROSSINGS BY DIRAC DELTA FUNCTION

Kac [97] used a narrow "testfunction" $\psi_\varepsilon(t) = 1$ for $|t| < \varepsilon$ and zero otherwise to identify the zeros of a real function. We will let $\psi_\varepsilon(t)/2\varepsilon \to \delta(t)$, the Dirac delta function, as $\varepsilon \to 0$.

We consider a differentiable random process $X(t)$ and count the number of crossings of level $u$ by $X(t)$ from below, in a time interval of duration $T$, taken to be $[0,T]$. We assume that the probability is zero that there are any $t$ for which simultaneously $X(t) = u$ and $X'(t) = 0$. Let the number of $u$-level upcrossings be denoted $N_T^+(u)$.

To establish the relation between the process $X(t), t \in [0,T]$ and $N_T^+(u)$, we define a new process,

$$Z(t) = U[X(t) - u], \tag{2.68}$$

where, $U[\cdot]$ is the Heaviside function and takes the value of unity when its argument is positive and zero otherwise. The process $Z(t)$ is therefore a zero-one process, which takes value of unity for the time duration $X(t)$ spends above level $u$. Note that the time interval spent by a segment of $X(t)$ above level $u$ is initiated by an upcrossing, when $X(t)$ crosses $u$ from below and is followed by a downcrossing, when $X(t)$ crosses $u$ from above.

Mathematically, these crossings are identified by differentiating $Z(t)$ with respect to $t$, which gives

$$Z'(t) = \delta(X(t) - u)X'(t), \tag{2.69}$$

where $\delta(\cdot)$ is the Dirac-delta function, acting as the derivative of the Heaviside function. Note the simple rule $\int_{-\infty}^{\infty} f(t)\delta(t-a)\,\mathrm{d}t = f(a)$ if $f(\cdot)$ is continuous at $t = a$.

We have now use for the following counting rule of the delta function.

**Lemma 2.1** (Counting with the delta function). *a) If the function $f$ is continuous and $g$ is differentiable then, if $g(t) \neq u, t = a = b$,*

$$\int_a^b f(t)\delta(g(t) - u)\,\mathrm{d}t = \sum \frac{f(t_i)}{|g'(t_i)|}, \tag{2.70}$$

*where the sum extends over all solutions $t_i$ to $g(t) = u$.*

*b) With* $f(t) = X'(t)U[X'(t) - 0]$ *and* $g(t) = X(t)$*, the conditions in part (a) are satisfied and one can express the count as*

$$N_T^+(u) = \int_0^T X'(t)U[X'(t) - 0]\,\delta(X(t) - u)\,\mathrm{d}t = \sum_{t_i \in [0,T]} 1. \tag{2.71}$$

Note that the direction of the arrows (upwards or downwards) depends on the sign of $X'(t)$, as clearly, if $X'(t) < 0$ a downcrossing of level $\alpha$ occurs, while an upcrossing occurs if $X'(t) > 0$. The total number of upcrossings in the time duration $T$, given by $N_X^+(u)$, can therefore be expressed as the total number of upward pointing arrows in Figure 2.8.

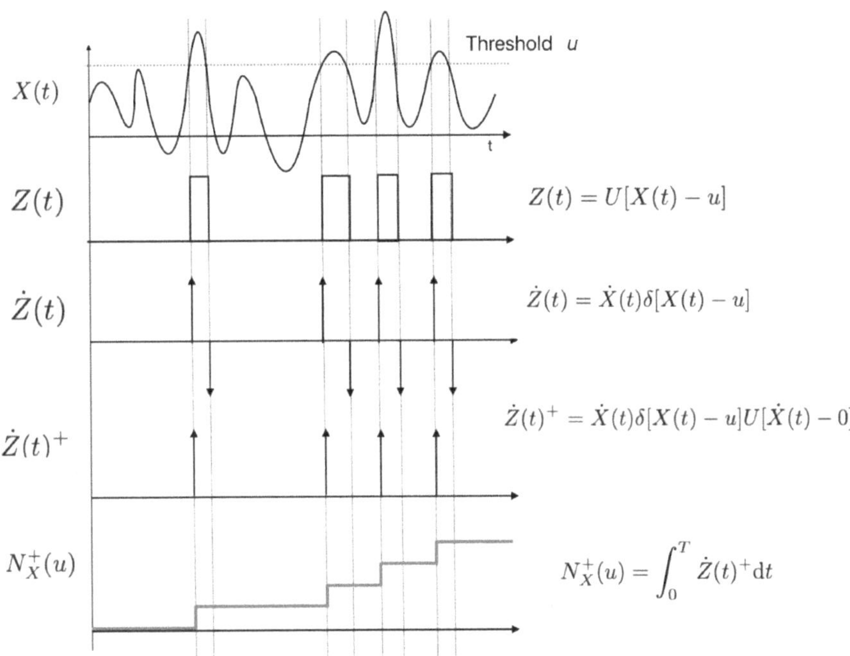


**FIGURE 2.8** Schematic for counting process.

Equation (2.71) establishes a linear relation between $N_T^+(u)$ and the process $X(t)$ as well as its instantaneous time derivative $X'(t)$. Characterising the pdf of $N_T^+(u)$, however, depends on the simultaneous joint pdf of $X(t), X'(t)$ for all $t$ in the interval. Computing the pdf for $N_X^+(\alpha, T)$ from the relation in (2.71) is therefore not possible. A considerably simpler task is to compute the statistical expectation of $N_T^+(u)$, from the joint pdf $p_{X(t),X'(t)}(x,x')$.

Noting that $x'U[x'-0] = x'$ for $x' > 0$ and zero otherwise, the expectation we seek is represented as

$$\mathsf{E}[N_T^+(u)] = \int_{x'=0}^{\infty}\int_{x=-\infty}^{\infty} N_T^+(u)\, p_{X(t),X'(t)}(x,x')\, \mathrm{d}x'\, \mathrm{d}x \tag{2.72}$$

$$= \int_{x'=0}^{\infty}\int_{x=-\infty}^{\infty}\int_{t=0}^{T} x'\,\delta(x-u)\, p_{X(t),X'(t)}(x,x')\, \mathrm{d}t\, \mathrm{d}x\, \mathrm{d}x'$$

$$= \int_{t=0}^{T}\int_{x'=0}^{\infty} x' \int_{x=-\infty}^{\infty} \delta(x-u)\, p_{X(t),X'(t)}(x,x')\, \mathrm{d}x\, \mathrm{d}x'\, \mathrm{d}t$$

$$= \int_{t=0}^{T}\int_{x'=0}^{\infty} x'\, p_{X(t),X'(t)}(u,x')\, \mathrm{d}x'\, \mathrm{d}t. \tag{2.73}$$

Thus, we have arrived at the non-stationary form of Rice's formula,

$$\mathsf{E}[N_T^+(u)] = \int_{t=0}^{T}\int_{x'=0}^{\infty} x'\, p_{X(t),X'(t)}(u,x')\, \mathrm{d}x'\, \mathrm{d}t. \tag{2.74}$$

**Remark 2.6** (Vertical window conditioning)**.** *Note that Equation (2.72) is a three-dimensional integral, which can be reduced to the double integral form of (2.73) by making use of the property* $\int_{-\infty}^{\infty} f(x)\delta(x-a)\, \mathrm{d}x = f(a)$*. This was possible due to the "vertical window" interpretation of the pdf* $p_{X(t)}(u)$*: The probability that the process* $X(\tau)$ *passes through a small vertical window* $[u, u+\Delta u]$ *at time* $t$ *is approximately* $\Delta u\, p_{X(t)}(u)$*. In the counting step in Lemma 2.1, we used the "horizontal window" interpretation: The probability that the process* $X(\tau)$ *upcrosses the level* $u$ *through a small horizontal window* $[t, t+\Delta t]$ *near time* $t$ *is approximately* $\Delta t\, x' p_{X(t),X'(t)}(x,x')$*. The distinction between the two aspects was clarified by Kac and Slepian 1959 [99].*

### 2.3.4 RICE'S FORMULA AND BANACH'S THEOREM ON TOTAL VARIATION

In the coming chapters on fatigue damage and load cycles, we will have use for the following relation between level crossings and total variation.

Let $f(t)$ be a differentiable function defined on an interval $I = [a,b]$ and define the total variation

$$V_f(I) = \sup \sum |f(t_{k+1}) - f(t_k)|,$$

where the supremum is taken over all subdivisions $a \le t_0 < t_1 < \cdots < t_n \le b$. Let $N_I(u)$ be the number of $u$-crossings by $f(t)$ for $t \in I$. Then, the following holds.

**Lemma 2.2** (Banach)**.** *(a)* $\int_{-\infty}^{\infty} N_I(u)\, \mathrm{d}u = \int_a^b |f'(t)|\, \mathrm{d}t = V_f(I)$*. (b) Let* $\chi_A(u)$ *be the indicator function for the set* $A$*, equal to one if* $u \in A$*, and zero otherwise. Then, for any arbitrary (measurable) set* $A \subset \mathbb{R}$*,*

$$\int_{-\infty}^{\infty} \chi_A(u) N_I(u)\, \mathrm{d}u = \int_a^b \chi_A(f(t))|f'(t)|\, \mathrm{d}t. \tag{2.75}$$

Part (b) of Banach's theorem can be used to prove Rice's formula for the expected number of level crossings by a stationary, differentiable process $X(t)$; see [125, Section 8.1].

*Proof of (2.60):* We prove that

$$\mu(u) = \mathsf{E}[N_{[0,1]}(u)] = f_{X(0)}(u)\,\mathsf{E}[|X'(0)| \mid X(0) = u)] \tag{2.76}$$

$$= \int_{-\infty}^{\infty} |x'| f_{X(0),X'(0)}(u,x')\, \mathrm{d}x'. \tag{2.77}$$

Taking expectation, changing the order of integration and expectation in (2.75) and using stationarity, we obtain

$$\int_{u\in A} \mu(u)\,\mathrm{d}u = \int_{-\infty}^{\infty} \chi_A(u)\mathsf{E}[N_{[0,1]}(u)]\,\mathrm{d}u = \mathsf{E}\left[\int_0^1 \chi_A(X(t))\,|X'(t)|\,\mathrm{d}t\right]$$
$$= \mathsf{E}\left[\chi_A(X(0))\,|X'(0)|\right] = \int_{u\in A} f_{X(0)}(u)\mathsf{E}[|X'(0)| \mid X(0) = u]\,\mathrm{d}u.$$

Since $A$ is an arbitrary measurable set, the two integrands in the left- and right-hand sides of this equation are equal (for "almost all" $u$) which shows (2.76). The form (2.77) is the form often used for Rice's formula, also called "the Rice integral". □

## 2.4 SOME SPECIAL TOPICS

### 2.4.1 STATIONARY TIME SERIES

A random time series is a sequence of random variables. It can be the result of sampling a process with continuous time, but it can also be a sequence of naturally aggregated data, e.g., a monthly economic series. We consider here only stationary time series, i.e. the statistical properties do not change with time. We use the integers $t \in \mathbb{Z}$ or $\mathbb{N}$ to index the time series $X_t$.

The moment functions are defined as in continuous time,

$$r_X(\tau) = \mathsf{Cov}[X_t, X_{t+\tau}],$$

but for the energy spectrum, there is room for a modification.

### 2.4.2 SAMPLING AND ALIASING

If the time series $X_t$ is obtained by sampling a continuous process $Y(t)$ at unit time steps, $X_t = Y(t)$, $t \in \mathbb{Z}$, then obviously (2.46) still holds,

$$r_X(\tau) = r_Y(\tau) = \int_{-\infty}^{\infty} e^{i\omega\tau}\, \widetilde{S}_Y(\omega)\,\mathrm{d}\omega, \quad \text{for } \tau = 0, \pm 1, \pm 2, \ldots. \tag{2.78}$$

Using that $e^{it}$ is a periodic function, $e^{i\omega\tau} = e^{i(\omega+2k\pi)\tau}$ for integer $k, \tau$, we can rewrite (2.78) as

$$r_X(\tau) = \sum_{k=-\infty}^{\infty} e^{i\omega\tau} \int_{k2\pi-\pi}^{k2\pi+\pi} \widetilde{S}_Y(\omega)\,\mathrm{d}\omega$$
$$= \int_{-\pi}^{\pi} e^{i\omega\tau} \left\{ \sum_{k=-\infty}^{\infty} \widetilde{S}_Y(\omega+2k\pi) \right\} \mathrm{d}\omega = \int_{-\pi}^{\pi} e^{i\omega\tau}\, \widetilde{S}_X(\omega)\,\mathrm{d}\omega.$$

The two-sided spectrum for $Y(t)$ has been *folded* around $\omega = 0$, and all energy in frequencies $\omega \pm 2k\pi$ has been *aliased* with a single frequency $-\pi < \omega \leq \pi$. The one-sided spectrum $S_X(\omega) = 2\widetilde{S}(\omega)$ is limited to $(0, \pi]$,

$$r_X(\tau) = \int_0^{\pi} \cos(\omega\tau)\, S_X(\omega)\,\mathrm{d}\omega. \tag{2.79}$$

Sampling with a different time resolution $dt \neq 1$ gives a folded spectrum up to the *Nyquist frequency* $\pi/dt$, and distortion will occur if the original signal contains significant energy above that frequency.

### 2.4.3 AUTO-REGRESSION AND MOVING AVERAGE

Let the sequence $e_n$ be white noise, i.e. a sequence of uncorrelated variables with mean zero and variance $\mathsf{Var}[e_n] = \sigma^2 < \infty$.

**Definition 2.6** (Autoregression). *A time series $X_i$ is called an autoregression of order p, AR(p), if it is stationary, has mean zero and finite variance, and there are constants $a_k, k = 1, 2, \ldots, p$, such that the sequence $e_n$, given by*

$$X_n + a_1 X_{n-1} + \cdots + a_p X_{n-p} = e_n, \tag{2.80}$$

*is a white noise sequence, such that all $e_n$ are uncorrelated with all previous $X_{n-1}, X_{n-2}, \ldots$. The variables $e_n$ are innovations that introduce new energy in the sequence.*

*The polynomial (with $a_0 = 1$)*

$$A(z) = a_0 + a_1 z + \cdots + a_p z^p$$

*is called the generating polynomial for the AR(p)-process, and the equation*

$$z^p A(z^{-1}) = a_0 z^p + a_1 z^{p-1} + \ldots a_p = 0$$

*is the characteristic equation. The roots of the characteristic equation must lie inside the unit circle for a stationary solution to* (2.80) *to exist.*

The spectral density of the AR(p)-process is

$$S(\omega) = \frac{\sigma^2}{|A(e^{-i\omega})|^2},$$

where $\sigma^2 = \mathsf{Var}[e_n]$.

**Example 2.10.** *The two recursions*

$$A: X_n = -0.5X_{n-1} + e_n, \quad B: Y_n = -2Y_{n-1} + e_n$$

*have characteristic equations with root inside and outside the unit circle:* $z+0.5=0$ *and* $z+2=0$*, respectively. Obviously, for recursion A,* $\mathsf{Var}[X_n] = 0.25\mathsf{Var}[X_{n-1}] + \mathsf{Var}[e_n]$ *which allows a stationary time series solution* $X_n$*. For recursion B,* $\mathsf{Var}[Y_n] = 2\mathsf{Var}[Y_{n-1}] + \mathsf{Var}[e_n]$*, and no stationary* $Y_n$ *exists.*

**Definition 2.7** (Moving average). *The process*

$$X_n = e_n + c_1 e_{n-1} + \ldots c_q e_{n-q} \tag{2.81}$$

*is called a moving average of order q, MA(q), with generating polynomial* ($c_0 = 1$)

$$C(z) = c_0 + c_1 z + \cdots + c_q z^q.$$

*The MA(q)-process has spectral density*

$$S(\omega) = \sigma^2 |C(e^{-i\omega})|^2.$$

For more details about AR- and MA-time series, see [130, Ch. 7].

### 2.4.4 WIENER PROCESS AND ORNSTEIN-UHLENBECK PROCESS

The Wiener process is a fundamental ingredient in random process theory and application. It was used as a test object in the development of the mathematical theory of random processes, provided one of the key pieces is in the physical puzzle of proof of the existence of molecules, and is in everyday use in control systems, economic decisions and weather forecasts. Its most remarkable property is that a Wiener process sample is nowhere differentiable, and nevertheless is its "derivative" a most useful offspring under the name of "white noise".

Of all definitions of a Wiener-process, we chose this as the shortest.

**Definition 2.8** (Wiener-process). *A Wiener-process* $W(t), t \geq 0$*, is a Gaussian process, starting at* $W(0) = 0$*, with mean zero* $\mathsf{E}[W(t)] = 0$*, and with covariance function* $r_W(s,t) = \mathsf{Cov}[W(s), W(t)] = \sigma^2 \min(s,t)$*. To deserve the name Wiener-process, we require its sample functions to be continuous; see note 2 in this chapter on page 51.*

From the Gaussianess and the covariance function, one can deduce that the Wiener-process has independent and normally distributed increments, i.e. for $0 = t_0 < t_1 < \cdots < t_n$,

$$W(t_1) - W(t_0), \quad W(t_2) - W(t_1), \ldots, W(t_n) - W(t_{n-1})$$

are independent and jointly normal, and that

$$\mathsf{Var}[W(t_{j+1}) - W(t_j)] = \sigma^2 (t_{j+1} - t_j).$$

Furthermore, $\text{Var}[(W(t+h)-W(t))/h] = \sigma^2 \cdot h/h^2 \to \infty$ as $h \to 0$, so the process is not differentiable, which is also obvious from the fact that the increment $W(t+h) - W(t)$ is independent of $W(u)$ for all $u < t$. The process simply "does not remember" how it arrived at $W(t)$.

**Definition 2.9** (Ornstein-Uhlenbeck). *An Ornstein-Uhlenbech process $X(t), t \in \mathbb{R}$, is a time-space deformed Wiener-process (with the constant $\alpha > 0$),*

$$X(t) = e^{-\alpha t} W\left(e^{2\alpha t}\right).$$

*It is a stationary Gaussian process with mean zero and covariance function and spectral density, respectively,*

$$r_X(t) = \sigma^2 e^{-\alpha|t|}, \quad S_X(\omega) = \frac{\sigma^2}{2\pi} \frac{2\alpha}{\alpha^2 + \omega^2}. \tag{2.82}$$

*It has continuous but non-differentiable paths.*

The historical origin of the Ornstein-Uhlenbeck process is as the solution to the differential equation

$$\alpha x(t) + x'(t) = \sqrt{2\alpha}\sigma w'(t), \tag{2.83}$$

where $w'(t)$ is "standard white noise". Formal integration rules give

$$x(t) = \sqrt{2\alpha}\sigma \int_{-\infty}^{t} e^{-\alpha(t-s)} w'(s)\, ds,$$

expressing $x(t)$ as an exponential smoothing of the white noise $w'(t)$. According to Section 2.2.3.3, the relation between the spectral densities is

$$S_x(\omega) = \frac{2\sigma^2}{\alpha} \left| \frac{\alpha}{\alpha + i\omega} \right|^2 S_{w'}(\omega). \tag{2.84}$$

This reveals one of the secrets of Gaussian standard white noise: In a linear filter, it can be treated as a stationary process with constant spectral density $S_{w'}(\omega) = 1/2\pi$. Inserting this into (2.84) yields the spectral density in (2.82) for the Ornstein-Uhlebeck process.

### 2.4.5 ERGODIC TIME SERIES AND PROCESSES

In this monograph, we use random processes as "models" for what we expect to meet and measure when making experiments with an irregular, "random" load over a time span of, say, one hour. The measurements may result in a single value, such as an average or a standard deviation, or in more detailed statistics, such as a table of extreme loading values or recorded load cycle ranges. If the experiment is repeated, the results will probably be different, and taking the average of many replicates may increase precision.

When we use a random process as model for this experiment, we prefer a model that can be run for infinitely many hours and in a single run produce all the possible results our one-hour experiment can expect to produce, when repeated many times. Ergodicity is the technical term for the described property.

**Definition 2.10** (Non-technical interpretation). *A stationary random process $X(t), t \geq 0$, is ergodic if every realisation, if observed long enough, will represent all possible (finite-dimensional) outcomes of the process with the correct proportions.*

**Example 2.11** (Law of large numbers). *The Law of large numbers, Theorem 2.3, can be rephrased as follows. For a stationary time series $X_n, n = 0, 1, \ldots$, we want to estimate its expected value, its mean value, $m = \mathsf{E}[X_n]$, the same for all n. The LLN says that one can make N independent realisations of the time series at time $t = 1$ and average the N outcomes, $x_1^{(1)}, x_1^{(2)}, \ldots, x_1^{(N)}$. If the time series is ergodic, it suffices to take one single realisation and average the N first values $x_1, \ldots, x_N$. Both estimates*

$$\widehat{m} = \frac{1}{N}\sum_{k=1}^{N} x_1^{(k)}, \quad \widehat{\widehat{m}} = \frac{1}{N}\sum_{k=1}^{N} x_k,$$

*will tend to the expected value m when N increases, but the rate with which is does so is usually different.*

For a Gaussian stationary time series or process $X(t)$ with covariance function $r(t)$, there are simple criteria for ergodicity:

(a) If $r(t) \to 0$ as $t \to \infty$, then $X(t)$ is ergodic.
(b) If $X(t)$ has a spectral density $S(\omega)$, then it is ergodic.
(c) If the spectrum has a discrete component, then the process is not ergodic.

Note that (a) implies (b) and that (c) says that the Monte Carlo simulated approximation $X^{appr}(t)$ in (2.57) is not ergodic.

## Notes

[1]If the mean is constant and the covariance function depends only on $t-s$, the process is called weakly stationary.

[2]To be mathematically correct, the guarantee only says that *there exists a process* whose sample functions are continuous, differentiable, etc.; see [125, Chapter 2].

# 3 Reliability Estimation

## 3.1 INTRODUCTION

It is the task of the structural designer to balance the competing needs of accuracy in the analysis and the feasibility of the associated computational costs. This requires often making simplifying assumptions on the mathematical modelling of the loadings, the system configuration, the material behaviour of the system (e.g., elastic behaviour or elasto-plastic behaviour and taking into account the effects of material non-linearities), the parameters of the material and their interaction effects with the loading and on the mathematical theories that are to be used to analyse the system response. These assumptions are based on the prevalent physical theories, current state-of-art knowledge, availability of data from laboratory or field measurements and experience and expertise of the modeller. These assumptions in the development of the mathematical model imply that the model is only an approximation of the physical system, and therefore the predictions obtained from the mathematical model are not expected to exactly reproduce the real-world system behaviour.

While physical experimentation usually is more trustworthy in the process of analysis and design, it is virtually impossible to carry out actual testing of the system at the operating conditions and the various possible scenarios considered during analysis and design. Moreover, physical experimentation is time-consuming and financially expensive. Therefore, in most cases during design, usually one resorts to computational analysis, which is significantly cheaper and quicker. Physical experiments are carried out usually to benchmark the mathematical model and validate a few test cases. However, this implies that there is a need to quantify the confidence in the computational predictions that have been obtained from the mathematical models. This can be carried out using a *posteriori* approach that involves comparing the outcomes from the mathematical model and the physical experiments. The alternative approach involves modelling the uncertainties in the mathematical models and carrying out an uncertainty quantification (UQ) analysis to quantify the propagation of the uncertainties in the input and in the mathematical model into the response predictions. This typically involves adopting probabilistic approaches that use the theories of probability, random variables and random processes. It is pertinent to note that uncertainties exist in the physical experiments as well, both in setting up the physical tests and in measurements and their acquisition. Therefore, it is increasingly being recognised that uncertainty quantification is an important step in analysis and design.

Uncertainties in mathematical modelling of structural systems are due to imprecise or insufficient knowledge about the loadings expected in its lifetime, the material behaviour over a period of time taking into account complexities arising due to ageing effects, simplifying assumptions in imposing the boundary conditions, availability of incomplete or corrupt data sets from measurements, the use of


DOI: 10.1201/9781003019947-3

simplified theories in the analysis and unknown errors induced in the analysis due to simplifying assumptions. These uncertainties are broadly classified as:

(a) uncertainties that arise due to the inherent randomness that exists in nature, such as in the specification of wind loads, material inhomogeneities at the micro-structural scales and the effect of changes in temperature on the material properties, and are therefore irreducible. These are referred to as aleatory uncertainties,
(b) uncertainties that arise due to incomplete information or lack of knowledge and therefore potentially can be reduced with increased knowledge. These uncertainties are referred to in the literature as epistemic uncertainties.

For reasonably well-constructed mathematical models, the effects of these uncertainties do not lead to any bias (or systematic errors) and lead to predictions that on the average lead to observations that are at consonance with those obtained from physical measurements. The effects of these uncertainties in the mathematical models lead to numerical predictions which enable quantifying the scatter that is usually observed in repeated so-called identical physical experiments. The effects of these variations are not so important if one is interested in the *on the average* behaviour of the system. The effects of uncertainties however become significant in black swan [200] type events – events that are rare but whose effects can be catastrophic. Uncertainty quantification is therefore important and critical in the context of assessing structural safety and evolving design procedures that meet specified reliability levels.

The presence of uncertainties in mathematical modelling, analysis and design has long been recognised in the literature. This had led to the development of the "factor of safety" approach in design [118]. Typically, this involves multiplying the demand on the structure and dividing the capacity of the structure to withstand the demand by factors greater than unity and adopting a design approach to meet these requirements. The use of these factors of safety are an implicit admission about the uncertainties in estimating the designed demand and the capacity of a structure. The selection of these factors of safety is usually subjective, depends on the opinion of "experts", lacks scientific rigour and leads to sub-optimal designs that make it impossible to estimate the probability of failure, a metric that is increasingly demanded by regulatory bodies and insurance organisations. Improvements on the factor of safety approaches constitute the so-called limit state design [39] approaches that have their underpinning on statistics of available data. This helps in a better identification of the design loads and capacities but does not help in assessing the failure probability of the designed structures against specific case scenarios.

A step towards probabilistic structure analysis was taken in the 1960s by the American Concrete Institute that resulted in the Cornell safety index, A. Cornell [41], relating resistance $R$ and loading $S$ to the probability of failure by means of the (inverse) coefficient of variation for their difference $R-S$,

$$\beta_C = \frac{m_R - m_S}{\sqrt{\sigma_R^2 + \sigma_S^2}}.$$

The rationale is that when $R$ and $S$ are Gaussian and independent, the failure probability is

$$P_f = \mathsf{P}(R < S) = \Phi(-\beta_C),$$

where $\Phi(\cdot)$ is the standard Gaussian probability distribution function. Its statistical sense was easy to comprehend, and its obvious deficiencies spurred the development of more flexible safety index. In Chapter 11, we will explore other forms of the coefficient of variation to follow time variable fatigue damage.

More modern approaches to accounting for the effects of uncertainties in analysis and design is by adopting probability-based theories for quantifying the uncertainties in a complex system and how these propagate through the system and get manifested into the response. These approaches also enable arriving at design procedures that meet target reliability levels. A key step here involves formulating the mathematical statement for computing reliability, or its complement, the failure probability. As is evident from the terminology, this is a measure of safety (or its complement, failure) in specific scenarios that depend on the loading, the system and its behaviour under the applied loadings and is formally discussed in the following section.

## 3.2 LIMIT STATE AND FAILURE PROBABILITY

Typically, the system equations are mathematically modelled in terms of a set of differential equations and are functions of the quantities that are modelled as uncertain. For spatially distributed uncertainties, these are modelled as random fields. If the correlation length scales of the random fields are greater than the physical dimensions of the system being analysed such that the effect of spatial fluctuations can be neglected, the uncertain quantities can be modelled as random variables. For properties modelled as random processes (or random fields), these are represented in the weak form as an equivalent vector of correlated random variables $\mathbf{Y}$. Several techniques have been discussed in the literature for the weak form representation of random fields and are the subject of stochastic finite element methods [70, 73, 76, 77, 103, 139].

For a particular problem, there could be uncertainties from various sources – the loading and different parameters associated with the system – each leading to vectors of correlated random variables. These can be collectively represented in terms of an $n$-dimensional vector of random variables $\mathbf{Y}$, where a point in the space spanned by $\mathbf{Y}$ represents one realisation of the uncertainties associated with the problem. The vector $\mathbf{Y}$ will contain both resistance and load variables.

Complete characterisation of $\mathbf{Y}$ is achieved through the $n$-dimensional joint probability density function (pdf) $p_{\mathbf{Y}}(\mathbf{y})$. It is easy to appreciate that there would exist some set of points in the $\mathbf{Y}$-space, which would cause the system to behave in a manner beyond that for which it was designed, leading to failure. Thus, the $\mathbf{Y}$-space can be demarcated into safe and failure regimes; see Figure 3.1. The boundary of these regimes can be mathematically represented as the function $g(\mathbf{Y}) = 0$ and is defined as the limit state function (alternatively, the failure state function or the performance function). It can be continuous, differentiable, piece-wise differentiable or

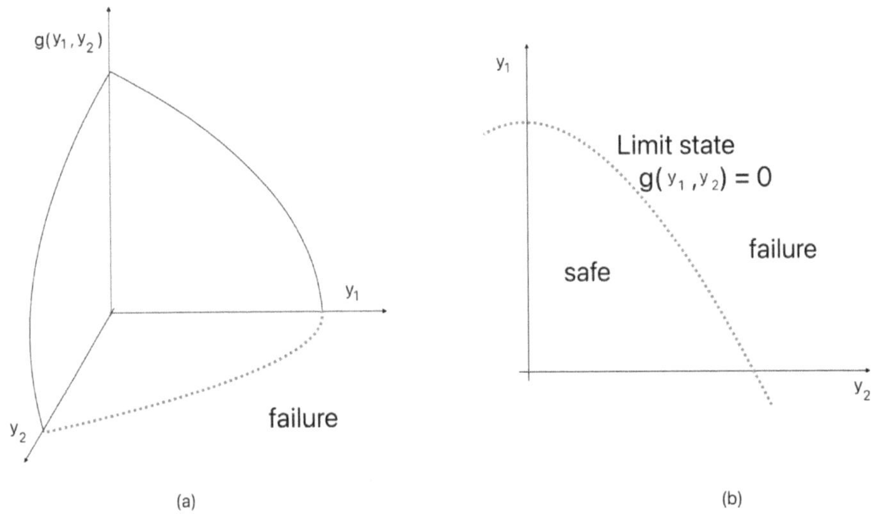


**FIGURE 3.1** (a) Two-dimensional function $g(y_1, y_2)$ where $Y_1$ and $Y_2$ are random variables. (b) Projection on the $Y_1 - Y_2$ plane corresponding to $g(Y_1, Y_2) = 0$; the dotted curve represents the limit surface, points to the right of the curve represent the failure domain and, those on the left represent the safe domain.

it can have discontinuities, depending on the nature of the problem being analysed. Typically, the safe regime is defined as $g(\mathbf{Y}) < 0$, and $g(\mathbf{Y}) > 0$ represents the failure domain in the **Y**-space.

As each point in the **Y**-space is characterised in terms of the pdf $p_{\mathbf{Y}}(\mathbf{y})$, the total integral probability measure across the entire domain is unity. It follows that an estimate of the failure probability is obtained by integrating the pdf over the failure domain and is mathematically given by

$$P_f = \int_{g(\mathbf{Y})>0} p_{\mathbf{Y}}(\mathbf{y})\, d\mathbf{y}. \tag{3.1}$$

Equation (3.1) is a multi-dimensional integral whose dimension is equal to the size of the vector **Y**. For industrial problems, especially those related to the aerospace industry, the dimension $n$ of **Y** could be in the order of millions. This implies that the dimensionality of the integral in (3.1) is also of the order of millions. It is therefore obvious that estimating $P_f$ is not a trivial task.

There are additional challenges in estimating $P_f$ using Eq. (3.1), the primary difficulty being the availability of the knowledge of the joint pdf $p_{\mathbf{Y}}(\mathbf{y})$. In general, information about the uncertainties is available in the form of measurement data. For certain parameters, the size of this data set may be small and contain spurious information. The dependence between different classes of data may be limited to the first few moments, from which it may be difficult to estimate the joint pdf $p_{\mathbf{Y}}(\mathbf{y})$, especially when the random variables are non-Gaussian. The failure surface function $g(\mathbf{Y}) = 0$ may also not be available in explicit form, especially when the governing equations are available only in implicit form through finite element models. This has led to the development of several approximate methods for estimating the failure probability, and these are discussed in the following sections. This chapter is not intended to provide a comprehensive discussion on all these methods but is focused on introducing the reader to the basic principles of these methods. For more detailed discussions, the interested reader is encouraged to look into the many textbooks, monographs and review papers that are devoted to the subject [73, 138, 146].

## 3.3 ANALYTICAL APPROXIMATE METHODS

Analytical estimates for $P_f$ from Eq. (3.1) are possible under certain specific conditions. If the vector of state variables $\mathbf{Y}$ and its partially known pdf with reasonable realism can be transformed into a vector $\mathbf{X}$ of independent standard normal variables, then the same transformation can be applied to the limit state function and the failure surface.

If the $n$-dimensional $\mathbf{Y}$-space is mapped to the space spanned by the vector of standard Gaussian random variables $\mathbf{X}$, the corresponding pdf $p_{\mathbf{X}}(\mathbf{x})$ is rotationally symmetric about the origin. As a result, the contour map of $p_{\mathbf{X}}(\mathbf{x})$ in the $\mathbf{X}$-space represents a set of concentric circles, with each circle of increasing radii enclosing higher probability measure, calculated as the integral of the pdf $p_{\mathbf{X}}(\mathbf{x})$ over the domain of these circles. Thus, the radii of these circles serve as indicators of the probability measure. The radius of the smallest circle that intersects the failure surface is therefore indicative of the failure probability $P_f$ of the system; see Figure 3.2. The point in the state space where the probability contour touches the limit state is defined as the most likely failure point. Note that reliability $R = 1 - P_f$.

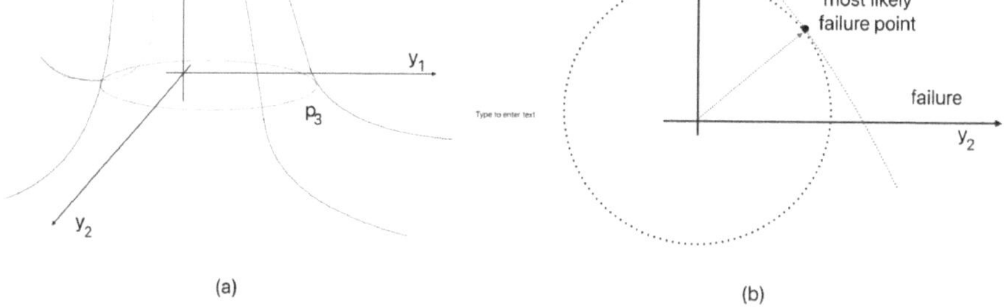


**FIGURE 3.2** (a) Two-dimensional joint pdf $p_{Y_1Y_2}(y_1, y_2)$; the dotted circles represent the contours levels $p_i$. (b) Superposition of the probability contours on the $Y_1 - Y_2$ plane; the radius of the smallest probability contour that intersects the limit surface is indicative of $P_f$.

Thus, the shortest distance from the origin to the failure surface is often used as a measure to compare the reliability of different systems and is referred to in the literature as the Hasofer-Lind reliability index, $\beta_{HL}$, [80]. An estimate of the failure probability is given by

$$P_f = \Phi(-\beta_{HL}). \tag{3.2}$$

Equation (3.2) is exact only when the failure surface $g(\mathbf{X}) = 0$ is linear. Mathematically, the Hasofer-Lind reliability index is given by

$$\beta_{HL} = \min_{g(\mathbf{X})=0} \sqrt{\mathbf{X}\mathbf{X}^t}.$$

The point in the $\mathbf{X}$-space that lies closest to the origin has importance and is referred to as the design point $\mathbf{x}^*$. The implication of the design point is that this combination of the realisations of the $n$-dimensional vector of random variables $\mathbf{X}$ leads to the most likely (highest probability) failure state, and hence the system is designed for these parameter values. It is worth emphasising here that this geometric view of Eq. (3.1) not only enables estimating the relative reliability of the system being analysed through the reliability indices but also enables identifying the design point. When the failure surface in the $\mathbf{X}$-space is not linear, the equality in Eq. (3.2) does not hold. This has led to the development of approximate analytical techniques, such as the first-order reliability method (FORM), second-order reliability method (SORM), and so on.

The FORM enables approximating $P_f$ from Eq. (3.1), when the failure surface, when transformed to the $\mathbf{X}$-space, is non-linear. The underlying principle in FORM is approximating the failure surface through a first-order Taylor series expansion that enables its linearisation and through an optimisation algorithm identifies the design point $\mathbf{x}^*$ – the point on the failure surface which is closest to the origin. This is typically achieved through constrained non-linear optimisation algorithms [38, 133, 162, 188, 217]. FORM enables a simple and efficient approach to identifying the design point in the $\mathbf{X}$-space (also referred to as the most probable failure point), and in turn, the reliability index. However, the reliability index cannot be mapped to a failure probability and can only be used to assess the relative reliability of different systems. A large number of studies have investigated the use of FORM and suggested developments, a review of which is available in [138].

One of the key criticisms of FORM is that it assigns the same reliability index for failure surfaces irrespective of the curvature in the vicinity of the design point. This is primarily because FORM enables a reasonable approximation to $P_f$ if $g(\mathbf{X} = 0)$ is flat near the design point, but in case of high curvatures, the errors can be quite large. This drawback can be somewhat mitigated by including the second-order terms in the Taylor series expansion, leading to the class of methods referred to as SORM. This approximates the failure surface near the design point as a quadratic surface and is achieved either by matching the principal curvatures or by point fitting strategies [25, 26, 27, 28, 83]. The introduction of the second-order terms in the Taylor series expansion however requires computing the Hessian matrix, comprising the second-order derivatives in an $n \times n$ matrix, where $n$ is the dimension of $\mathbf{X}$, and is not only computationally expensive when analytical forms of the failure surface are not available, but also introduces larger computational errors [51]. Strategies have been proposed to reduce these errors by adopting strategies that bypass the computation of the Hessian through analytical approximations [29, 35, 160, 161], iterative algorithms [52, 108] or point fitting strategies [85, 220, 221, 222].

In Chapter 10, we will use an asymptotic SORM-technique by Breitung [26], to approximate, in Section 10.3.2.4, the risk for high rainflow fatigue damage. The method is expanded, in Appendix 10.A, to an asymptotic $n$-dimensional Rice's formula.

## 3.4 TIME VARIANT RELIABILITY

Quantifying the failure probability in structural systems subjected to time varying loadings is fundamentally and significantly more challenging. Here, the loading is typically modelled as a vector of random processes in time, and the system response can be viewed as filtered random processes, where the structure behaves like a filter. Mathematically, the statement for failure probability is expressed in the general form as

$$P_f = 1 - \mathsf{P}(X(t) \leq \alpha \text{ for all } t \in (t_o, t_o + T)), \tag{3.3}$$

where $X(t)$ is the filtered random process in time representing the system response, $\alpha$ is the safe threshold, the exceedance of which by $X(t)$ constitutes a failure, $T$ is the time duration of interest and is often described as the designed structure lifetime and $t_o$ is initial time, which can be taken to be zero without loss of generality. Here, the exceedance of $X(t)$ across the safe threshold $\alpha$ need not define a structural or catastrophic failure but could also represent design failure. The threshold level $\alpha$ could also be a function of time, though generally, the time scales of its variations could be significantly larger than the time scales of variations of the structure response $X(t)$.

It is obvious that failure probability estimation from Eq. (3.3) involves quantifying the probability of exceedance at all time instants in the uncountably infinite time interval $(t_o, t_o + T)$ and consequently involves evaluating an infinite dimensional integral. This is neither practical nor feasible. An alternative approach is to recast the problem by defining a random variable

$$X_m = \max_{t_o \leq t \leq t_o + T} X(t), \tag{3.4}$$

which defines the maximum of the process $X(t)$ in the time interval $(t_o, t_o + T)$. This enables rewriting Eq. (3.3) in the equivalent form as

$$P_f = 1 - \mathsf{P}(X_m \leq \alpha) = 1 - F_{X_m}(\alpha), \tag{3.5}$$

where $F_{X_m}(\alpha)$ is the cumulative distribution function (cdf) of the variable $X_m$. As can be seen, Eq. (3.5) is now expressed in the time invariant format and the problem of evaluating $P_f$ essentially boils down to estimating the probability distribution of the extreme value random variable $X_m$. The possible distributions $F_{X_m}(\alpha)$ are referred to in the literature as extreme value distributions (evd).

### 3.4.1 EXTREME VALUE DISTRIBUTIONS

The problem of estimating extreme value distributions has received wide attention in the literature, for a wide range of problems involving disciplines as diverse as physics, statistics, mechanics, meteorology, social networks, etc. [37, 64, 72]. The fundamental question here lies in estimating the distribution of the maxima (or minima) from a given sample of observations/data. For a sequence of independent and identically distributed (iid) random variables $X_i$, $i = 1, \dots, n$, only three types of limiting distributions are possible for the maximum over an increasing number of observations,

$$X_m = \max_{i=1,\dots,m} X_i, \qquad \text{as } m \to \infty.$$

It was shown by Fisher and Tippett (1928) [59] that if, for sequences of location $a_m$ and scale parameters $b_m$,

$$\mathsf{P}\left(\frac{X_m - a_m}{b_m} \leq x\right) \to F(x),$$

for some non-degenerate distribution function $F$, then $F$ is one of the extreme value distributions, the Weibull probability distribution, [210], given by

$$\mathsf{P}(X \leq x) = \begin{cases} \exp[-(-x)^{\alpha}], & \text{for } x < 0, \\ 1, & \text{for } x \geq 0, \end{cases}$$

the Fréchet probability distribution, Fréchet (1927) [61],

$$\mathsf{P}(X \leq x) = \begin{cases} 0, & \text{for } x \leq 0, \\ \exp[-x^{-\alpha}], & \text{for } x > 0, \end{cases}$$

and the Gumbel probability distribution, named after the wide spread book by Gumbel, Statistics of Extremes (1958) [72],

$$\mathsf{P}(X \leq x) = \exp[-\exp^{-x}], \text{ for } -\infty < x < \infty.$$

Here, $\alpha > 0$.

These three limiting distributions can be represented through the generalised extreme value (gev) distribution function (also known as the von Mises extreme value distribution, [71]), of the form

$$\mathsf{P}(X \leq x) = \exp\left\{-\left[1 + \zeta\left(\frac{x-\mu}{\sigma}\right)\right]^{-\frac{1}{\zeta}}\right\},$$

where, $1 + \zeta(\frac{x-\mu}{\sigma}) > 0$, with $\mu, \sigma > 0$ and $-\infty < \zeta < \infty$, being referred to as, respectively, the location, scale and shape parameters. The generalised extreme value distribution leads to the Fréchet distribution for $\zeta > 0$, the Weibull distribution when $\zeta < 0$ and the Gumbel distribution at $\zeta = 0$. For details and proofs, see [60, 79, 115, 137].

Estimating extreme value distributions for a sequence of mutually dependent random variables is, however, more complicated. This can be attributed to the fact that in constructing the extreme value distribution, information on the joint distribution of the sequence of dependent random variables is necessary. Unlike the case of iid random variables, no unique limiting distributions exist for a sequence of dependent random variables.

However, under certain conditions on the dependence structure, a stationary sequence of random variables can be shown to have identical limiting extreme value distributions as a sequence of iid random variables [115, Chapter 3]. These conditions prevent large values to "cluster", and further they impose restrictions on the dependence between large maxima over intervals $I, J$ far apart,

$$X_I = \max_{t \in I} X_t, \quad X_J = \max_{t \in J} X_t.$$

Studies in the literature have shown that slight variations in the model accounting for mutual dependencies of the random variables or their marginal density functions can lead to dramatically different models for extremes, and hence caution must be exercised on the use of extreme value theory for predicting extremes. Statistical diagnostics for analysis of extremes in dependent sequences can be found in, among others [8, 37, 40, 48, 91].

For a sequence of vector random variables, such as $\{X_i\}_{i=1}^k$ and $\{Y_i\}_{i=1}^k$, where the $X_i$s and the $Y_i$s constitute a sequence of iid random variables but there exists a dependence between the $X_i$s and the $Y_i$s, it is obvious that this dependency will be propagated to the extreme values $X_m = \max\{X_i\}_{i=1}^k$ and $Y_m = \max\{Y_i\}_{i=1}^k$. A complete characterisation of this dependency is possible through the joint extreme value distribution, given by

$$P_{X_m Y_m}(\alpha, \beta) = \mathsf{P}[(X_m < \alpha) \cap (Y_m \leq \beta)]. \tag{3.6}$$

Equation (3.6) represents a bivariate form of the extreme value distributions and can be considered a special case of $k$-dimensional multivariate extreme value distributions. However, unlike the univariate case, there are no unique limiting extreme value distributions. The limiting extreme value distributions are influenced by the dependency function [49] that exists between the vector of random variables and can be expressed through the generic functional form [90, 91] as

$$\mathbf{D_P}(P_1(x_1), \ldots, P_k(x_k)) = \mathbf{P}(x_1, \ldots, x_k),$$

where $X_1, \ldots, X_k$ are $n$-dimensional vectors of random variables and $P_k(x)$ is the parent probability distribution function of these random variables. As is obvious, the dependency function has two limiting cases corresponding to the cases of asymptotic total independence and asymptotic complete dependence between the componentwise extremes leading to the limiting forms of the multi-variate extreme value distributions, given by

$$\mathbf{P}(x_1, \ldots, x_k) = P_1(x_1) \ldots P_k(x_k)$$

for the former and

$$\mathbf{P}(x_1,\ldots,x_k) = \min[P_1(x_1),\ldots,P_k(x_k)]$$

for the latter. Here, $P_j(x_j)$ represents the component asymptotic extreme value distributions.

Several theorems have been proposed in the literature allowing for the characterisation of limit distributions and their domains of attraction, depending on the distributions of the parent variables and the nature of the dependency functions [105]. Under special conditions, the limiting distributions for the iid cases still hold for dependent sequences, the necessary and sufficient conditions for which have been formulated. There have been other significant mathematical developments in the field of extreme value theories, especially in higher dimensions; see [8, 40, 48].

### 3.4.2 RICE'S FORMULA

In the context of time variant reliability, systems are subjected to random loadings that are modelled as random processes in time, and the response is also a random process, whose characteristics have been transformed with the system behaving like a filter. Thus, for example, if the loading to a system is modelled as a random process $Y(t)$, the response can be expressed as

$$X(t) = g(Y(t)),$$

where the function $g(\cdot)$ depends on the system and behaves as a filter. The failure probability for the system can be expressed as Eq. (3.3), which can be reformulated in the time invariant format of Eq. (3.5) by introducing the extreme value random variable as in Eq. (3.4). Thus, it can be seen that the extreme value distribution plays a central role in time variant reliability problems.

Following the definition in Eq. (3.3), a no-failure condition implies that the process $X(t)$ undergoes no crossings of level $\alpha$ in the time duration $(t_0, t_0+T)$. If $X(t)$ crosses the threshold $\alpha$, the times at which these crossings occur are random, and these events form a discrete process in time. The first occurrence of a crossing is defined as the first passage time, denoted by $T_f$, and is a random variable. Clearly, if the extreme value of the process $X_m \leq \alpha$, it follows that $T_f > T$. Hence, the extreme value distribution and the distribution of the first passage time are related as

$$\begin{aligned} F_{X_m}(\alpha) &= \mathsf{P}(X_m \leq \alpha) = \mathsf{P}(T_f > T) \\ &= 1 - \mathsf{P}(T_f \leq T) = 1 - F_{T_f}(T), \end{aligned}$$

where $F_{T_f}(T)$ is the cumulative distribution function for the first passage time. In the case of $X_m < \alpha$, there is no failure and there are no crossings of level $\alpha$ by $X(t)$ from below. It is therefore clear that there exists a connection with the crossings of the processes and the extreme value distribution.

A first step involves defining the number of crossings of level $\alpha$ by $X(t)$, from below, in a time duration of $T$. Let this be denoted as $N_X^+(\alpha, T)$. It is a random variable, whose characteristics depend on the level $\alpha$ and duration $T$. The subscript denotes the random associated process $X(t)$, while the superscript $+$ indicates that only the upcrossings by $X(t)$ of level $\alpha$ from below are being considered. To completely characterise $N_X^+(\alpha,T)$, one needs to estimate its probability density function. Since $N_X^+(\alpha,T)$ is a function of $X(t)$, for $t$ in the whole interval $[t_o, t_o+T]$, it is clear that the relationship between these two quantities is complex.

In fact, while a full description of the statistical relation seems to be beyond reach, numerical algorithms and approximations based on low-order moments often give good results. For safety analysis, the expected number of upcrossings $\mathsf{E}[N_X^+(\alpha,T)]$ is the fundamental parameter. It can be expressed as

$$\mathsf{E}[N_X^+(\alpha,T)] = \int_0^T \int_0^\infty x' p_{X(t),X'(t)}(\alpha,x')\,\mathrm{d}x'\,\mathrm{d}t, \tag{3.7}$$

where $p_{X(t),X'(t)}(\alpha,x')$ is the joint pdf of $X(t)$ and its instantaneous time derivative $X'(t)$.

Further simplification is possible if $X(t)$ is assumed to be a stationary process, indicating that its statistical properties are invariant with time. This enables expressing the mean number of upcrossings as

$$\mathsf{E}[N_X^+(\alpha,T)] = \mu_X^+(\alpha)T, \tag{3.8}$$

where $\mu_X^+(\alpha)$ is the mean number of upcrossing per time unit across level $\alpha$.

For a non-stationary process, one can use

$$\mu_X^+(\alpha,t) = \int_0^\infty x' p_{X(t),X'(t)}(\alpha,x')\,\mathrm{d}x', \tag{3.9}$$

as the local intensity of $u$-upcrossings at time $t$.

Note that in general, this joint pdf is not available explicitly, and one has to adopt approximations for the joint pdf to evaluate the mean crossing rate. These approximations will be discussed later in this monograph.

It is worthwhile to note that similar arguments can be used to derive the downcrossing statistics for the process $X(t)$ across level $-\alpha$. The corresponding equivalent expression for Eq. (2.72) is given by

$$\mathsf{E}[N_X^-(\alpha,T)] = -\int_0^T \int_{-\infty}^0 x' p_{X(t),X'(t)}(\alpha,x')\,\mathrm{d}x'\,\mathrm{d}t, \tag{3.10}$$

and for the stationary case, the mean downcrossing rate is

$$\mu_X^-(\alpha) = -\int_{-\infty}^0 x' p_{X(t),X'(t)}(\alpha,x')\,\mathrm{d}x'. \tag{3.11}$$

The total crossing rate – which includes both upcrossings and downcrossings – is

$$\mu_X(\alpha,t) = \int_{-\infty}^\infty |x'| p_{X(t),X'(t)}(\alpha,x')\,\mathrm{d}x'. \tag{3.12}$$

It is important to note here that Equations (3.9, 3.11, 3.12) are the most general forms for the crossings of a random process across level $\alpha$ and were originally developed by Rice [167, 168] and are referred to as Rice's formula, alternatively the Rice's integrals or Kac-Rice formula. History and formal proofs are given in Section 2.3.

### 3.4.3 HIGH-LEVEL CROSSINGS AS A POISSON POINT PROCESS

One aspect of the extreme value theory for dependent sequences and processes is the crossings as a point process. Under certain conditions, it can be shown that these enable approximating the probability distribution of the extreme values of a process as defined in Eq. (3.4). A failure can be defined to be an event such that an up-crossing of level $\alpha$ occurs at least once in the time interval

$$\begin{aligned}
P_f &= P_f(0) + \mathsf{P}(N_X^+(\alpha,T) \geq 1 \mid \text{no failure at } t=0)(1-P_f(0)) \\
&= P_f(0) + (1-P_f(0)) \sum_{k=1}^{\infty} \mathsf{P}(N_X^+(\alpha,T)=k) \\
&\leq P_f(0) + (1-P_f(0)) \sum_{k=1}^{\infty} k\mathsf{P}(N_X^+(\alpha,T)=k) \\
&= P_f(0) + (1-P_f(0))\mathsf{E}[N_X^+(\alpha,T)]. \qquad (3.13)
\end{aligned}$$

Here, $P_f(0)$ is the probability of an up-crossing (failure) at $t=0$.

For processes for which high-level crossings have no tendency to cluster, it can be assumed that the crossings of very high levels can be modelled as a Poisson point process after a suitable time normalisation [43, 44]. The conditions are similar to those for the stability of asymptotic extreme value distributions. This enables representing the probability of $k$ crossings in time $[0,T]$ as

$$\mathsf{P}(N_X^+(\alpha,T)=k) = \frac{\lambda^k}{k!}\exp(-\lambda), \qquad (3.14)$$

where $\lambda$ is the parameter for the Poisson distribution. Note that for the Poisson distributed random variable, $\lambda$ also represents the mean and hence

$$\mathsf{E}[N_X^+(\alpha,T)] = \lambda. \qquad (3.15)$$

Comparing with Eq. (2.73), it follows that

$$\lambda = \int_0^T \int_0^\infty x' p_{X(t),X'(t)}(\alpha,x')\,\mathrm{d}x'\,\mathrm{d}t.$$

From Eq. (3.15), it follows that for zero-crossings of the level $\alpha$

$$\mathsf{P}(N_X^+(\alpha,T)=0) = \exp(-\lambda).$$

It therefore follows that

$$P_f = P_f(0) + [1-P_f(0)]\{1-\exp[-N_X^+(\alpha,T)]\}. \qquad (3.16)$$

Equation (3.16) can be further simplified by assuming that $P_f(0)=0$ to the form

$$\begin{aligned}
P_f &= 1-\exp[-\int_0^T \int_0^\infty x' p_{X(t),X'(t)}(\alpha,x')\,\mathrm{d}x'\,\mathrm{d}t] \\
&= 1-\exp[-\int_0^T \mu_X^+(\alpha,t)\,\mathrm{d}t], \qquad (3.17)
\end{aligned}$$

where $\mu_X^+(\alpha,t)$ is the time variant mean upcrossing rate across level $\alpha$.

Comparing Eq. (3.17) with Eq. (3.8), it follows that the extreme value distribution for $X_m$ defined in Eq. (2.73) can be expressed as

$$P_{X_m}(\alpha) = \exp[-\int_0^T \mu_X^+(\alpha,t)\,\mathrm{d}t], \tag{3.18}$$

thereby establishing the relationship between the extreme value distribution of a random process and its mean crossing statistic.

### 3.4.4 FIRST PASSAGE TIMES

The first passage time is defined as the time a random process $X(t)$ undergoes an excursion of the safe threshold limit $\alpha$ for the first time. This time is denoted by $T_f$ and is obviously a random variable for a random process $X(t)$. The excursion of a process $X(t)$ across the safe threshold $\alpha$ is an overloading type of failure and though may not lead to a structural collapse, can indicate serviceability criteria failure. Examples of such failures could be excessive deflections in a bridge due to vehicular movement or excessive deflections or accelerations in the top floors of a tall building under wind loads. Compliance with the serviceability criteria would imply that the random process $X(t)$, which could represent for example, the bridge deflections due to vehicular movement, would be below the safe threshold $\alpha$ for all time $t \in [0,T]$. Here, $T$ denotes the structure lifetime.

In the context of first passage times, a structure would be deemed to be safe if the first passage time $T_f > T$. In other words, the first excursion of the safe threshold $\alpha$ by the process $X(t)$ does not occur within the designed structure lifetime $T$. This is equivalent to the situation that the extreme value $X_m$ of the process $X(t)$, defined as in Eq. (3.4), is less than the threshold $\alpha$. Since both $T_f$ and $X_m$ are random variables, it follows that

$$\mathsf{P}(T_f > T) = \mathsf{P}(X_m \leq \alpha) = P_{X_m}(\alpha). \tag{3.19}$$

In other words,

$$P_{T_f}(t) = \mathsf{P}(T_f \leq t) = 1 - P_{X_m}(\alpha), \tag{3.20}$$

i.e. the probability distribution function $P_{T_f}(t)$ is the complement of the extreme value distribution function of the process $X(t)$.

# 4 Cycle Count and Expected Damage, Basic Concepts

## 4.1 INTRODUCTION

In this chapter, we define the rainflow cycle counting algorithm and the simpler min-max cycle count. A damage function is defined that gives the partial fatigue damage for each load cycle. The statistical distribution of the individual load cycles is here condensed to the expected damage, leading to the concept of damage intensity, i.e. the expected damage per time unit. In later chapters, we will present tools to calculate the full statistical distribution of load cycles, including the rainflow cycles, for different classes of useful load process models.

Industrial on-field applications geared towards quantifying fatigue damage and estimating the residual life time of structural components are based on available measurements in terms of time histories of the loadings (or strain time histories measured through sensors) and the available S-N data obtained for the material in question. As was discussed in Chapter 1, the S-N diagrams are empirically obtained data from laboratory experiments on standard specimens of the material when subjected to uniform cycles of loading, corresponding to specified stress levels. A regression analysis is typically carried out to estimate the best fit curve for the data generated, which is subsequently used to estimate the number of cycles to failure for stress loadings for which data are not available.

The key feature in the use of S-N curves is that all the loadings are assumed to be composed of identical amplitude cycles. However, in real life, often the loading is an irregular time history. If the same S-N diagrams are to be used to estimate the fatigue damage, the crux lies in decomposing the irregular stress amplitude time history into an equivalent number of cycles, with each cycle comprising a peak and a valley of equal height and calculating the fractional fatigue damage from the S-N curves valid for each of these cycles. As is therefore quite obvious, it is essential to devise cycle counting techniques for irregular random loads that combine empirical fidelity with reasonable simplicity in data acquisition and analysis. This chapter is devoted to a discussion of the more popular cycle counting methods available in the literature. In Appendix 7.C in Chapter 7, we have collected a list of abilities to predict fatigue damage for the rainflow count and for 10 of these other methods.

An important aspect of fatigue life prediction by means of cycle counts is the statistical uncertainty in the calculated damage when the load is an irregular random load. In this book, we will use several different stochastic models, i.e. random processes, to derive the theoretical uncertainties of derived cycle counts and their effect of the fatigue life predictions.

DOI: 10.1201/9781003019947-4

Two models are particularly useful for statistical analysis of the different cycle count methods, the so-called Markov chain of turning points and the Gaussian load models, which are smooth functions of time. The link between the statistics of Markov chain of turning points and Gaussian processes, on the one hand, and the statistical properties of the rainflow cycle count, on the other hand, is a challenging problem.

The key to the solution was presented by Rychlik in the PhD thesis 1986 [169]. It consists of a local description of the rainflow cycles in terms of local extremes of the load process, described in Section 4.2.3.3, combined with an exact analysis of its statistical properties for the two types of data, the Gaussian process and the Markov chain of turning points. The separate parts were published in [62, 131, 170, 171].

Chapters 6–8 are devoted to cycle counts in Gaussian and Markov loads. Both models use the concept *oscillation across intervals* as a link between the distribution of rainflow cycles and load distribution; Section 4.8.

### 4.1.1 ROADMAP TO RANDOM FATIGUE DAMAGE CALCULATION

We identify the following steps to reliable random fatigue damage analysis, based on accepted damage rules.

a. Define possible load cycles and identify possible biases and weaknesses.
b. Define a damage function that relates individual cycles to the damage they cause, allowing for material properties and physical knowledge.
c. Identify what stochastic load models can be realistically fitted and used for random fatigue loads – we treat general time series loads, Gaussian loads, Markov loads, and Laplace loads.
d. Derive what statistical properties of the different models that are relevant for fatigue life.
e. Identify how the models can be modified to meet empirical deviations from ideal statistical assumptions.
f. Evaluate variability and confidence in fatigue damage prediction models.

### 4.1.2 ILLUSTRATING CYCLE COUNTS

The first stage in the roadmap is the choice of cycle definition. A cycle is a pair of a local maximum $M_i$ and a local minimum $m_{k(i)}$, selected by the cycle algorithm. For example the min-Max count, when $m_{k(i)}$ is the minimum immediately before $M_i$, and the rainflow count.

Figure 4.1 illustrates the typical difference between the min-Max cycle count and the rainflow cycle count. Data come from simulation of a Gaussian sea wave load with Pierson-Moskowitz spectrum. Figure 5.3 shows more examples of the great variety of cycle patterns.

The cycle data observed in an experiment (or from a theoretical model) are collected in a *cycle matrix*, obtained as follows.

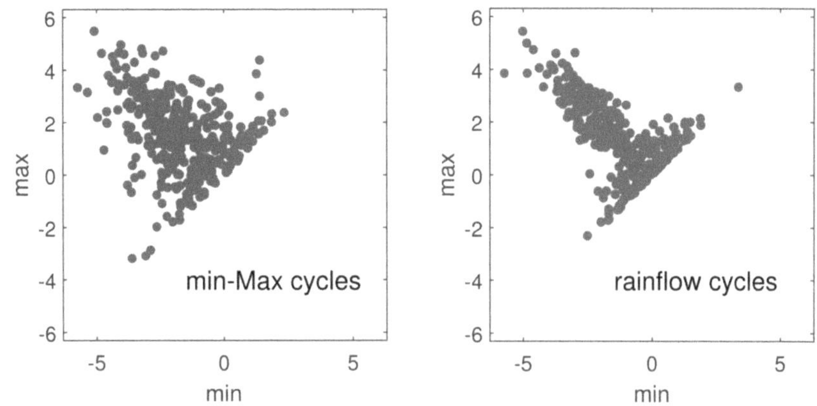


**FIGURE 4.1** min-Max and rainflow cycle counts from sea wave simulations with Pierson-Moskowitz spectrum. A cycle matrix is the histogram of the cycle count.

**Definition 4.1** (Cycle matrix). *Let $x_1 > x_2 > \ldots > x_J$ be an equidistant partition of the load range. For the observed cycles $\{(m_{k(i)}, M_i); i = 1, \ldots, n\}$, the cycle matrix is a two-dimensional histogram, a $J \times J$-dimensional matrix $X$ with elements $x_{jl}$ equal to the number of $i$ such that $x_{l+1} \leq m_{k(i)} < x_l$ and $x_{j+1} \leq M_i < x_j$.*

*For the min-Max and rainflow counts, we use the notation from [177],*

- *Markov matrix: $F^{mM}$,*
- *Rainflow matrix: $F^{rfc}$.*

*A count matrix can be empirical (absolute or normalised) or represent the model probability distribution of $(m_{k(i)}, M_i)$.*

## 4.2 CYCLE COUNTS

For a smooth stress function $x(\tau)$, $\tau \in [0,t]$, having a finite number of local extremes, consider $\tau_i$ to be the time when the $i$th local maximum occurs. Denote the range of a cycle by $H = M_i - m_{k(i)}$ and let its amplitude be $S = H/2$. Further, let $m_i$ be the height of the preceding minimum. If there are no local minima in $[0, \tau_1]$, then $m_1 = M_1$. Let the number of local maxima in $x(\tau)$ be denoted by $n_t$. The sequence $m_1, M_1, m_2, \ldots, M_{n_t}$ is called *the sequence of turning points* and is denoted by $T_P(x)$.

It is obvious that rate-independent damage $D(t)$ must be a function of the sequence of turning points of the stress function. For a stiff structure, one can extract the sequence of turning points from the applied load time history by scaling it by some constant $c$. In the discussions that follow, no distinction is made between loads and stresses, and it is assumed that the damage process depends only on the sequence of turning points of $x(\tau)$.

**Definition 4.2.** *Let $T_P = (m_1, M_1, \ldots, m_n, M_n)$ be a sequence of turning points and let $k(i)$ denote a permutation of the indices $1, 2, \ldots, n$, i.e. any reordering of the indices.*

*For the sequence $T_P$, the permutation $k(i)$ defines a set of pairs $\{(m_{k(i)}, M_i) : i = 1, \dots, n\}$ that defines the cycles. Note that not all permutations are practically useful, e.g., one requires that $M_i \geq m_{k(i)}$ and that the damage is not decreasing when $n$ grows.*

**Remark 4.1.** *We defined the turning points as starting with a minimum, labelled $m_1$, ending with a maximum $M_n$. This choice is arbitrary and depends on context.*

A cycle count is defined to be a procedure of paring the local maxima and minima in the time history $x(t)$. Next, we will introduce three different important cycle counts, the min-Max that tends to underestimate damage, the crossing-cycle count that usually overestimates it, and rainflow counts, the most reliable one. The cycles will be marked by letters $mM$, $cc$ and $rfc$, respectively.

### 4.2.1 MIN-MAX CYCLE COUNT

For the sequence of turning points $T_P = (m_1, M_1, \dots, m_n, M_n)$, the simplest cycle count is the so-called *min-Max count*, when the local maximum $M_i$ is paired with the preceding minimum $m_i$. Obviously for that count, the permutation is the identity $k(i) = i$, and the min-Max cycle count is $\{(m_i, M_i) : i = 1, \dots, n\}$. It has been observed that the estimates of fatigue damage using the min-Max cycle count lead to the underestimation of the fatigue damage, and hence it is not used for risk estimation due to fatigue. However, the min-Max cycle count has interesting properties which will be investigated later in this chapter. The presence of many small range cycles is seen in Figure 4.2.

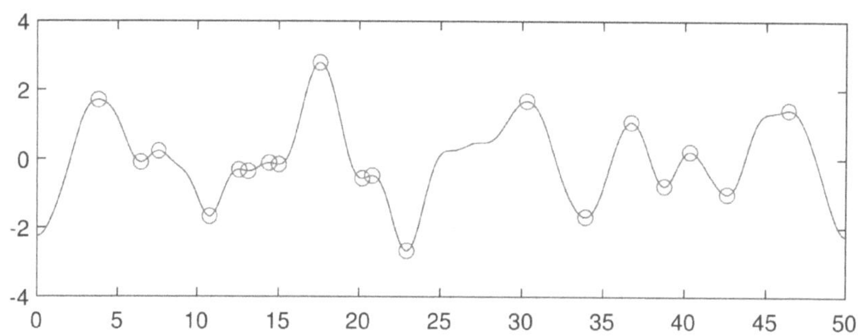


**FIGURE 4.2** True local minima and maxima.

### 4.2.2 CROSSING-CYCLE COUNT, CC COUNT

While the min-Max count underestimates fatigue damage intensity, the crossing-cycle count overestimates it. The method is based on the upcrossing spectrum $N^+(u)$, i.e. the observed number of upcrossings as a function of the level $u$, (2.58). Figure 4.3 illustrates the principle, applied to the load in Figure 4.2. The solid line is the upcrossing spectrum, the circles on the increasing sections represent local minima and circles of the decreasing sections represent maxima.

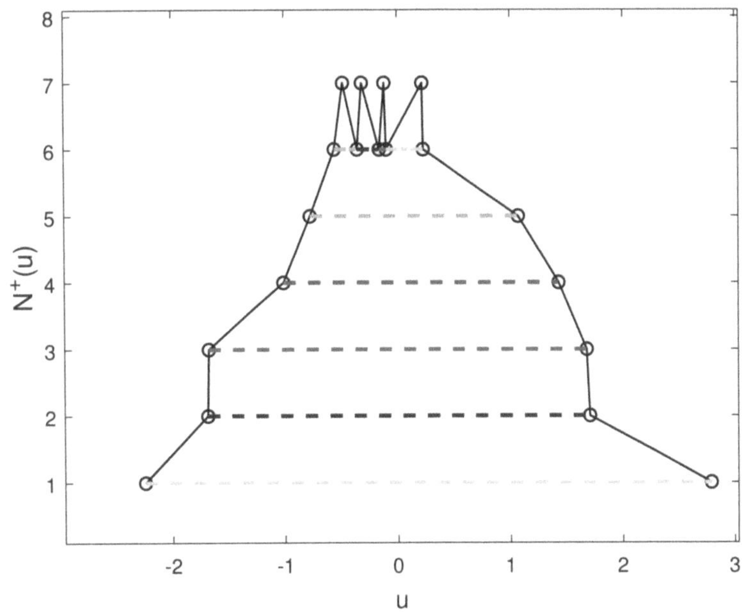


**FIGURE 4.3** Definition of crossing-cycle count.

The crossing-cycles are obtained by pairing local minima with the corresponding maxima, from the more extreme ones, to the less extreme, as illustrated by the horizontal dashed lines in the figure.

### 4.2.3 RAINFLOW CYCLE COUNT

The *rainflow cycle count* (rfc) was introduced by Endo [145] in the late 60s and is now the most frequently used cycle count to estimate fatigue damage from random/irregular time histories. The rainflow method was introduced in two papers presented at the Kyushu District Meeting of the Japanese Society of Mechanical Engineers in November 1967. The name rainflow count was first proposed in the master thesis by Matsuishi in March 1969, where the original rainflow algorithm to identify the closed hysteresis loops from the stress time history $s(\tau)$ was proposed. The algorithm identified the pairing of the maxima and minima from an irregular time history by following the path traversed by rain water as it flows in time, when the time history figure is rotated with time axis pointing downwards.

The underlying principle of the rainflow counting method is based on the hysteretic properties of the underlying material. Namely when studying jointly the strain $\varepsilon(\tau)$ and the local stress $s(\tau)$ time histories, the curve $(s(\tau), \varepsilon(\tau))$ forms hysteresis loops in the $s-\varepsilon$ plane. The topmost point of a hysteresis loop, when projected on the $s$-axis, is identified as one of the maxima $M_i$, while the projection of the lowermost point is denoted as one of the minima. The rainflow cycle permutation

$k(i)$ enables identifying which of these local minima forms the bottom of a hysteresis loop. However, some of the local maxima and minima do not form closed hysteresis loops; these are termed as the residuals. As these residuals do not form a complete cycle, while computing the fatigue damage $D(T)$, they are treated as a collection of half-cycles. Here, for mathematical convenience, we propose to count the min-Max cycles in the residual, to be subsequently added to those pairs that form closed hysteresis loops.

The original definition of the rainflow algorithm is presented next for completeness. However, it must be noted that there is a more efficient way to identify the rainflow cycles in the stress signal, the so-called 4-points algorithm, as has been proposed by de Jonge [50]. For mathematical computations, there is still another definition proposed by Rychlik [170]. This definition considers the more general case of the time history being available as a continuous function, having finite or infinite number of cycles in any interval $[0,t]$, and the sequence of turning point may not exist. It has been proved that this mathematical formulation leads to the same cycles as the original Endo algorithm for smooth loads with finite number of local extremes. This will be discussed later in this chapter.

Why is rainflow count so important? As has been shown by Dowling [56] and many others subsequently, the use of the rainflow cycle counting algorithm leads to conservative predictions of the fatigue damage and therefore leads to designs that are the safest (conservative), even though the rfc method may not always be the best or the simplest means to predict the fatigue life, if only constant amplitude experiment S-N curves are available.

#### 4.2.3.1 The Rainflow Algorithm by Endo

Find the sequence of turning points in the load and imagine that the time history is a template for a rigid sheet (pagoda roof). Then, turn the sheet clockwise 90° such that the origin of the time axis is at the top, as in Figure 4.4.

1. Each peak (local maximum) is imagined as a source of water that "drips" down the pagoda.
2. Count the number of half-cycles by looking for termination in the flow occurring when either:
   a) The rain drop reaches the end of the time history;
   b) It merges with a flow that started at an earlier peak;
   c) It flows opposite a peak of equal or greater magnitude.
3. Repeat steps 1,2 for troughs (local minima).
4. Assign a magnitude to each half-cycle equal to the stress difference between its start and termination.
5. Pair up half-cycles of identical magnitude (but opposite sense) to count the number of complete cycles.
6. Typically, there are some residual half-cycles left.

As we mentioned before, we propose to consider residual half cycles with positive slope as contributing to the accumulated damage in the same way as the pairs of

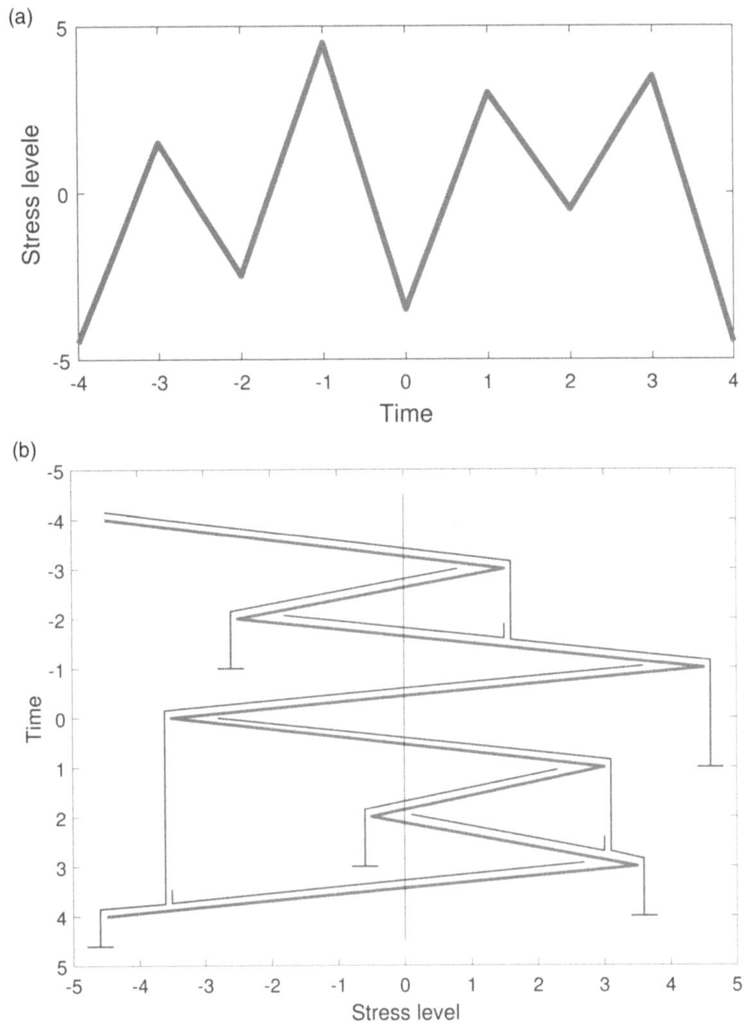


**FIGURE 4.4** The rainflow algorithm by Endo. (a) Stress sequence. (b) Water flowing down stress sequence turned to a pagoda roof.

minima and maxima that form close hysteresis loops. It is easy to see that now the rainflow cycles are defined by the water flow that originated at troughs and dropping to the right. Now, there are as many cycles as there are troughs in the load, and cycles starting at the troughs have the range equal to the length of the water flow started at the trough.

In the next subsection, we present the "Four-point recursive algorithm" to find the rainflow cycles in a measured load.

#### 4.2.3.2 Four-Point Algorithm by de Jonge

First, the sequence of turning points is extracted from the signal. Then, the algorithm scans the sequence and extracts a cycle as soon as three consecutive ranges satisfy a certain criterion:

1. Four successive points 1,2,3,4 are chosen from the sequence of turning points and their values $x_1, x_2, x_3, x_4$ recorded.
2. The three ranges, $\Delta x_1 = |x_1 - x_2|$, $\Delta x_2 = |x_2 - x_3|$, and $\Delta x_3 = |x_3 - x_4|$ are then computed.
3. If $\Delta x_2 \leq \Delta x_1$ and $\Delta x_2 \leq \Delta x_3$, then $(x_2, x_3)$, $x_2 < x_3$ (or $(x_3, x_2)$ if $x_2 > x_3$) forms a rainflow cycle. The two points $x_2$ and $x_3$ are removed, and the two next points are taken from the sequence of turning point to form the four points record. Then, start again with 1.
4. If $\Delta x_2 > \Delta x_1$ or $\Delta x_2 > \Delta x_3$, then point $x_1$ is disregarded, and the next point in the sequence added to form the four points. Then, one is going to 1.

The algorithm is repeated until all points in the sequence of turning points have been used. Once the last point has been reached, the not paired turning points form the *residuals.*

#### 4.2.3.3 The Local Definition by Rychlik

The de Jonge four-point algorithm is very efficient to extract rainflow cycles from measured signals, and to on-line cycle count in a field experiment, e.g. from stresses measured in an operating vehicle. The signals (loads) are recorded in "digital" with a given frequency in Hz and finite resolution which means that they are digitised, taking a finite number of values with a fixed resolution $\Delta x$, say. Then, the local maxima and minima are extracted to form the sequence of turning points. Finally, the algorithm presented above can be used to pair local minima and maxima that form closed hysteresis loops. The turning points that could not be paired form the residual. Observe that if the signal would be measured during a longer time interval, then some extremes in the residual could be paired with extremes observed before or after the actually used measured period. In the following, the residual will be transformed into cycles using the min-Max counting method, i.e. a maximum is paired with the previous minimum.

We turn next to a local definition of rainflow count, proposed by Rychlik, introduced in the PhD thesis 1986 [169] and published [170]. The idea, illustrated in Figure 4.5, is particularly useful for studying rainflow range distributions for random loads. The method is defined next.

**Local definition of rainflow cycles:** From each local maximum $M_i = x(\tau_i), \tau_i < t$, $(i < n)$, one should try to reach the same level, in the backward (left) and reach **above** the level in forward (right) directions, with an as small a downward excursion as possible. Let $m_i^-$ and $m_i^+$ denote the lowest points reached to the left and to the right of $\tau_i$, respectively.

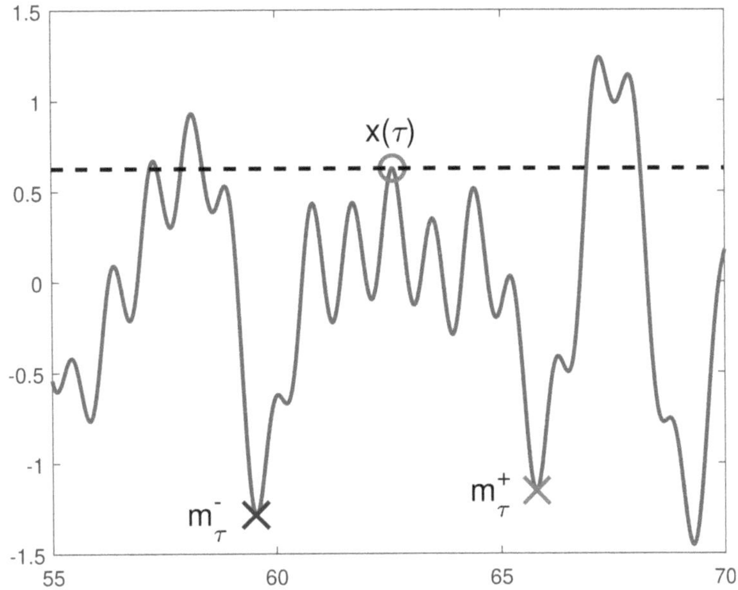


**FIGURE 4.5** Local definition of rainflow cycles.

Then, the maximum of the two minima $m_i^-$ and $m_i^+$ should be chosen to represent the smallest deviation from the maximum $M_i$. The corresponding rainflow minimum is then defined as $m_i^{rfc} = \max(m_i^-, m_i^+)$.

**Definition 4.3** (Rainflow cycle)**.** *The i:th rainflow cycle is defined as* $(m_i^{rfc}, M_i)$*. For the last maximum* $M_n$ *the rainflow minimum* $m_n^{rfc} = m_n$*, the minimum preceding the maximum* $M_n$*. The rainflow range is finally defined as*

$$H_i^{rfc} = M_i - m_i^{rfc}. \tag{4.1}$$

**A more general range definition:** For derivation of the rainflow range distribution for random loads, we will use the two "semi-ranges",

$$H_i^- = M_i - m_i^-, \qquad H_i^+ = M_i - m_i^+. \tag{4.2}$$

Obviously, it is possible to apply the definition of the minima $m_i^-$ and $m_i^+$ for a time point $\tau$ that is not the location of a local maximum, and define the two left and right minima $m_\tau^-$ and $m_\tau^+$. The "local semi-ranges"

$$H^-(\tau) = x(\tau) - m_\tau^-, \qquad H^+(\tau) = x(\tau) - m_\tau^+, \tag{4.3}$$

are now functions of time $\tau$. If $x(\tau)$ is a stationary process, both $H^-(\tau)$ and $H^+(\tau)$ are stationary. It is easy to see that one of the ranges $H^-(\tau)$, $H^+(\tau)$ is equal to zero when $\tau$ is not the location of a local maximum and that $\tau$ is the location of local maximum if

$$H^{rfc}(\tau) = \min(H^-(\tau), H^+(\tau)) > 0. \tag{4.4}$$

The distribution of $H^-(\tau)$ for a general fixed $\tau$ will be different from $H_i^-$, i.e. $H^-$ evaluated at the $i$th local maximum, see discussion of Rice formula in Section 6.4. We will show that both distributions of $H^-$ find applications in the evaluation of expected damage. We will meet the continuous time rainflow process $H^{rfc}(\tau)$ as we introduce the *rainflow count memory* in Section 4.5.

**Example 4.1** (Local definition). *Consider a load $x(t)$ starting at time zero at level zero, i.e. at point $(0,0)$ in $(t,x)$ plain, and then reaches the points $(1,2)$, $(2,1)$, $(3,2)$ and finishes at $(4,0)$. The Endo and de Jonge rainflow counting algorithms would find one closed hysteresis loop (1,2) and a residual 0,2,0. The local definition identifies two cycles $(m_i^{rfc}, M_i)$, viz. (0,2) and (1,2).*

*Note that it is the condition "one should try to reach above the same level, in the forward (right) direction" that caused that the local maximum at time 2 has been paired with the local minimum at time 0. Further, if one would replace the requirement "one should try to reach above the same level" with "one should try to reach the same level", then Rychlik's definition would identify two small cycles (1,2).*

## 4.3 DAMAGE FUNCTION

As we mentioned in the introduction, the load $x(\tau)$ is usually transformed into a sequence of turning points $T_P = (m_1, M_1, \ldots, m_n, M_n)$ from which a sequence of fatigue damaging cycles are extracted with the help of a suitable permutation $k(i)$ of the indices $1, 2, \ldots, n$. The pairs $(m_{k(i)}, M_i)$ such that $M_i \geq m_{k(i)}$ constitute the cycle count. This is step (a) in the Roadmap in Section 4.1.1 to fatigue damage calculation. We now turn to step (b), the damage caused by the load cycles.

In the additive fatigue damage theory, the severity of a cycle $(u, v)$ is measured by a damage function $f(u, v)$. A common and simple form is

$$f(u,v) = \alpha(v-u)^{\beta}, \tag{4.5}$$

for $v > u$ and zero otherwise. The parameters $\alpha, \beta$ are material dependent, empirically estimated constants. While $\alpha$ determines the absolute damage level, the constant $\beta$ determines the sensitivity to changes in stress variability, cycle range and amplitude. Typically, $\beta \geq 3$ is the *damage exponent*. When we in this work discuss the relative merits of different cycle counts and damage models, we set $\alpha = 1$.

The damage function (4.5) is related to the S-N curve (1.1). If a specimen can survive $N = \alpha S^{-\beta}$ cycles with amplitude $S$ before it breaks, the damage from each cycle with range $H = v - u = 2S$ is

$$N^{-1} = \alpha^{-1} 2^{-\beta} H^{\beta}.$$

Thus, the $\beta$ in (4.5) is inherited from the S-N curve and the $\alpha$ gets inverted, including a factor $2^{\beta}$.

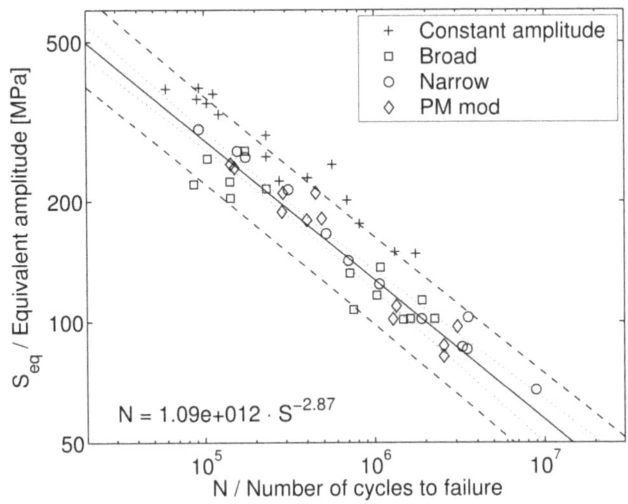


**FIGURE 4.6** S-N curve estimated from constant amplitude tests, with scatter bands of two standard deviations. The fatigue lives for broad, narrow, and P-M spectra are plotted versus equivalent loads calculated using the estimated $\beta$. (Reproduced with permission from [15]. Copyright © 2009 Elsevier).

The measured load is characterised by its *cycle spectrum*, which contains the information that there are $n_i$ cycles with amplitude (or range) $S_i$, obtained by a cycle counting procedure, e.g., the Rain-flow-Count algorithm. Another quantity of interest, besides the cycle spectrum, is the stress ratio $R$, defined as the ratio of the minimum stress to the maximum stress amplitude, counted from the average stress. The spectrum is used for prediction of fatigue life, which is the number of cycles to failure, which could be based on constant amplitude test, i.e. the S-N diagram that relates amplitude $S$ to the number of cycles to failure $N$. The S-N curves are derived for different stress ratios $R$, and the parameter $\alpha$ can vary with $R$. For variable amplitude load, the ratio $R$ may be different for the different cycles $S_i$ in the load spectrum, and the parameters $\alpha, \beta$ are only valid for the $R$-value used in the constant amplitude test. This can be addressed by using an empirical correction formula which will give the general form of damage function $f(u,v)$, or by using the global $R$-value for the variable amplitude load process, e.g., for the zero-mean Gaussian load the average (over the counted cycles $(v_i, u_i)$), or simply by neglecting the mean stress effect. Validation tests have shown that this procedure often gives poor predictions, and the use of the Palmgren-Miner hypothesis has been questioned.

An example of where the predictions are non-conservative is shown in Figure 4.6, where the data come from Agerskov [3].

There are several methods proposed to improve the prediction, see e.g. [3, 95] and references therein. However, this is not the subject of this book. Here, we will study properties of cycle counts for random loads which can be used to construct efficient methods to predict time to failure.

Here, the damage caused by a cycle $(u,v)$ will be measured by means of a damage function $f(u,v)$. For convenience, we require that $f$ is a non-negative, twice continuously differentiable function, and for simplicity of notation, we write

$$f_1(u,v) = \frac{\partial f(u,v)}{\partial u}, \qquad f_{12}(u,v) = \frac{\partial^2 f(u,v)}{\partial u \partial v}.$$

**Definition 4.4.** *The function $f$ will be called a damage function if*

$$f(u,u) = 0, \quad f_1(u,v) \le 0, \quad f_{12}(u,v) \le 0 \quad \text{for all } u < v. \tag{4.6}$$

*Note that* (4.6) *implies that* $f_2(u,v) \ge 0$.

The assumptions on the damage function imply that larger cycles will cause more damage. The $f(u,v)$ which will be most often used in this text is $f(u,v) = \alpha(v-u)^\beta$, $\beta \ge 1$, with the scaling constant $\alpha$ set to one.[1]

**Definition 4.5** (Total damage and damage intensity).
Total damage: *For a load $x(\tau)$, $0 \le \tau \le T$, first, the sequence of turning points $T_P(x)$ is extracted from $x$ and the cycle counts are found, i.e. the local maxima are paired with the local minima according to the chosen cycle definition. Finally, the severity of the load $x$ is measured by the total damage*

$$D(T) = \sum_{i=1}^{n} f(m_{k(i)}, M_i). \tag{4.7}$$

Damage intensity: *The damage intensity* (2.64) *is the expected rate of damage increase $\mathsf{E}[\dot{D}(t)]$, i.e. fatigue damage per time unit caused by the individual cycles, and it is expressed as the product of the cycle intensity $\mu$, the average number of cycles per time unit and the expected damage caused by an individual load cycle, for $f(u,v) = (v-u)^\beta = H^\beta$,*

$$d = \mu \mathsf{E}[H^\beta]. \tag{4.8}$$

For the min-Max and rainflow counts, we use the notations

$$D^{mM}(T) = \sum_{i=1}^{n} f(m_i, M_i), \qquad D^{rfc}(T) = \sum_{i=1}^{n} f(m_i^{rfc}, M_i). \tag{4.9}$$

In Section 4.6, we will prove some general properties of the total damage function that hold for any damage function $f(u,v)$. The most important is the relation

$$D^{mM}(T) \le D^{rfc}(T),$$

i.e. min-Max count predicts smaller damage that rainflow count.

If $f(u,v) = (v-u)^\beta$, the damage depends only on the cycle ranges. Further, if $f(u,v) = (v-u)$, then for any cycle count, e.g. permutation $k(i)$, the damage $D(T)$ is equal to $\sum_{i=1}^{n_T}(M_i - m_i)$, which is called the *positive variation* of the signal.

**Remark 4.2** (Scaling the random load). *When modelling random fatigue, it is common to scale the random load process in time and space. In view of relation* (4.8)*, we need symbols for time scale and space scale, in the specific application area. The most common ones used in this book are for time and space:*

*For time scale,* $\mu$*:*

- $f_z$*: apparent frequency, expected number of zero upcrossing cycles per time unit, with the inverse, zero-crossing period* $t_z = 1/f_z$*,*
- $f_{max}$*: expected number of maxima per time unit, equal to* $\alpha_2 f_z$*,*
- $f_p$*: peak frequency,* $\arg\max S(\omega)/2\pi$*, with peak period* $T_p = 1/f_p$*.*

*For space scale,* $\mathsf{E}[H^\beta]$*:*

- $\sigma$*: load standard deviation,*
- $H_s, h_s$*: significant wave-height or significant load-range defined as* $4\sigma$*, inherited from marine statistics.*

## 4.4 RAINFLOW DAMAGE FOR A RANDOM WALK

Often, environmental loads vary in an unpredictable way and are modelled by means of random processes. Important quantities are expected damage, rainflow range distribution and others. The distributions can be used to predict risk for fatigue failure, testing the strength of components, comparison of severity of different missions, planning routes, calculation of inspections periods, etc. This book will describe several type of random models of loads, e.g., Gaussian, Markov and Laplace. In many situations, it is convenient to fit models (random processes) having only a small number of parameters that describe the unpredictability of the load environment.

The separation of the type of random environment and the parameters in the chosen model often reduces the uncertainty in a reliability analysis. The model choice for the environment depends the type of mission under study, while the variability in the environment during the missions is a statistical problem based on empirical studies. Here, we give an introduction to these types of problems by showing how the rainflow range $H^{rfc}$ (or min-Max range $H^{mM}$) distribution can be found and then expected damage computed. The considered load is very simple so that explicit formulas for the distribution can be found.

We start with a short remark on how the expected damage can be evaluated in a stationary (ergodic) random load. The first approach would be to simulate long realisations (time series) of loads, extract the rainflow ranges and fit a distribution to the observed ranges $H$. In this way, we obtain the model of cycles range distribution. Then, the damage increase in unit time $d^{rfc}$ (or $d^{mM}$) is equal to

$$d^{rfc}\ (d^{mM}) = \mu\, \mathsf{E}[H^\beta], \tag{4.10}$$

where $\mu$ is the frequency of cycles, i.e. expected number of maxima in unit time. Then, for example, the expected total rainflow damage during a time $t$ is $\mathsf{E}[D^{rfc}(t)] = t\, d^{rfc}$.

The second approach is to directly compute $d^{rfc}$ which means the speed of damage growth from the load model variability. Here, all these issues will be illustrated using a random walk as a model for load variability.

### 4.4.1 RANDOM WALK

Random walk $\{Y_i\}$ is a simple model for a discretised load sequence. It takes $N+1$ values $0,1,\ldots,N$. The sequence is stationary, so all $Y_i$ have the same distribution, but there is a simple dependence between successive values that allows us to derive explicit formulas for quantities of fatigue interest. The example will also highlight the question how dependence between variables $Y_i$ may affect the damage. Random walk is a simple example of a Markov chain, when $Y_i$ depends only on the value of $Y_{i-1}$.

The random walk moves by one unit up or down, independently of its present value, with probability 1/2. More precisely, let $e_i$ be a sequence of independent random variables taking values $-1$ or 1 with probabilities $1/2$. (One can imagine it as a sequence of results of tossing a fair coin.) Then, define $Y_{i+1} = Y_i + e_i$. Two reflecting boundaries are added, at zero and $N$, such that whenever $Y_i = 0$, then with probability one $Y_{i+1} = 1$, and similarly, if $Y_i = N$, then with probability one $Y_{i+1} = N-1$. The load $Y_i$ will start at time zero from its stationary distribution, i.e. $\mathsf{P}(Y_0=0)=\mathsf{P}(Y_0=N)=1/2N$ and $\mathsf{P}(Y_0=n)=1/N$ for $n=1,\ldots,N-1$. For convenience, we extend the random walk in the negative negative direction by defining $Y_{-n-1}=Y_{-n}+e_{-i}$ and adding reflecting boundaries at 0 and $N$.

For this random model, one can give an explicit formula for the rainflow range probability function, see Appendix 4.A.1 for details, viz

$$\mathsf{P}(H^{rfc}=N)=\frac{2}{N(N+1)}, \quad \mathsf{P}(H^{rfc}=k)=\frac{4}{k(k+1)(k+2)}, \tag{4.11}$$

if $0<k<N$. As $N$ tends to infinity, i.e. the influence of the reflecting barrier at $N$ fades away and the probability function of the rainflow range becomes

$$\mathsf{P}(H^{rfc}=k)=4/k(k+1)(k+2),$$

for $k=1,2,\ldots$. This is a "Pareto" type distribution having finite mean, but infinite variance.

The frequency of cycles, i.e. of local maxima, is $\mu^{rfc}=(N+1)/4N$, and the expected damage of each cycle is

$$\mathsf{E}[(H^{rfc})^{\beta}]=\sum_{k=1}^{N} k^{\beta}\,\mathsf{P}(H^{rfc}=k).$$

Thus, the damage intensity (4.10), the speed of damage growth, is equal to

$$d^{rfc}=\frac{N+1}{4N}\,\mathsf{E}\left[(H^{rfc})^{\beta}\right]=\frac{1}{2}N^{\beta-2}+\frac{N+1}{N}\sum_{k=1}^{N-1}\frac{k^{\beta}}{k(k+1)(k+2)}. \tag{4.12}$$

Obviously, for $\beta \geq 2$, the damage intensity grows to infinity when $N$ increases without bound, which is not a realistic or interesting property of a random load model. Consequently, we will modify $Y_i$ in such a way so that the load will have bounded values but becomes more and more noisy with increasing $N$.

### 4.4.2 SCALED RANDOM WALK – A VERY NOISY LOAD

Again, let us consider random walk and scale it to have values in the interval $[0,1]$, i.e divide its values by $N$. The normalised random walk takes also values between 0 and 1 with step $1/N$. The scaled random walk has the rainflow damage intensity given by

$$d^{rfc} = \frac{N+1}{4N}\,\mathsf{E}\left[(H^{rfc}/N)^{\beta}\right] = \frac{1}{2}N^{-2} + \frac{N+1}{N^{\beta+1}}\sum_{k=1}^{N-1}\frac{k^{\beta}}{k(k+1)(k+2)}, \tag{4.13}$$

which for large $N$ is close to zero. The reason for this is that the normalised random walk model has a low frequency of cycles $\mu^{rfc} \approx 1/4$ and a lot of very small cycles.

The signal is noisy, but in the limit it is a constant. In order to get an interesting model to study, we need to increase the frequency of cycles $\mu^{rfc}$, e.g., from $1/4$ to $N^2/4$, which will give the following damage intensity,

$$d^{rfc} = \frac{1}{2} + N^{2-\beta}\sum_{k=1}^{N-1}\frac{k^{\beta}}{k(k+1)(k+2)} \approx \frac{1}{2} + \frac{1}{N^{\beta-2}}\sum_{k=1}^{N-1}k^{\beta-3} \approx \frac{1}{2} + \frac{1}{\beta-2},$$

with equality as $N$ tends to infinity and $\beta > 2$.

Note that when $\beta = 2$, the damage intensity grows as $\log N$, as $N$ increases without bound. An interesting observation is that the damage intensity due to the largest rainflow cycle, with range one, is equal to 0.5, while the damage intensity of all remaining cycles is smaller and decreases with increasing $\beta$. It is also easy to see that only a very small fraction of large ranges are responsible for the damage.

**Remark 4.3** (Rainflow filtering). *The scaled random walk illustrates a phenomenon often present in measured loads, namely the presence of high frequency/low energy noise, that results in a huge number of very small cycles. Those small cycles do not contribute much to the damage rate but are inconvenient when loads are used in testing rigs of fatigue strength of components. Consequently, this high frequency "noise" is usually removed by means of rainflow filtering, a technique to remove small cycles from the measured signal/model; see Figure 4.7.*

*Rainflow filtering is important in costly fatigue testing since it reduces the number of cycles in the sequence used to test a component. This leads to shorter sequences of turning points, and reduction up to 90% can be achieved. The rainflow filter just removes all cycles* $(m_{k(i)}, M_i)$ *such that* $h_i = M_i - m_{k(i)} < \varepsilon$ *for a suitably chosen threshold* $\varepsilon$.

## 4.5 THE RAINFLOW COUNT MEMORY

### 4.5.1 WHAT IS RAINFLOW MEMORY?

The damage $D(t)$ is a continuously increasing function. If there are a finite number of cycles, then it is almost everywhere differentiable and hence there is a function $\dot{D}(t)$ such that $D(T) = \int_0^T \dot{D}(t)\,\mathrm{d}t$. The derivative, called damage rate, can be used to compute $D(T)$, $\mathsf{E}[D(T)]$, $\mathsf{Var}[D(T)]$, or to compare damage intensities of different cycle

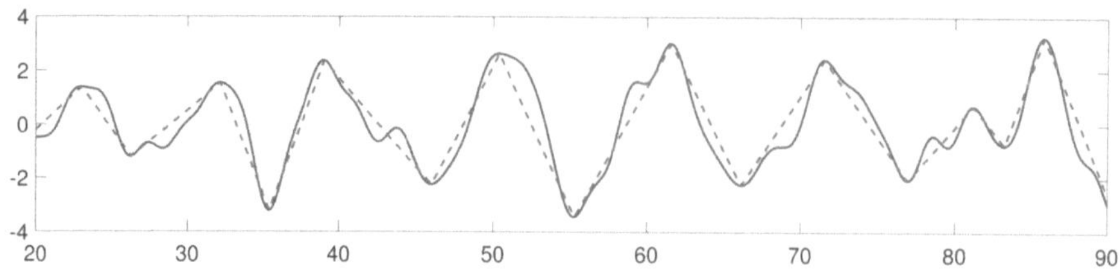


**FIGURE 4.7** Gaussian load and its rainflow filtered sequence of turning points. Filter threshold $\varepsilon = 0.6$.

counts. Note that if the process is not differentiable, like the Ornstein-Uhlenbeck process, then the rainflow damage intensity does not exist. However, the damage intensity can be evaluated using the cycle intensity, and the function $\mu(u,v)$ is defined in (4.27).

In this section, we define the rainflow damage rate $\dot{D}^{rfc}(t)$ which obviously depends only on the load history prior to time $t$ and actually on the residual rainflow count at time $t$. However, similarly like one can count rainflow cycles locally,

$$H^{rfc}(\tau) = \min(H^-(\tau), H^+(\tau)),$$

one can define a memory locally (see (4.17)), and prove that rainflow damage rate is a function of load value and its derivative and the variable $H^-(t) = x(t) - x^-(t)$, see (4.16).

In Section 4.4.1, we introduced a random walk process to illustrate some properties of the rainflow counting method. That model is a time series taking finitely many discrete values. Registrations of continuous load signals are most often sampled and discretised, and then the random walk could be used as a model of the measured load. The accumulated damage is also sampled and increases only at the sampling points.

If the random load model $X(\tau)$, $0 \le \tau \le t$, is a differentiable process (e.g. a differentiable Gaussian process), then a more general rainflow range is everywhere defined, as in (4.4),

$$H^{rfc}(\tau) = \min(H^-(\tau), H^+(\tau)).$$

The accumulated damage $D^{rfc}(t)$ is a sum of $H^{rfc}(\tau_i)^\beta$ over all $\tau_i$, $0 < \tau_i \le t$, such that $H^{rfc}(\tau_i) > 0$. For the damage function $f(u,v) = (v-u)^\beta$, the following result is valid,

$$\frac{dD^{rfc}(t)}{dt} = \dot{X}(t)^+ \beta H^-(t)^{\beta-1}, \tag{4.14}$$

where $\dot{X}(t)^+ = \max(\dot{X}(t), 0)$.

Using (4.14), one can show that for a random load, the damage rate $d^{rfc}(t)$ can be evaluated if the joint distribution of $\dot{X}(t), H^-(t)$ is known. The average speed of rainflow damage growth is

$$d^{rfc}(t) = \mathsf{E}[\dot{X}(t)^+ \beta H^-(t)^{\beta-1}]. \tag{4.15}$$

For a stationary load process $X(t)$, $d^{rfc}(t) = d^{rfc}(0)$.

For a general damage function $f(u,v)$, a similar formula is valid, and one can easily demonstrate, see e.g., Rychlik (1993) [173], that the time derivative of the damage is

$$\frac{dD^{rfc}(t)}{dt} = \dot{X}(t)^{+} \cdot f_2(x(t) - H^{-}(t), x(t)). \tag{4.16}$$

Here, $f_2(u,v)$ denotes the partial derivative of the damage function $f(u,v)$ on $v$.

**Rainflow memory:** Let us introduce the term *rainflow memory* for the left local semi-range (4.3),

$$H^{-}(\tau) = X(\tau) - X^{-}(\tau), \tag{4.17}$$

where

$$X^{-}(\tau) = \min_{\tau^{-} \leq t \leq \tau} X(t), \qquad \tau^{-} = \sup\{t < \tau; X(t) > X(\tau)\}. \tag{4.18}$$

The reason for this name is the way the original definition of rainflow cycle count was proposed. It is the semi-range that is not yet used to produce a full rainflow cycle – it is kept in memory until the value of $H^{+}(\tau)$ is known.

The original Endo's definition of rainflow count used imaginary water as it flows down a roof, starting at valleys and peaks (local minima, maxima) of a load. Let us consider the water that started at the valleys. For any time point $t$ where the load is growing (positive derivative), there is water that reaches that point. The valley from which it originated has height $X^{-}(t)$. The pair $(X^{-}(t), X(t))$ was a half-cycle that was paired with a half-cycle started at local maximum. The half-cycles that could not be paired formed a residual, which served as a memory needed if one wished to online identify cycles and save them. (Note that if the load derivative is negative, such point is "dry" and hence $X^{-}(t) = X(t)$.)

The practical application of formula (4.16) is in the computation of expected damage increase under different operation conditions. For example, the sea wave environment encountered by a sailing vessel changes with time. Often, it is assumed that for a period $T$ of half an hour, it can be described by means of a stationary Gaussian random process. Then, the damage accumulated during the period is equal to $T \cdot \mathsf{E}[\dot{X}^{+}(0) \cdot f_2(X^{-}(0), X(0))]$, by the assumed stationarity of sea condition. In Chapter 6, we shall present means to evaluate the expected damage intensity for Gaussian loads. (Here, we see that the Slepian model for $X$ around crossing will be needed, while for computation of rainflow range distribution, the Slepian model around local maximum is required.)

**Another application of the rainflow memory** is in estimation of crack growth by means of cycle by cycle methods. Here, we use the simple model offered by the Paris-Erdogan law (1.5–1.7). In the case of constant amplitude loading, the Paris-Erdogan equations postulate that the crack increment during the $n$:th load cycle with range $H_n$ is given by

$$a_n - a_{n-1} = \varepsilon a_{n-1}^{\beta/2} H_n^{\beta},$$

where $\varepsilon$ is a small constant that may depend on a slowly growing $a$. Here, $H_n$ is usually the range of a min-Max cycle of the far stress $x(t)$.

Let $t_0 = 0$ and $t_n$, $n = 1,\dots,N$, be the locations of the local maxima in the time interval $[0,T]$. Then the $n$:th range is

$$H_n^\beta = x(t_n) - \min_{t_{n-1}\le t\le t_n} x(t),$$

e.g. the range of $n$:th min-Max cycle. In fatigue failure prediction for variable range loads, one prefers to use rainflow cycles ranges $H_n^{rfc}$ instead of min-Max ranges, and one may ask why not use them in prediction of crack grow, e.g. in Paris-Ergodan's formula. There are several indications that it should be a better method. Experience with variable amplitude crack growth indicates that consideration must be given to the sequencing of the cycles. Crack growth during a load cycle will depend not only on its magnitude but also on the magnitude of the preceding cycles. In addition, there are experiments by Socier and Kurath [193] that show that crack growth prediction using rainflow cycles gives more accurate results than min-Max ranges.

However, the rainflow counting is not causal, since some hysteresis loops may close after time $T$ and some ranges $H_n^{rfc}$ may change with $T$. Using the negative rainflow ranges $H_i^-$ can solve the problem, since these are causal.

The rainflow damage $D^{rfc}(T)$ can be evaluated using rainflow cycle or damage rate $\dot{D}^{rfc}(t)$, viz.

$$D_T^{rfc} = \sum_{n=1}^{N} (H_n^{rfc})^\beta = \int_0^T \dot{D}^{rfc}(t)\,\mathrm{d}t = \sum_{i=1}^{N} \int_{t_{n-1}}^{t_n} \dot{D}^{rfc}(t)\,\mathrm{d}t.$$

Obviously,

$$\int_{t_{n-1}}^{t_n} \dot{D}^{rfc}(t)\,\mathrm{d}t = D^{rfc}(t_n) - D^{rfc}(t_{n-1}),$$

where, for a fixed time $T$,

$$D^{rfc}(t_n) - D^{rfc}(t_{n-1}) = (H_n^{rfc})^\beta.$$

However, if $T$ changes, the range of the n:th rainflow cycle may change too, and this is why direct use of the rainflow ranges is not recommendable, since one would need to alter the crack increment $a_n - a_{n-1}$. A solution for this inconvenience was proposed in the Appendix in [183] by using rainflow memory (the local semi-range $H^-(t)$ defined in (4.3)) instead of rainflow cycle ranges. One employs the following formula for the rainflow damage derivative for the damage function $f(u,v) = (v - u)^\beta$, viz.

$$\dot{D}^{rfc}(t) = \dot{x}(t)^+ \beta (H^-(t))^{\beta-1},$$

where $x^+ = \max(0,x)$. Note that for a stationary process, the derivative of the damage is stationary too. Hence

$$\int_{t_{n-1}}^{t_n} \dot{D}^{rfc}(t)\,\mathrm{d}t = \int_{t_{n-1}}^{t_n} \dot{x}(t)^+ \beta (H^-(t))^{\beta-1}\,\mathrm{d}t = H_n^\beta,$$

which is independent on $T$ as long as $t_n \le T$.

Next, we turn to replace the incremental formula for crack growth $a_{n+1} - a_n$ by the differential equation

$$\mathrm{d}a(t) = \varepsilon\,\beta a(t)^{\beta/2}\dot{x}(t)^{+}H^{-}(t)^{\beta-1}\,\mathrm{d}t.$$

This simple equation has an explicit solution. Namely, there exists a deterministic function $G(\cdot)$, which depends only on $\beta$ and $a_0$, such that

$$a(t) = G\left(\varepsilon \int_0^t \beta\,\dot{x}(s)^{+}H^{-}(s)^{\beta-1}\,\mathrm{d}s\right),$$

and, for $t = t_n$,

$$a_n = a(t_n) = G(D_n^{rfc}),$$

where $D_n^{rfc}$ is the rainflow damage accumulated in the first $n$ rainflow cycles, see Appendix in [183] for more details.

### 4.5.2 RANDOM WALK, CONTINUED

Here, we again consider the random walk process to illustrate the relation between rainflow memory

$$H^{-}(\tau) = X(\tau) - X^{-}(\tau)$$

and rainflow range $H^{rfc}$. We will compute the rainflow damage intensity, i.e. the average increase of the damage in unit time, using rainflow memory and rainflow range distributions. The continuous curve in Figure 4.8 illustrates the random walk $Y(t)$ and the broken curve is its rainflow memory.

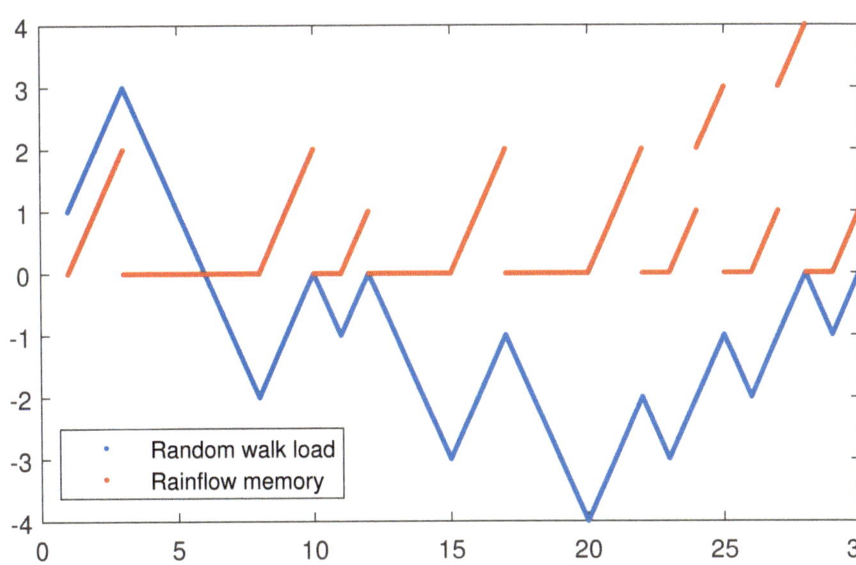


**FIGURE 4.8** A random walk load $\{Y(t)\}$ and its rainflow memory $\{H^{-}(t)\}$. Note that the memory is zero when the load decreases.

Again let $\ldots Y_{-1}, Y_0, Y_1, \ldots$ be a stationary random walk model. Since the rainflow memory is defined for a continuous function, we shall introduce a random process $Y(t)$ such that $Y(i) = Y_i$, $Y(i+1) = Y_{i+1}$ and linearly interpolated for $s$ between $i$ and $i+1$, e.g. $Y(s) = Y_i + (Y_{i+1} - Y_i)(s-i)$.

Now, we are able to evaluate the derivative of the rainflow damage accumulated for $Y(t)$, given in (4.14). Since $\dot{Y}(t)^+ = 1$, we have that

$$\frac{dD^{rfc}}{dt}(t) = \beta H^-(t)^{\beta-1}.$$

In Appendix 4.A.3, we will derive the formula for the random walk rainflow memory distribution, viz.

$$\mathsf{P}(H^- = k) = \frac{N+1}{2Nk(k+1)} \tag{4.19}$$

for all $k$, $0 < k \leq n$.

Now, it is easy to demonstrate that for random walk

$$\mathsf{E}[(H^{rfc})^\beta] = \frac{4N}{N+1}\left(\mathsf{E}[(H^-)^\beta] - \mathsf{E}[(H^- - 1)^\beta]\right), \tag{4.20}$$

where $H^- = H^-(0)$.

*Proof.* If $H^-(i+1) > 0$, then for all $s$, $0 \leq s < 1$, we have that $H^-(i+1-s) = H^-(i+1) - s$ and hence

$$D^{rfc}(i+1) - D^{rfc}(i) = \int_i^{i+1} \beta H^-(s)^{\beta-1}\,\mathrm{d}s = H^-(i+1)^\beta - (H^-(i+1)-1)^\beta,$$

since if $H^-(i+1) = 0$, then $H^-(i+s) = 0$, too. Consequently, the damage intensity is

$$d^{rfc}(i) = \mathsf{E}[D^{rfc}(i+1) - D^{rfc}(i)] = \mathsf{E}[H^-(0)^\beta] - \mathsf{E}[(H^-(0)-1)^\beta], \tag{4.21}$$

since the sequence $H^-(i)$ is a stationary sequence. □

## 4.6 CUMULATIVE CYCLE COUNTS

The empirical variability of fatigue damaging load cycles can be described by means of a two-dimensional histogram of $(u_{k(i)}, v_i)$, i.e. of cycle maxima $v_i$ and their corresponding cycle minima $u_{k(i)}$. An alternative is to use the ranges $h_i = v_i - u_{k(i)}$, disregarding the absolute load level. In both cases, a normalised histogram is an estimate of a probability density function, and it can be used to estimate the damage rate, i.e. the average damage increase per cycle, or time unit.

The damage function (4.4), $f(u,v) = \alpha(v-u)^\beta$, is often used in applications. It is a function of the cycle range $h_i = v_i - u_{k(i)}$, viz. $D(t) = \alpha \sum h_i^\beta$. This is simple if one has measured loads acting on a component. However, at the design stage, uncertainty in loading conditions are described by means of random models, defined by some parametric model, while the variability of operations (missions) is described by letting parameters of those models be random variables. In such a case, in order to estimate damage rates, the probability distribution of cycle ranges has to be computed for the relevant load model.

### 4.6.1 AN INTRODUCTION TO CYCLE COUNT DISTRIBUTIONS

In the following chapters, we will present different types of stochastic models and methods how to compute the expected damage intensity, mostly using the rainflow cycles counting method. It means that we wish to find the probability distribution $F(h) = \mathsf{P}(H \le h)$ or the probability density function (pdf) $f_H(h)$ of the cycle ranges $h_i$. However, for most of the models, one needs to first find the two-dimensional probability density of local maxima and the attached local minima, i.e. top and bottom of a cycle. For example, one starts with estimating the conditional distribution $F(h \mid v) = \mathsf{P}(H \le h \mid Max = v)$, and then deriving the probability density of the cycle ranges by integration, since $\mathsf{P}(H \le h) = \int_{-\infty}^{\infty} \mathsf{P}(H \le h \mid Max = v) f(v)\, \mathrm{d}v$. For this reason, we will put most effort on methods to estimate the two-dimensional distributions (histograms) of $(u_{k(i)}, v_i)$.

Throughout this section, we assume we have observed a load process $x(t)$ over a time interval $[0,T]$; the time $T$ is often suppressed in the notation.

#### 4.6.1.1 Cumulative Range Count

We now introduce some empirical measures for damage-causing properties of a load. First, we introduce the **cumulative range count**,

$$N^h(s) = \text{ number of ranges } h_i \text{ exceeding } s. \tag{4.22}$$

If the load is modelled by a random process, then $N^h(h)/N^h(0)$ is an estimate of the probability that a cycle range exceeds $h$. In the following, we shall use the notations $N^{rfc}$ and $N^{mM}$ for cumulative rainflow and min-Max range counts, respectively. Obviously, it is possible to recover all ranges $h_i$ from the cumulative range count by finding locations and sizes of jumps in $N(s)$. (However, the order those occurred is lost.) Most importantly, the total damage $D(T)$, defined by the damage function $f(v-u) = (v-u)^\beta$, can be evaluated by means of

$$D(T) = \sum_{i=1}^{N} h_i^\beta = \beta \int_0^\infty s^{\beta-1} N^h(s)\, \mathrm{d}s, \tag{4.23}$$

where $h_i = v_i - u_{k(i)}$ and $N = N^h(0)$.

**Positive peak count** is a range count with $h_i = 2v_i^+$ and $N^{pp}(s)$ is the number of positive maxima with $2v_i^+ > s$, and by (4.23),

$$D^{pp}(T) = \beta \int_0^\infty s^{\beta-1} N^{pp}(s)\, \mathrm{d}s. \tag{4.24}$$

Note that formally positive peak count is not a cycle count as defined in this chapter, since the method is not paring local maxima with local minima. Note that only positive peaks are used. It is a convenient model since for many random loads, the probability $\mathsf{P}(H^{pp} > h)$ can be efficiently computed from random load parameters. The disadvantage of the model is that the positive peak damage is always higher than the rainflow damage. The method is suitable for loads with narrow-band spectrum.

#### 4.6.1.2 Cumulative Cycle Count

We first recall the definition of **[up]crossing spectrum** defined by (2.58),

$$N^+(u) = \text{ number of upcrossings of level } u \text{ for } u \in \mathbb{R}, t \in [0,T]. \tag{4.25}$$

For cycle analysis, we can use the two-dimensional cumulative cycle count[2],

$$N(u,v) = \text{ number of cycles } (u_{k(i)}, v_i) \text{ such that } v_i > v \text{ and } u_{k(i)} < u, \tag{4.26}$$

and the corresponding **cycle intensity**, i.e. the expected number of cycles per time unit,

$$\mu(u,v) = \mathsf{E}[N(u,v)]/T. \tag{4.27}$$

We will use the notation $N^{rfc}(u,v)$, $N^{mM}(u,v)$ or $N^{pp}(u,v)$ for the cumulative cycle counts of rainflow, min-Max and positive peak counts.

There exist some simple properties of cumulative counts:

$$N(u_1,v_1) \le N(u_2,v_2), \quad \text{if } u_1 \le u_2 \le v_2 \le v_1, \tag{4.28}$$

$$N(u,u) = N^+(u) \quad \text{for all cycle counts}, \tag{4.29}$$

$$N^{mM}(u,v) \le N^{rfc}(u,v). \tag{4.30}$$

Note that (4.29) does not hold for positive peak count, since it is not a cycle count according to our definition.

Figure 4.9 illustrates the fundamental difference between rainflow cycles and Max-min cycles. The figure shows the theoretical statistical distribution in the form of level curves for a “probability density” of cycle maxima and minima. Each curve encloses a well-defined proportion of cycles, e.g. only 0.1% of all cycles will fall outside the outer curves. The load process is generated as a Gaussian load process with two dominating frequencies.

#### 4.6.1.3 Summary of Crossing and Counting Variables

For a load or load model $x(t)$ observed over a time interval $0 \le t \le T$, we have introduced some principles in notation.

**Counts:** Counts $N$ during interval $[0,T]$.

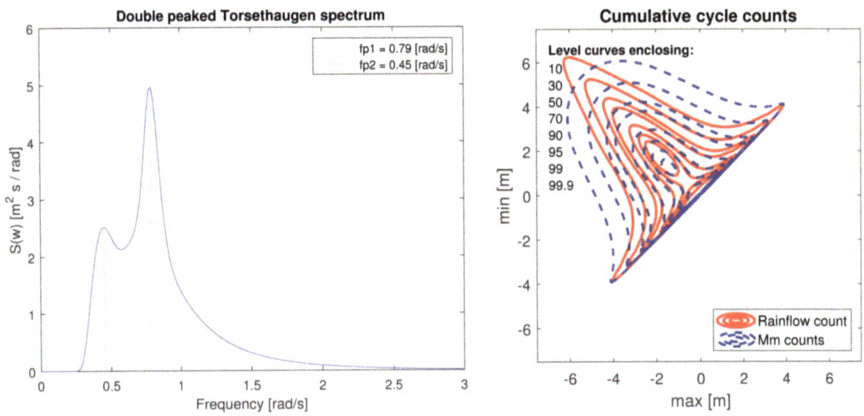


**FIGURE 4.9** Double peaked Torsethaugen seastate spectrum and its cumulative counts of rainflow (solid, red) and Max-min (dashed, blue) cycles.

- $N^+(u)$: number of upcrossings of level $u$ during time $T$, crossing spectrum.
- $N(u,v)$: number of counted cycles $(u_i, v_i)$ with $u_i < u < v < v_i$. Superscript indicates cycle type.
- $N^+(u,v) = N^{osc}(u,v)$: number of oscillations across interval $[u,v]$.

**Expectation:** $\mu_T(N) = \mathsf{E}[N]$, total expected number of counts.

**Intensity:** $\mu = \mathsf{E}[N]/T$, expected number of counts per time unit. For example,

- $\mu^+(u) = \mathsf{E}[N^+(u)]/T$, expected number of $u$-upcrossings per time unit, mean crossing spectrum.
- $k^+(u,v) = min_{u \le z \le v} \mu^+(z)$, minimal upcrossing spectrum.
- $\mu^{osc}(u,v) = \mathsf{E}[N^{osc}(u,v)]/T$, expected number of $(u,v)$-oscillations per time unit.

**Damage and damage bounds:**

- $D(T)$ = total accumulated damage; $\dot{D}(t)$ = damage increase at time $t$.
- $\mathsf{E}[D(T)]$ = total expected damage; $d, d(t) = \mathsf{E}[\dot{D}(t)]$ = damage intensity.
- $d^{nb}$ = narrow-band damage approximation/bound, Definition 4.6.
- $d^+$ = general damage bound, Definition 4.7.

### 4.6.2 A MASTER-FORMULA FOR TOTAL DAMAGE

The cumulative cycle count can be used to evaluate the damage $D(T)$, accumulated at time $T$, for any cycle count by means of the following *master-formula*,

$$D(T) = -\int_{-\infty}^{\infty}\int_{-\infty}^{v} f_{12}(u,v)N(u,v)\,\mathrm{d}u\,\mathrm{d}v + \int_{-\infty}^{\infty} f_2(v,v)N(v,v)\,\mathrm{d}v. \tag{4.31}$$

Formulas (4.23) and (4.31) are particularly useful tools to evaluate expected damage for random processes. For discretised loads that take only a finite number of values, one can retrieve the cycle count from the cumulative cycle count and compute damage by summing damages of individual cycles.

For example, consider a discretised load, which takes values $1,\dots,N$. Then, the cumulative rainflow count $N^{rfc}$ is an $N$ by $N$ matrix with elements $N^{rfc}(i,j)$. (For $i>j$, $N^{rfc}(i,j)=0$.) Now, the number of cycles $(u_{k(n)},v_n)=(i,j)$, denoted by $N_{ij}$, say, can be computed from the matrix $N^{rfc}$ as follows:

$$N_{ij}=N^{rfc}(i+1,j-1)+N^{rfc}(i,j)-N^{rfc}(i,j-1)-N^{rfc}(i+1,j), \qquad (4.32)$$

leading to the following discrete master-formula for the accumulated damage,

$$D^{rfc}=\sum_{ij}N_{ij}f(i,j). \qquad (4.33)$$

## 4.7 THE NARROW-BAND BOUND FOR RAINFLOW DAMAGE

### 4.7.1 BENDAT'S NARROW-BAND APPROXIMATION

**Definition 4.6** (Bendat's narrow-band damage approximation). *For a stationary load with mean zero and variance $\sigma^2$ and with apparent frequency $f_z$, the Bendat narrow-band approximation of expected damage per time unit for damage function $(v-u)^\beta$ is defined as*

$$d^{nb}=f_z\sigma^\beta\,2^{3\beta/2}\Gamma(1+\beta/2). \qquad (4.34)$$

*Here, $f_z$ is the average number of upcrossings of the mean level per time unit.*

Bendat (1964) [12] considered zero-mean stationary Gaussian loads with spectrum concentrated around a mean frequency. Essentially, such loads have one positive peak $H_i$ between the $i$th zero level upcrossing and the following downcrossing. In [12, Eq. (23)], Bendat gave the following formula for the peak distribution,

$$\mathsf{P}(H>u)=\frac{\mu^+(u)}{\mu^+(0)},$$

where $\mu^+(u)$ is the frequency of upcrossing of the level $u$. Using Rice's formula for Gaussian loads (2.63), he concluded that $H$ has a Rayleigh distribution, i.e. $H=\sqrt{\omega_0}R=\sigma R$, where $\mathsf{P}(R\geq r)=e^{-r^2/2}$. Then, with $\mu^+(0)$ cycles per time unit and damage per cycle equal $(2H)^\beta$, the expected fatigue damage in time $T$ could be approximated by

$$\mathsf{E}[D(T)]=T\mu^+(0)\mathsf{E}[(2H)^\beta]=T\mu^+(0)\mathsf{E}[(2\sigma R)^\beta]. \qquad (4.35)$$

The narrow-band approximation is expressed in terms of "apparent frequency" $f_z=\mu^+(0)$, (for a Gaussian load $\frac{1}{2\pi}\sqrt{\omega_2/\omega_0}$), load standard deviation $\sigma=\sqrt{\omega_0}$, and the $\beta$-moment of the Rayleigh distribution, $\mathsf{E}[R^\beta]=2^{\beta/2}\Gamma(1+\beta/2)$.[3] Then, the

**narrow-band approximation** for a stationary zero-mean Gaussian load of expected total damage is

$$\mathsf{E}[D(T)] = T f_z \sigma^\beta 2^{3\beta/2}\Gamma(1+\beta/2) = D_T^{nb} = T d^{nb}. \tag{4.36}$$

This approximation is often used in offshore engineering for wave loads even if the wave spectra are not narrow. In the literature, one remarked that the narrow-band approximation "seams" to give conservative estimate of the expected rainflow damage and proposed different modifications in order to limit the overestimation.

**Remark 4.4** (What is the distribution of a local peak?). *The narrow-band approximation was motivated by the observation that a zero-mean stationary narrow-band Gaussian load has, essentially, one and only one local maximum per zero upcrossing, and that its random amplitude A is Rayleigh distributed. In fact, Rice, in [168, Eqn. 3.6–3.9], derived the exact distribution for the height of local maxima in Gaussian noise. The height of a randomly chosen maximum M is distributed as*[4] *the sum of a Rayleigh variable R and a normal variable U,* (6.10),

$$M \overset{\mathscr{L}}{=} \sigma\left(\alpha_2 R + \sqrt{1-\alpha_2^2}U\right), \tag{4.37}$$

*where* $\alpha_2 = \omega_2/\sqrt{\omega_0\omega_4}$ *is a measure of bandwidth;* $\alpha_2$ *near one indicates narrow spectrum and a near Rayleigh maximum distribution.*

### 4.7.2 THE NARROW-BAND AND A GENERAL DAMAGE BOUND

The narrow-band damage approximation is based on the crossing spectrum, $N^+(u)$, or the crossing intensity $\mu^+(u) = \mathsf{E}[N^+(u)]/T$. The specific form in Definition 4.6 derives from Rice's formula for $\mu^+(u)$, valid only for Gaussian loads. It can be refined to an upper bound for damage under some simple restrictions on the load and on the definition of the counting permutation; see the definition of cycle counts, Definition 4.2.

Actually, as we show below, the narrow-band approximation is a bound for the expected damage for any stationary symmetrical load ($\mu^+(-u) = \mu^+(u)$), e.g. Gaussian or Laplace with symmetrical kernels, and for any correctly defined cycle count.

First, we conclude from properties (4.28) and (4.29) that for any cycle count

$$N(u,v) \le N(z,z) = N^+(z), \qquad \text{for any } z,\ u \le z \le v, \tag{4.38}$$

and hence the cumulative cycle intensities,

$$\mathsf{E}[N(u,v)]/T \le \min_{u\le z\le v} \mathsf{E}[N^+(z)]/T = \min_{u\le z\le v} \mu^+(z) = k^+(u,v). \tag{4.39}$$

Thus, the narrow-band approximation is conservative whenever $\min_{u\le z\le v}\mu^+(z) = \mu^+(z)$ where $z = \max(|u|,|v|)$.

The minimum expression in (4.39) appears in many damage bounds in this and later chapters. It was defined in Definition 2.5 and called the *minimal upcrossing spectrum*,

$$k^+(u,v) = \min_{u \le z \le v} \mu^+(z). \tag{4.40}$$

Finally, the master-formula (4.31) gives a bound for expected damage per time unit valid for any cycle count,

$$\mathsf{E}[D(T)]/T \le -\int_{-\infty}^{\infty}\int_{-\infty}^{v} f_{12}(u,v)k^+(u,v)\,du\,dv + \int_{-\infty}^{\infty} f_2(v,v)\,\mu^+(v)\,dv. \tag{4.41}$$

The reader is reminded that the positive-peak count is not a true cycle count and that the formulas given above are not useful in that case.

**Definition 4.7** (A general damage bound $d^+$). *The right hand side in* (4.41) *is called the general damage intensity bound and is denoted* $d^+$. *For a damage function* $f(u,v)$ *with* $f_2(v,v) = 0$, *it is given as*

$$d^+ = -\int_{-\infty}^{\infty}\int_{-\infty}^{v} f_{12}(u,v)k^+(u,v)\,du\,dv. \tag{4.42}$$

**Remark 4.5.** *The master-formula (4.31) consists of two integrals. In the second, one is integrating* $f_2(u,u)N(u,u)$*, and this integral is often equal to zero, e.g., if* $f(u,v) = (v-u)^\beta$ *and* $\beta > 1$. *For* $\beta = 1$*, the second-order partial derivative* $f_{12}(u,v) = 0$*, and we have the result that*

$$\sum (v_i - u_i) = \int N(u,u)\,du = \int N^+(u)\,du.$$

*The sum is equal to the total distance the load is moving upwards, also called the positive variation of the load, and it is equal to an integral of the crossing spectrum. (This result is also valid for very irregular loads having an infinite number of cycles in finite interval. This follows from Banach's theorem on total variation.) Note that for any cycle count,* $\sum(v_i - u_{k(i)}) = \sum(v_i - u_i)$.

The crossing spectrum $N^+(u)$ is often used in fatigue analysis to describe load variability and was introduced by Gaßner 1954, in German [67] and in English 1956 [68], for zero-mean and narrow-band Gaussian loads.

**Example 4.2** (Gaussian load). *Consider the damage function* $f(u,v) = (v-u)^\beta$ *and let* $X(t)$ *be a Gaussian load with mean zero and variance one and assume time is scaled, so the variance of the derivative* $\dot{X}(t)$ *is one. Then, Rice's formula for the expected number of crossings, introduced in Chapter 2, (2.63), gives*

$$\mathsf{E}[N^+(u)] = \frac{1}{2\pi}\exp(-u^2/2),$$

*and for $\beta \geq 2$, we have that*

$$d^{rfc} \leq d^{+} = \int_{-\infty}^{\infty} \int_{-\infty}^{v} \beta(\beta-1)(v-u)^{\beta-2} k^{+}(u,v)\, du\, dv. \tag{4.43}$$

*Using symmetry of $k^{+}(u,v)$, we can rewrite this integral in the following way,*

$$\begin{aligned} d^{rfc} \leq & 2\int_{0}^{\infty} \int_{-v}^{v} \beta(\beta-1)(v-u)^{\beta-2} \frac{1}{2\pi} e^{-v^2/2}\, du\, dv \\ = & \frac{1}{2\pi} \int_{0}^{\infty} \beta(2v)^{\beta-1} e^{-v^2/2}\, dv = \frac{1}{2\pi} \int_{0}^{\infty} (2r)^{\beta} e^{-r^2/2}\, r\, dr \\ = & \frac{1}{2\pi} 2^{3\beta/2} \Gamma(1+\beta/2). \end{aligned} \tag{4.44}$$

Take a look at formula (4.44). The term $1/2\pi$ is the frequency of mean level upcrossings by the Gaussian process ($\mathsf{E}[N^{+}(0)]$), while the integral is equal to $\mathsf{E}[(2R)^{\beta}]$ where $R$ is a Rayleigh distributed variable $\mathsf{P}(R > r) = \exp(-r^2/2)$. This is what Bendat proposed 1964 [12] to estimate fatigue damage caused by a Gaussian amplitude load at a time when rainflow cycle counting was not yet available. It was named narrow-band approximation since it is exact for the simplest Gaussian process consisting of a single harmonic having Rayleigh distributed amplitude. Here, it was demonstrated that Equation (4.44) actually is a bound for the rainflow damage under stationary Gaussian loading.

## 4.8 CROSSING OF INTERVALS, OSCILLATION INTENSITY

### 4.8.1 MOTIVATION AND DEFINITION

The local definition of rainflow cycles in Section 4.2.3.3 is easy to understand and easy to transform into an algorithm for online load analysis. The theoretical analysis of the statistical properties leading to assessment of the fatigue damage is more difficult.

Level upcrossings with the crossing spectrum $N^{+}(u)$ defined in Equation (4.25) and the upcrossing intensity $\mu^{+}(u)$ have an interesting and very useful generalisation in *crossings of intervals* also called *oscillations across intervals*. As will be shown in this section, counting oscillations across intervals will give the same information as the cumulative rainflow cycle count, at a lower computational cost. In the following chapters, we will present dedicated exact and approximate methods for oscillation analysis, developed for different stochastic load models.

Figure 4.10 shows a sample of a continuous load that combines regular low frequency cycles of high amplitude with high frequency components of small amplitude. The oscillation between the lower region $U = \{x < u\}$ and the upper region $V = \{x > v\}$ is clearly identifiable even if the high and low extreme parts are "cluttered" by high frequency cycles.

In this section, we make a formal analysis of this visual impression by defining the notion *[up]crossings of intervals* $N^{+}(u,v)$, which will turn out to contain the same information about load variability as the cumulative rainflow count. In fact,

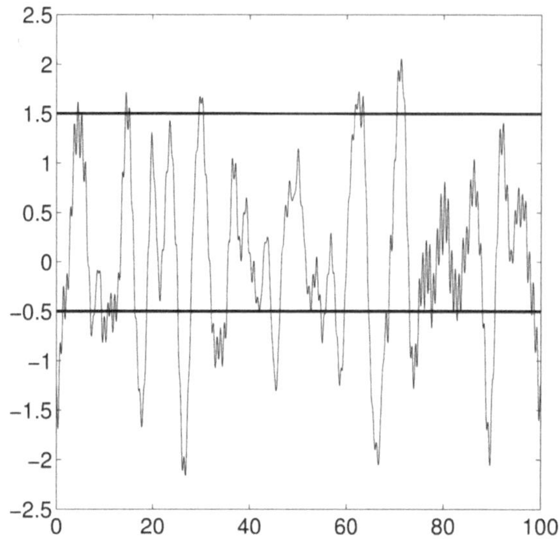


**FIGURE 4.10** Crossings of interval $[u, v]$, $u = -0.5$ and $v = 1.5$, $N^+(u, v) = 5$.

for a load $x(t)$, $0 \le t \le T$, having finitely many local extremes, i.e. the sequence of turning points $T_P$ is finite, there is an equality (with a possible difference $\pm 1$ at the ends of the observation interval),

$$N^{rfc}(u, v) = N^+(u, v). \tag{4.45}$$

For a stochastic stationary load process, $X(t), t \in \mathbb{R}$, under mild regularity conditions, the two intensities are equal,

$$\mu^{rfc}(u, v) = \mu^+(u, v). \tag{4.46}$$

**Definition 4.8** (Interval oscillation). *Let $x(\tau).0 \le \tau \le T$ be a continuous load function and let $u < v$ be real numbers. a) An upcrossing (up-oscillation) of the interval $[u, v]$ is said to take place between times $s$ and time $t$ if*

$$x(s) = u, x(t) = v, \text{ and } u < x(\tau) < v \text{ for all } \tau, s < \tau < t.$$

*b) The number of completed up-oscillations in $[0, T]$ is the oscillation count, denoted $N^+(u, v)$ or $N^{osc}(u, v)$.*
*c) For a stationary load process, the oscillation intensity is denoted*

$$\mu^{osc}(u, v) = \mathsf{E}[N^+(u, v)]/T.$$

In engineering applications, $N^+(u, v)$ and $N^{rfc}(u, v)$ are equivalent. However, the interval crossings can be finite even for very irregular loads having an infinite number of rainflow cycles. In the following chapters, we will see that the evaluation of $\mathsf{E}[N^+(u, v)]$ is easier than $\mathsf{E}[N^{rfc}(u, v)]$. Later in this section, we will use the expected number of interval crossings to derive the distribution of rainflow cycle ranges and rainflow memory for the random walk model.

### 4.8.2 OSCILLATION INTENSITY = RAINFLOW CYCLE INTENSITY

As mentioned, for many load models, it is easier to compute the distribution of the oscillation count $N^{osc}(u,v)$ than for the rainflow count $N^{rfc}(u,v)$. But the equivalence between the two is not obvious at first glance. Figure 4.11 should clarify the matter.

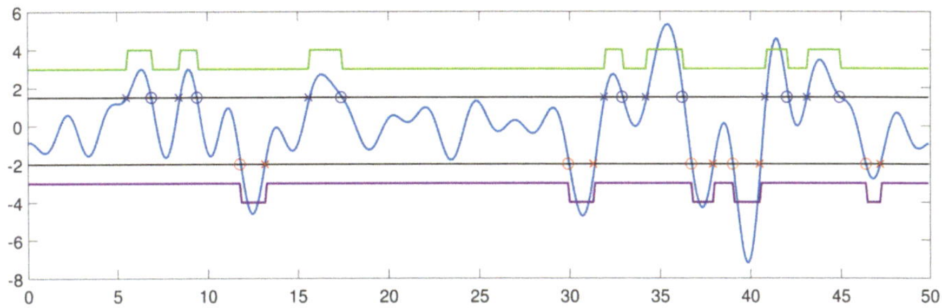


**FIGURE 4.11** Upwards and downwards interval oscillations across $[-2,1.5]$. The upper green and lower purple tooth curves show the locations of visits outside the interval. An oscillation according to the definition is counted from an upper(lower) tooth only if the next tooth is of type lower(upper). Only the last tooth in a group of similar teeth connects to the opposite tooth.

**Theorem 4.1.** *If the random load function $X(t), t \in \mathbb{R}$ is stationary and fulfils the conditions for Bulinskaya's Lemma 6.5, the cumulative rainflow cycle intensity and the interval crossing intensity coincide,*

$$\mu^{rfc}(u,v) = \mu^{osc}(u,v),$$

*and the master-equation for rainflow damage intensity reads*

$$d^{rfc} = -\int_{-\infty}^{\infty}\int_{-\infty}^{v} f_{12}(u,v)\mu^{osc}(u,v)\,\mathrm{d}u\,\mathrm{d}v + \int_{-\infty}^{\infty} f_2(u,u)\mu^{+}(u)\,\mathrm{d}u. \tag{4.47}$$

*Proof.* We show that each rainflow cycle $(m_i^{rfc}, M_i)$ with $m_i^{rfc} < u$ and $M_i > v$ can be coupled to exactly one upward oscillation over $[u,v]$ and vice versa, i.e. exact one upward oscillation over $[u,v]$ can be coupled to exactly one such rainflow cycle outside the bounds. We assume that all rainflow cycles are complete. Note that Bulinskaya's Lemma assures that no turning point falls exactly on any of the border lines.

First, we observe that up- and down-oscillations alternate and that between two adjacent oscillations, there can be a variable number of boundary crossings. For example, in Figure 4.11, there is the sequence: "v-upcrossing, 2 extra v-crossings, down-oscillation, up-oscillation, down-oscillation, up-oscillation, 2 extra v-crossings, down-oscillation, 2 extra u-crossings, up-oscillation, 2 extra v-crossings, down-oscillation, u-upcrossing". We consider what happens between an up-oscillation starting at $t_u^-$, ending at $t_v^-$ and the following down-oscillation starting at $t_v^+$, ending at $t_u^+$, say, see Figure 4.12.

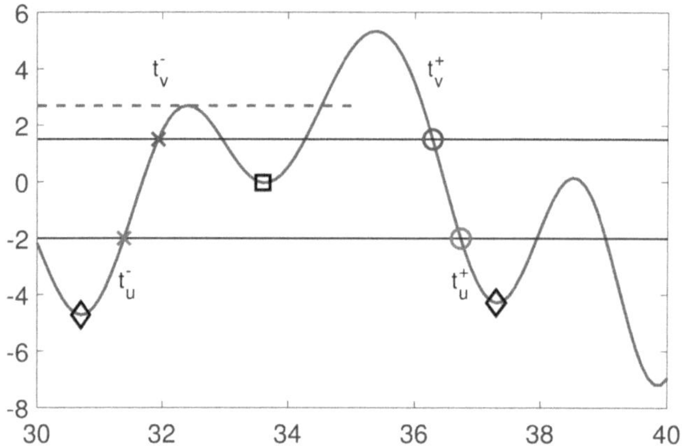


**FIGURE 4.12** Detail of alternating up-oscillations and down-oscillations.

If there is only one local maximum $M$ in the interval $[t_v^-, t_v^+]$, it acts as a rainflow maximum. Its closest local minima lie outside the interval $[t_u^-, t_u^+]$, like the two diamonds in the figure. These are contenders of being the rainflow minimum to the single local maximum. There may be more local minima to be considered, but the smallest minima on either side before the cycle can be closed are at least as small as the two closest.

With more than one local maximum in the interval $[t_v^-, t_v^+]$, one has the situation in Figure 4.12. Maxima lower than the largest will be stopped within the interval, and there will be a local minimum, like the square in the figure, with height above $u$. Only the largest maximum has a rainflow cycle spanning $[u, v]$.

We have shown that between an up-oscillation and the next down-oscillation, there is exactly one rainflow cycle that spans $[u, v]$. A little reflection shows that after a down-oscillation, there must be an up-oscillation before there can be any local maximum above $v$, which implies that in the space between a down-oscillation and the following up-oscillation, there cannot be any rainflow cycle that spans $[u, v]$. This completes the proof of (4.46). □

**Remark 4.6** (Relation to rainflow memory). *During the proof, we have the following expression as a by-product:*

$$N_T^{rfc}(u, v) = \#\{\tau : x(\tau) = v, \dot{x}(\tau) > 0, x_\tau^- < u\}. \tag{4.48}$$

**Remark 4.7** (Discretised loads). *Before we start with calculations of* $\mathsf{E}[N^+(u, v)]$, *we need to mention that there are technical/methodological differences when the random load is discretised, i.e. data is a time series* $X_n$ *taking enumerable values, sampled from a smooth continuous random functions* $X(t)$. *The reason for this difference is that for stationary, smooth random loads (under general conditions, see*

*Bulinskaya's Lemma [33]), the probability that there is a maximum with height exactly u, or minimum with height v, in $X(t)$ is zero. Consequently*

$$\mathsf{E}[N^+(u,v)] = \mathsf{P}(H^- \geq v-u \mid X(0)=v)\mathsf{E}[N^+(v)].$$

*The situation is different when the load is a time series having discretised with fixed steps, e.g., all values of $X_n$ are integers and hence*

$$\mathsf{E}[N^+(i,j)] = \mathsf{P}(H^- \geq k+1 \mid Y_0 = j+1)\mathsf{P}(Y_0 = j+1), \tag{4.49}$$

*where $k = j-i$.*

### 4.8.3 ON COMPUTATION AND GENERALISATIONS

Knowing the interval crossing intensities, one can evaluate the expected rainflow damage both for discrete and for continuous loads. Unfortunately, efficient means to derive the interval crossing intensities are rare. Hence, one tries to find approximate but accurate bounds. (Obviously, Monte Carlo methods are always available but can be inconvenient.) A notable exception is the very broad class of Markov loads, for which efficient and exact means to evaluate $\mathsf{E}[N^+(u,v)]$ exist. Those loads will be described in a separate chapter. Here, we will introduce this model by considering the random walk, which is a Markov chain.

In this section, we give an algorithm to evaluate $\mathsf{E}[N^+(u,v)]$ for a stationary, time-reversible Markov chain, such that $\mathsf{P}(X_n = X_{n+1}) = 0$. Obviously, if $X_n = X_{n+1}$, then the rainflow memory $H^-_{n+1} = 0$, so no damage is accumulated at time $n+1$.

The presented approach employs a somewhat unexpected relation between the distribution of oscillations in a Markov chain and the currents in an electric circuit. This will be briefly presented in the following subsection. The reason for presenting this "network" method here is that it is a convenient method to evaluate the variability of multi-axial loads by studying the so-called oscillations between sets. Note that crossing of intervals $N^+(u,v)$ is also equal to the number of oscillations between the sets $A = (-\infty,u)$ and $B = (v,\infty)$, denoted by $N^{osc}(A,B)$.

This method was introduced in Beste et al. 1992 [16] to study rainflow counts for linear combinations of stress vectors, and by this also to define the rainflow method for multi-axial loads. The network approach to rainflow cycles distribution was presented by Rychlik 1999, as "Rainflow cycles, Markov chain, and electrical circuits", [178].

### 4.8.4 MARKOV CHAIN AND ELECTRIC CIRCUITS

Consider a Markov chain $Y_n$. The chain is defined by a two-dimensional probability function $F_{kl} = \mathsf{P}(Y_0 = k, Y_1 = l)$. Here, we consider only reversible and stationary Markov chain $Y_n$, with $N$ states. This means that

$$F_{kl} = \mathsf{P}(Y_0 = k, Y_1 = l) = \mathsf{P}(Y_0 = l, Y_1 = k) = F_{lk}.$$

The assumed time reversibility means that the Markov chain "looks the same" regardless of whether we run it forwards in time or backwards in time. We also assume that $F_{kk} = 0$. For the Markov chain defined by the matrix $F = [F_{kl}]$, we introduce an electric circuit defined by a conductance matrix $C_{kl} = F_{kl}$. Obviously, the resistance matrix has elements $R_{kl} = 1/F_{kl}$.

Now, for the above-introduced Markov chain, defined by the transition matrix $[F_{kl}]$, the upcrossing intensity $\mathsf{E}[N^+(i,j)]$ can be found by studying currents flowing in the electrical circuit defined by the conductance matrix $C_{kl} = F_{kl}$. Next, we will demonstrate how to use the electrical circuit model to evaluate the interval upcrossing intensities $\mu^+(i,j) = \mathsf{E}[N^+(i,j)]$ for the Markov chain model.

Denote by $S$ the collection of terminals, in our example $S = \{0,1,2,\dots,N\}$, which are the labels attached to the terminals. Often, for Markov chain load models, the labels correspond to the values taken by the load, but also more complex networks can be interesting for fatigue investigations.

We wish to study the frequency of upcrossings of intervals $[i,j]$, i.e. oscillations of the Markov chain between two disjoint subsets of $S$, $U = 0,\dots,i-1$ and $V = j+1,\dots,N$. Assume that we have an outside source with two terminals at voltage 0 and 1, respectively. Connect all terminals in $U$ to the 1 voltage of the source and all terminals in $V$ to the voltage 0. Then, the current $I_{U,V}$ flowing out of the source is equal to the interval upcrossing intensity $\mathsf{E}[N^+(U,V)]$. The current can be found by means of Kirchhoff's Law.

We finish this section by proving that for random walk

$$\mathsf{E}[N^+(i,j)] = \frac{1}{2N(j-i+2)}. \tag{4.50}$$

*Proof* of (4.50). For the random walk, the chain has $N+1$ states $0,1,\dots,N$ with $F_{ij} = 1/2N$ if $|i-j| = 1$, $1 < i \le j < N+1$ and zero otherwise. Consequently, the Markov chain defines an electrical network of resistors (all having resistance $R = 2N$ Ohm) connected in series. For fixed values of $i$ and $j$, the system contains $j-i+2$ resistors and hence the current flowing out of the source is $1/2N(j-i+2)$. □

In the following appendix, we demonstrate how one can derive a formula for rainflow range distribution (4.11) when one knows $\mathsf{E}[N^+(i,j)]$ and prove (4.11) from (4.50).

## APPENDIX

## 4.A SOME PROOFS

### 4.A.1 PROOF OF FORMULA (4.11) FOR RAINFLOW RANGE DISTRIBUTION

In this section, we will prove (4.11), viz.

$$\mathsf{P}(H^{rfc} = N) = \frac{2}{N(N+1)}, \quad \mathsf{P}(H^{rfc} = k) = \frac{4}{k(k+1)(k+2)}.$$

Let us consider a random walk with states $0,1\ldots,N$. By Equation (4.32), the frequency $N_{ij}$ of rainflow cycles having top $j$ and bottom $i$, $0 \le i < j \le N$, is

$$N_{ij} = N^+(i+1,j-1) + N^+(i,j) - N^+(i,j-1) - N^+(i+1,j).$$

Note that if $i=0$ or $j=N$, then $N^+(i,j)=0$.

We first compute the frequency of rainflow cycles with range $N$, i.e. $\mathsf{E}[N_{0N}]$. Formula (4.50) gives

$$\mathsf{E}[N_{0N}] = \mathsf{E}[N^+(1,N-1)] = \frac{1}{2N^2}.$$

Now, $\mathsf{P}(H^{rfc}=N)$ is equal $E[N_{0N}]/\mu^{max}$, where $\mu^{max}$ is the frequency of rainflow cycles. For random walk, $\mu^{max}=(N+1)/4N$ and hence $\mathsf{P}(H^{rfc}=N)=2/N(N+1)$.

Next, we evaluate $\mathsf{P}(H^{rfc}=k)$ for $k<N$. Let $k=j-i$ and $0<i<j<N$. Then

$$\mathsf{E}[N_{ij}] = \frac{1}{2N}\left(\frac{1}{k} - \frac{1}{k+1} + \frac{1}{k+2} - \frac{1}{k+1}\right) = \frac{1}{Nk(k+1)(k+2)}.$$

Consider the remaining two cases, $j-i=k$ and $i=0$ or $j=N$. Similar calculations give $\mathsf{E}[N_{0k}] = 1/2Nk(k+1)$ and $\mathsf{E}[N_{(N-k)N}] = 1/2Nk(k+1)$. Now

$$\sum_{j-i=k} \mathsf{E}[N_{ij}] = \frac{N-k+1}{Nk(k+1)(k+2)} + \frac{1}{Nk(k+1)} = \frac{N+1}{Nk(k+1)(k+2)},$$

which finishes the proof since

$$\mathsf{P}(H^{rfc}=k) = \sum_{j-i=k} \mathsf{E}[N_{ij}]/\mu^{max} = \frac{4}{k(k+1)(k+2)}.$$

### 4.A.2 PROOF OF FORMULA (4.12) FOR DAMAGE INTENSITY OF RANDOM WALK

We evaluate $d^{rfc}$ using (4.21), i.e. employing the probability distribution of $H^-$,

$$\begin{aligned} d^{rfc} &= \mathsf{E}[(H^-)^\beta] - \mathsf{E}[(H^- - 1)^\beta] \\ &= \sum_{k=1}^{N} k^\beta \frac{N+1}{2Nk(k+1)} - \sum_{k=1}^{N} (k-1)^\beta \frac{N+1}{2Nk(k+1)}. \end{aligned} \tag{4.51}$$

We turn to the evaluation of the average damage increase in unit time,

$$\begin{aligned} d^{rfc} &= \mathsf{E}[(H^-)^\beta - (H^- - 1)^\beta] = \mathsf{E}[(H^-)^\beta] - \mathsf{E}[(H^- - 1)^\beta] \\ &= \sum_{k=1}^{N} k^\beta \frac{N+1}{2Nk(k+1)} - \sum_{k=1}^{N} (k-1)^\beta \frac{N+1}{2Nk(k+1)}. \end{aligned} \tag{4.52}$$

Now

$$\mathsf{E}[(H^- - 1)^\beta] = \sum_{k=1}^{N} (k-1)^\beta \frac{N+1}{2Nk(k+1)} = \sum_{m=1}^{N-1} m^\beta \frac{N+1}{2N(m+1)(m+2)},$$

by variable change $k-1=m$. Now, we can write the final result,

$$\begin{aligned}
d^{rfc} &= \mathsf{E}[(H^-)^\beta - (H^- - 1)^\beta] \\
&= \sum_{k=1}^{N-1} k^\beta \frac{N+1}{2N}\left(\frac{1}{k(k+1)} - \frac{1}{(k+1)(k+2)}\right) + N^\beta \frac{N+1}{2N}\frac{1}{N(N+1)} \\
&= \frac{1}{2}N^{\beta-2} + \frac{N+1}{N}\sum_{k=1}^{N-1} k^\beta \frac{1}{k(k+1)(k+2)} \\
&= \frac{N+1}{4N}\mathsf{E}[(H^{rfc})^\beta],
\end{aligned} \tag{4.53}$$

and we proved (4.12).

### 4.A.3 PROOF OF FORMULA (4.19) FOR RAINFLOW MEMORY DISTRIBUTION IN RANDOM WALK

Using formulas (4.49) and (4.50), we obtain that for all $k \leq n$,

$$\mathsf{P}(X^- \geq k, Y_0 = n) = \mathsf{E}[N^+(n-k-1, n+1)] = 1/2Nk, \tag{4.54}$$

while $\mathsf{P}(H^- \geq n+1, Y_0 = n) = 0$. It is also obvious that $\mathsf{P}(X^- = 0, Y_0 = N) = 0$, $\mathsf{P}(H_0^- = 0, Y_0 = 0) = 1/2N$ for all $k < N$. Now, since

$$\mathsf{P}(H_0^- = k, Y_0 = n) = P(H_0^- \geq k, Y_0 = n) - \mathsf{P}(H_0^- \geq k+1, Y_0 = n),$$

we obtain that

$$\mathsf{P}(H^- = k) = \frac{N-k}{2Nk(k+1)} + \frac{1}{2Nk} = \frac{N+1}{2Nk(k+1)}.$$

for all $k$, $0 < k \leq n$. Here, we wrote $H^-$ for $H_0^-$.

### 4.A.4 PROOF OF THE MASTER-FORMULA (4.31)

Before we consider a general cycle count, suppose that one has only one cycle $(u,v)$; then $D(t) = f(u,v)$. The following integral representation of $f(u,v)$ can be obtained by integration by parts applied to the damage function and is also easily verified by carrying out the double integration:

$$f(u,v) = -\int_u^v \int_u^y f_{12}(x,y)\,\mathrm{d}x\,\mathrm{d}y + \int_u^v f_2(y,y)\,\mathrm{d}y.$$

Using the indicator function $1_A(x) = 1$ if $x \in A$ and zero otherwise, the last equation can be written

$$f(u,v) = -\int_{-\infty}^{\infty}\int_{-\infty}^{v} f_{12}(x,y) 1_{(u,v)}(x) 1_{(u,v)}(y)\,\mathrm{d}x\,\mathrm{d}y + \int_{-\infty}^{\infty} f_2(y,y) 1_{(u,v)}(y)\,\mathrm{d}y.$$

Finally, consider the cycle count $(u_i, v_i)$, $i = 1, \ldots, N$, and the damage $D(t) = \sum_{i=1}^{N} f(u_i, v_i)$. Exchanging summation and integration (allowed by means of Fubini's theorem), one gets

$$D(t) = -\int_{-\infty}^{\infty}\int_{-\infty}^{v} f_{12}(x,y)\sum_{i=1}^{N} 1_{(u_i,v_i)}(x)1_{(u_i,v_i)}(y)\,\mathrm{d}x\,\mathrm{d}y$$
$$+\int_{-\infty}^{\infty} f_2(y,y)\sum_{i=1}^{N} 1_{(u_i,v_i)}(y)\,\mathrm{d}y,$$

and (4.31) is proved, since it is easy to see that

$$N(x,y) = \sum_{i=1}^{N} 1_{(u_i,v_i)}(x)1_{(u_i,v_i)}(y), \quad N(y,y) = \sum_{i=1}^{N} 1_{(u_i,v_i)}(y).$$

**Corollary 4.1.** *Let $f(u,v) = v - u$ be a damage function defined in Definition 4.4. For any cycle count with counting distribution $N(x,y)$, the damage is*

$$\sum(v_i - u_i) = \int_{-\infty}^{\infty} N(x,x)\,\mathrm{d}x. \tag{4.55}$$

*Proof:* The corollary is a simple consequence of (4.31), since for $f(u,v) = v - u$, $\frac{\partial^2 f(x,y)}{\partial x \partial y} = 0$, while $\frac{\partial f(x,y)}{\partial y} = 1$.

**Remark 4.8.** *Note that for $f(u,v) = (v-u)^\beta$, $\beta > 1$ the second integral in (4.31) vanishes, while for $\beta = 1$ the first integral is equal zero.*

*In addition, the formula can also be written in an alternative way*

$$D(t) = -\int_{-\infty}^{\infty}\int_{-\infty}^{v} f_{12}(u,v)N(u,v)\,\mathrm{d}u\,\mathrm{d}v - \int_{-\infty}^{\infty} f_1(u,u)N(u,u)\,\mathrm{d}u. \tag{4.56}$$

However, those are not equivalent in the sense that those give the same result for any function $f(u,v)$. For example, consider the positive peak count and damage function $f(u,v) = (2v)^\beta$, then formula (4.31) would give correct positive peak damage (4.24), while formula (4.56) would give damage 0.

## Notes

[1]However, we could use also $\beta > 0$, in the limit as $\beta$ tends to zero, the damage will be equal to the number of cycles.

[2]The terms "cumulative range count" and "cumulative cycle count" allude to cumulative distribution functions in probability theory.

[3]$\Gamma(x) = \int_0^\infty y^{(x-1)} \exp(-y)\,\mathrm{d}y$ is the gamma function.

[4]The symbol $X \stackrel{\mathscr{L}}{=} Y$ means that the random variables $X$ and $Y$ have the same distribution.

# 5 Sampled Loads – Time Series

## 5.1 INTRODUCTION

In the previous chapter, we used a simple random load, the random walk, to illustrate the concept and elementary tools to estimate rainflow damage. The result was explicit due to the Markov property of the walk. The loads were discretised and sampled, and the electric circuits analogy was used to get the results.

This chapter is an introduction to rainflow analysis for the most often used random load models, the stationary Gaussian time series. We will start with Gaussian loads that are also Markov, the discrete time AR-model and the continuous time Ornstein-Uhlenbeck (O-U) process, presented in Sections 2.4.1 and 2.4.4, respectively. We will work with an explicit equation for the interval oscillation intensity from Section 4.8 and investigate the accuracy of the numerical solution. In principle, the techniques of the previous section could also be used and result in an integral equation that has to be solved numerically. The numerical errors can be controlled to get very accurate approximations of rainflow damage intensities.

More details on Gaussian and Markov load models will be presented in the following three chapters, devoted to Gaussian loads and Markov loads.

Although the stress function $X(t)$ is a continuous function, it is often measured sampled at a suitable frequency. Set $X_i = X(i\,dt)$ where $dt$ is the sampling step. Crossings, local extremes and oscillations in $X$ will be approximated by similar quantities counted in the time series $X_i$. If the sampling frequency is high and $X$ smooth, then there is no need to distinguish between cycle counting in $X(t)$ and $X_i$. However, sometimes the signal is sampled less dense so that the damage intensity computed or estimated for the series $X_i$ will be smaller than that for the process $X(t)$.

Usually, the time series $X_i$ is used to estimate the basic structure of the continuous model $X(t)$, e.g., its mean and power spectrum. Further, if a statistical test accepts that $X_i$ is stationary Gaussian, then we can use a stationary Gaussian model also for $X(t)$. We will focus our attention on analysis of the oscillation intensity of $X$ and the resulting estimated fatigue damage.

The sequence of turning points $T_P(X)$ is a time series that alternates between two types of states, Max and min. It is used to compress the measured stress to what is essential for fatigue. It is then used to test the strength of components for fatigue under variable amplitude loading. The advantage is that the sequence is much shorter than the sampled load $X_i$, but the damage intensity derived for $X$ and $T_P(X)$ are identical. Having a sequence of turning points $T_P(X)$, it is convenient to describe the dependency between extremes by means of Markov chain even if the time series $X_i$ is Gaussian. The sequence of turning points can be the result of onboard measured

 DOI: 10.1201/9781003019947-5

signals but also a result of mathematical derivations based on models of a variable environment where a structure is used and of its dynamical properties.

Mathematically, it is not an easy task to use the two models $X_i$ and $T_P(X)$ simultaneously, i.e. for the same measured signal. The problem is that even for most used standard random load models, i.e. the stationary Gaussian process, the sequence of turning points $T_P(X)$ is usually a non-stationary time series. (We will discuss this issue in Section 6.5 about Slepian models.) The dilemma can be overcome by numerically calculating the joint density $f_{mM}$ of a local minimum and the following maximum in a Gaussian process. Then, the sequence of turning points $T_P(X)$ in a Gaussian load can be approximated by a Markov chain with transition matrix derived from $f_{mM}$. In many cases, it is an accurate approximation, convenient to use for testing the strength of components which may be exposed to Gaussian loads.

In case of non-stationary but slowly changing Gaussian loads, a useful approximation is the narrow-band bound. Simply, locally one can bound the damage intensity $d^{rfc}$ by $d^{nb}$. Then, the total damage for longer service times, when stationarity can not be assumed, can be estimated (bound) by summing the local damages $D^{nb}$, or constructing a random model for the variability of the loading environment.

Since this can be a quite conservative approach, we will give here another approximation that combines Markov chain modelling of $T_P(X)$ and narrow-band bounding. This will be done by use of the upcrossing intensity $\mu^+(u)$ to approximate the $f_{mM}$ kernel. We will call this method the narrow-band Markov chain model. The important property of this approximation is that it will have the same crossing intensity as the load $X(t)$. Consequently, the expected damage of such a sequence will be smaller than the narrow-band bound $d^{nb}$. Such sequences could be used in tests of components for random load sequences. The method can be applied also to non-Gaussian loads.

In this chapter, we present a mathematical tool for numerical computation of the oscillation intensity, the Fredholm integral equation. In Sections 5.2 and 5.3, we apply it to simple Markov chain loads and to Markov chain of turning points. The accuracy of the numerical solutions are tested in Section 5.4 on Gaussian AR(1) load and the O-U process.

## 5.2 CROSSINGS OF INTERVALS BY SIMPLE TIME SERIES

In this section, we consider infinite sequences of random variables $X_i$, $i = 0, 1, \ldots$; often, the time series is also defined for negative $i$. Here, we consider time series that possess a probability density function (pdf), which we also denote $f(x)$ for simplicity. An important case is Gaussian loads. One reason for this limitation is that in such a time series, there are no two local maxima that have the same height from which it follows that for any $u \le v$, there are no maximum and minimum that have values $u, v$. For such time series, it is convenient to derive damage intensity by first estimating the intensity of interval upcrossings $\mu^{osc}(u, v)$ and then using master-equation (4.47) to evaluate the rainflow damage intensity.

### 5.2.1 FREQUENCY OF INTERVAL UPCROSSINGS $\mu^{osc}(u,v)$

This section is somewhat technical and will give means to evaluate $\mu^{osc}(u,v)$ for specific load models possessing some Markov properties.

For the time series $X_i, i \in \mathbb{N}$, an upcrossing of level $u$ is counted at time $j$ if $X_j < u$ and $X_{j+1} > u$. As before, let $N_t^+(u)$ be the number of $u$-upcrossings by $X_i$, $i = 0,\ldots,t-1$. An upcrossing of the interval $[u,v]$ is counted at $j$ if $X_j < u$ and $X_{j+1} > v$ or $X_j < u$ and $u < X_{j+1} < v$ and the sequence $X_k$, $k = j+2,\ldots$, gets a value above $v$ before a value below $u$. (This is a discrete time version of the continuous time Definition 4.8.) We denote that event $A_{j+2}(u,v)$. For simplicity of notation, we will write $A_i$ instead of $A_i(u,v)$ whenever there is no risk for confusion. Figure 5.1 illustrates the event $A_2 = A_{j+2}$, for an interval upcrossing starting at $j = 0$.

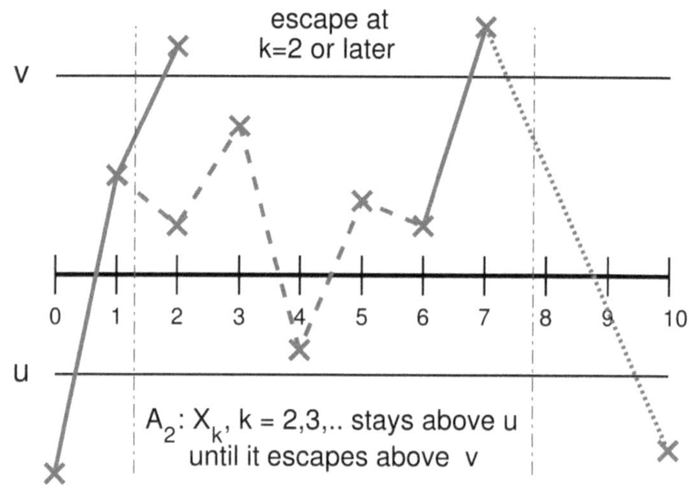


**FIGURE 5.1** Definition of event $A_2$. An oscillation from below $u$ to above $v$ starting at 0 with "miss" in first try.

Using this notation, we can write the expected number of interval upcrossings as

$$\mathsf{E}[N_t^{osc}(u,v)] = \sum_{j=0}^{t-1} \mathsf{P}(X_j < u, X_{j+1} > v) + \sum_{j=0}^{t-1} \mathsf{P}(X_j < u, u \le X_{j+1} < v, A_{j+2}).$$

If the sequence $X_j$ is stationarity and ergodic,

$$\begin{aligned}\mu^{osc}(u,v) &= \mathsf{E}[N_t^{osc}(u,v)]/t \\ &= \mathsf{P}(X_0 < u, X_1 > v) + \mathsf{P}(X_0 < u, u < X_1 < v, A_2).\end{aligned} \tag{5.1}$$

Computation of $\mathsf{P}(X_0 < u, u < X_1 < v, A_2)$ is in general very difficult, except for some degenerated cases, e.g. if $X_i$ is a sequence of iid variables, see the following example.

**Example 5.1** ($X_i$ are iid). *Let us consider the sequence of independent, identically distributed random variables. Then,* $\mathsf{P}(X_0 < u, u < X_1 < v, A_2)$ *simplifies and is equal to* $\mathsf{P}(X_0 < u)\mathsf{P}(u < X_1 < v)\mathsf{P}(A_2)$. *Further* $\mathsf{P}(A_2)$ *satisfies the following recursion,*

$$\mathsf{P}(A_2) = \mathsf{P}(X_2 > v) + \mathsf{P}(A_3)\mathsf{P}(u \leq X_2 \leq v).$$

*Since for all i,* $\mathsf{P}(A_i) = \mathsf{P}(A_0)$, *for fixed* $u < v$,[1]

$$\mathsf{P}(A_0) = \frac{\mathsf{P}(X_0 > v)}{\mathsf{P}(X_0 < u) + \mathsf{P}(X_0 > v)}. \tag{5.2}$$

*Finally, by (5.1),*

$$\begin{aligned}\mu^{osc}(u,v) &= \mathsf{P}(X_0 < u)\mathsf{P}(X_0 > v) + \mathsf{P}(X_0 < u)\mathsf{P}(u < X_0 < v)\mathsf{P}(A_0) \\ &= \frac{\mathsf{P}(X_0 < u)\mathsf{P}(X_0 > v)}{1 - \mathsf{P}(u \leq X_0 \leq v)}. \end{aligned} \tag{5.3}$$

### 5.2.2 ONE-STEP MARKOV CHAIN

For dependent variables, the simple argument behind (5.2) does not hold, and one has to take account for what has happened before time $j$.

#### 5.2.2.1 A Fredholm Integral Equation for the Oscillation Intensity

Suppose $X_i$ is a stationary Markov chain with continuous state space, i.e. for any $n$ the joint density of $X_0, X_1, \dots, X_n$ is given by

$$\begin{aligned}&f_{X_n,\dots,X_0}(x_n, x_{n-1}, \dots, x_0) \\ &\quad = f(x_n|x_{n-1}) \cdot f(x_{n-1}|x_{n-2}) \cdot \ldots \cdot f(x_1|x_0)f(x_0),\end{aligned} \tag{5.4}$$

where $f(y|x) = f_{X_1,X_0}(y,x)/f_{X_0}(x)$ is the the conditional probability density of a step from $x$ to $y$. Then, (5.1) can be somewhat simplified using the Markov property, viz. the conditional probability of $A_2$ given $X_0, X_1$ only depends on the latest $X$,

$$\mathsf{P}(A_2 \mid X_1 = y, X_0 = x) = \mathsf{P}(A_2 \mid X_1 = y), \quad u \leq y \leq v.$$

Then, expanding (5.1) by integrating the density of $(X_0, X_1)$, we get

$$\begin{aligned}\mu^{osc}(u,v) &= \mathsf{P}(X_0 < u, X_1 > v) + \int_{-\infty}^{u}\int_{u}^{v} \mathsf{P}(A_2 \mid X_1 = y, X_0 = x) f_{X_0,X_1}(x,y)\, \mathrm{d}y\, \mathrm{d}x \\ &= \mathsf{P}(X_0 < u, X_1 > v) + \int_{u}^{v} \mathsf{P}(A_2 \mid X_1 = y) q(y)\, \mathrm{d}y,\end{aligned} \tag{5.5}$$

where $q(y) = \int_{-\infty}^{u} f_{X_0,X_1}(x,y)\, \mathrm{d}x$.

What remains is to compute $\mathsf{P}(A_2 \mid X_1 = y)$. Again, using the Markov property, we can write an integral equation for $\mathsf{P}(A_2 \mid X_1 = y)$, $u \leq y \leq v$, viz.

$$\begin{aligned}\mathsf{P}(A_2 \mid X_1 = y) &= \mathsf{P}(X_2 > v \mid X_1 = y) + \int_{u}^{v} \mathsf{P}(A_3 \mid X_2 = z) f(z|y)\, \mathrm{d}z \\ &= \mathsf{P}(X_2 > v \mid X_1 = y) + \int_{u}^{v} \mathsf{P}(A_2 \mid X_1 = z) f(z|y)\, \mathrm{d}z.\end{aligned} \tag{5.6}$$

We conclude that in order to evaluate $\mu^{osc}(u,v)$ using (5.5), one needs to solve the integral equation (5.6) for each pair $u < v$ to get $\mathsf{P}(A_2 \mid X_1 = y)$ needed in (5.5). The questions are as follows: Is there any solution and is the solution unique? The equation is called a Fredholm integral equation of the second type, and the answer to both questions is, in our case, "Yes". We deal with this in the next section.

#### 5.2.2.2 Approximate Solution of (5.6)

A Fredholm integral equation like (5.6) has the form

$$f(y) = g(y) + \int_a^b k(y,z) f(z)\, \mathrm{d}z, \quad \text{for } y \in [a,b], \tag{5.7}$$

where $g$ and $k$ are given functions. The problem is to find a function $f$ that satisfies (5.7) for all $y$. The mathematical answer is as follows: A unique and continuous solution exists if $g$ and $k$ are continuous and $\sup_{a \le y \le b} \int_a^b |k(y,z)|\, \mathrm{d}z < 1$.

In most cases, explicit solution of the Fredholm integral equation is not possible. The equation has to be solved numerically and there are several procedures proposed in the literature. Here, we shall present a simple method to find $f(y) = \mathsf{P}(A_2 \mid X_1 = y)$ for a discrete set of $y$-values which employs quadrature of integrals.

For fixed $u, v$, let $(y_i, h_i)$, $i = 1, \ldots, n$, be a quadrature of an integral in (5.6) leading to the following approximation of the equation

$$\mathsf{P}(A_2 \mid y_i) \approx \mathsf{P}(X_2 > v \mid X_1 = y_i) + \frac{1}{f_{X_1}(y_i)} \sum_{j=1}^{n} \mathsf{P}(A_2 \mid y_j) f_{X_1,X_2}(y_i, y_j)\, h_j. \tag{5.8}$$

Note that one needs to solve one equation system (5.8) for each pair of levels $(u,v)$. The method approximates the integral equation by a system of linear equations with unknowns $\mathsf{P}(A_2 \mid y_i)$.

A particularly simple quadrature is obtained by using the equidistant grid $u_1 > u_2 > \ldots > u_N$, with $\delta u = u_i - u_{i+1}$ constant, letting $y_i$ take values from the grid and $h_i = \delta u$. The algorithm is simple and may work accurately when $N$ is large and the density $f_{X_0,X_1}(x,y)$ is smooth.

The method has a very appealing probabilistic interpretation, viz. the numerical solution is the same as the exact solution of (5.6) for the discrete valued Markov chain taking values $u_i$. Simply, one is approximating the continuous valued chain by a discreet chain having a finite number of states. Since the time when this method was proposed for about 40 years ago, the speed of computations has increased dramatically, and there is no real need to solve problems using the theory presented above. One can just approximate the Markov time series by a discrete Markov chain and use means later presented in Chapter 8. In order to support this claim, we will investigate the accuracy of this approach by considering an AR(1) process, see Section 5.4.

## 5.3 TIME SERIES OF TURNING POINTS

### 5.3.1 TRANSITION PROBABILITIES AND STATIONARITY

In fatigue analysis, one often saves only the heights of the local maxima and minima, called the sequence of turning points, and in this context defined as

$$T_P(x) = (M_0, m_0, M_1 \ldots).$$

The rainflow cycles can be found from $T_P(x)$ as well as rainflow damage and the interval upcrossing intensity $\mu^{osc}(u,v)$. The oscillation intensity is one of the important measures of load variability, and it can be computed from the probabilistic description of load variability, i.e. a random load model $X(\tau)$ or a random time series $T_P(X)$ of turning points. It is convenient to use models for $T_P(X)$ possessing Markov property. Here, we present the most important model, the one-step alternating Markov chain with continuous state space.

The one-step Markov chain of turning points is defined by the transition kernels, i.e. the bivariate densities $f_{m_0,M_0}$ and $f_{M_0,m_1}$. For simplicity, we assume that the chain is *time-reversible*. More precisely, we assume that at time zero, we have a local maximum $M_0$ with probability density function $f_{M_0}$ followed by a local minimum with pdf $f_{m_0}$. and that the transitions between local maxima and minima have a density satisfying

$$f_{M_0,m_0}(x_0,x_1) = f_{m_0,M_1}(x_1,x_0). \tag{5.9}$$

The transition probabilities can be estimated from a measured signal, e.g. using the so-called Markov matrices, or estimated from a model for the load $X(\tau)$. Here, we have some marginal problem which we just mention for completeness.

Suppose that we have a random load $X(\tau)$, e.g. a stationary Gaussian process. Then, the sampled time series $X_i = X(idt)$ is stationary, and $X_i$ has the same distribution as $X(0)$. On the contrary, if we consider a sequence of turning points $T_P(X)$ with $M_0$ being the first maximum in $X(\tau)$ after zero, then the second local maximum $M_1$ will have a different distribution than $M_0$. In other words, the sequence of turning points $M_j > m_j$, $j = 0,1,\ldots$ is not stationary. This is because first local maximum occurs at a random time. Practically, it is not so important since one is interested in the expected rainflow damage for a long sequence of turning points, not just the first few ones. This problem is resolved by computing the long-run expected damage of a rainflow cycle $(M_i, m_i^{rfc})$, it means for a rainflow cycle taken at random from the sequence. The reason for this choice is that we have tools for finding such distributions, e.g., in Gaussian loads. This involves the so-called Slepian model processes, see Lindgren and Rychlik (1991) [132] and Section 6.5 in this book.

For Gaussian (and related) loads, one can numerically compute the transition kernel $f_{M_0,m_0}$ for a pair of local maximum and the following minimum. Then, the Markov method to model the load is to approximate $T_P(X)$ by a time-reversible Markov chain with the kernel $f_{Mm}$. In the following, we assume that the kernel is given.

### 5.3.2 RAINFLOW MEMORY AND DAMAGE INTENSITY FOR A TIME-REVERSIBLE MARKOV CHAIN

In Section 5.2, we gave formulas for $\mu^{osc}(u,v)$ for a Markov chain $X_i$ for any fixed levels $u < v$. For the sequence of turning points, the problem of finding the oscillation intensity, alternatively the rainflow damage intensity, becomes somewhat simpler since we are working with local extremes of a process $X(\tau)$ and not its sampled values as in the previous section. We assume that the Markov chain is time-reversible and stationary.

We seek the rainflow damage intensity for the damage function $f(u,v) = (v-u)^\beta$, by (2.5) given by

$$d^{rfc} = \mu_{Max} \int_0^\infty \beta\, h^{\beta-1} \mathsf{P}(H^{rfc} > h)\, \mathrm{d}h.$$

Thus, we need the rainflow range distribution, averaging over the height of a maximum, which we can take as $M_0$ by the assumed stationarity,

$$\mathsf{P}(H_0^{rfc} > h) = \int_{-\infty}^{\infty} \mathsf{P}(H_0^{rfc} > h \mid M_0 = v) f_{M_0}(v)\, \mathrm{d}v.$$

Now, we can use the rainflow memories for the load and for the time-reversed load, defined by (4.17); $H_0^-$, for the load and $H_0^+$ for the reversed load. That will give a formula for $\mathsf{P}(H_0^{rfc} > h \mid M_0 = v)$ when $H_0^{rfc} = \min(H_0^-, H_0^+)$.[2]

By the assumed time-reversibility of the Markov chain (5.9), local maxima and minima occurring before and after $M_0$ are conditionally independent, given height $M_0$. Consequently

$$\begin{aligned} \mathsf{P}(H_0^{rfc} > h \mid M_0 = v) &= \mathsf{P}(H_0^- > h \mid M_0 = v)\, \mathsf{P}(H_0^+ > h \mid M_0 = v) \\ &= \mathsf{P}(H_0^+ > h \mid M_0 = v)^2. \end{aligned} \tag{5.10}$$

We will now derive a Fredholm integral equal equation, similar to that in Section 5.2.2.1, to compute the conditional probability that $H_0^+ > h$ given that maximum $M_0 = x \geq v$. Write $u = v - h$ and introduce the conditional probability of transitions from a maximum to the following maximum

$$f_{M_1|M_0}(y \mid x) = \int_{-\infty}^{x} f_{M_0,m_0,M_1}(x,z,y)/f_{M_0}(x)\, \mathrm{d}z,$$

and the transition probability with the restriction that $m_0$ is in the interval $[u,v]$, $q_{[u,v]}(y \mid x)$, say,

$$q_{[u,v]}(y \mid x) = \int_u^v f_{M_0,m_0,M_1}(x,z,y)/f_{M_0}(x)\, \mathrm{d}z. \tag{5.11}$$

Let us also introduce event $B_i$, which occurs if the sequence of turning points $(m_i, M_{i+1}, \ldots)$ down-crosses $u$ before it exceeds $v$ (similar to $A_i$ in Section 5.2.2.1).

Now, with $h = v - u$,

$$\begin{aligned} \mathsf{P}(H_0^+ > h \mid M_0 = v) &= \mathsf{P}(m_0 < u \mid M_0 = v) \\ &\quad + \int_u^v \mathsf{P}(B_1 \mid M_1 = y) q_{[u,v]}(y \mid v)\,\mathrm{d}y \\ &= \mathsf{P}(m_0 < u \mid M_0 = v) + \int_u^v \mathsf{P}(B_0 \mid M_0 = y) q_{[u,v]}(y \mid v)\,\mathrm{d}y, \end{aligned} \tag{5.12}$$

by the Markov property of turning points, $\mathsf{P}(B_1 \mid M_1 = y) = \mathsf{P}(B_0 \mid M_0 = y)$.

Now, we have found the Fredholm type integral equation that gives our missing probability, $\mathsf{P}(H_0^+ > h \mid M_0 = x)$ for $x = v$ and $u = v - h$,

$$\begin{aligned} \mathsf{P}(B_0 \mid M_0 = x) &= \mathsf{P}(m_0 \le u \mid M_0 = x) \\ &\quad + \int_u^v \mathsf{P}(B_0 \mid M_0 = y)\, q_{[u,v]}(y \mid x)\,\mathrm{d}y, \quad u \le x \le v. \end{aligned} \tag{5.13}$$

Equation (5.13) has to be solved numerically as presented before, e.g., by discretising the sequence of turning points, e.g. by using finite valued Markov chain of turning points.

The most difficult problem is to find the kernel $q_{[u,v]}$, or equivalently the density $f_{M_0,m_0}$. We will later discuss how to do it for a Gaussian model but for non-Gaussian models the kernel is much harder to evaluate. Note that for non-Gaussian loads, even computation/approximation of the upcrossing intensity $\mu^+(u)$ can be a difficult task. Having the intensity $\mu^+(u)$, one can employ the narrow-band bound $d^{nb}$ for estimation (an upper bound) of the rainflow damage; see Section 4.7.

In the following section, we propose a simple approximation of the kernel $f_{M_0,m_0}$, given $\mu^+(u)$. The Markov chain of turning points defined by such a kernel will have the upcrossing intensity $\mu^+(u)$, but the damage $d^{rfc}$ will be smaller than the narrow-band bound $d^{nb}$.

**Remark 5.1** (An asymptotic approximation)**.** *We finish this section with an asymptotic result valid for zero-mean stationary Gaussian loads. We consider extreme levels $u < 0 < v$. Then, a Poisson limit theorem can be used to estimate the oscillation intensity $\mu^{osc}(u,v)$, leading to*

$$\mu^{osc}(u,v) \approx \frac{\mu^+(u)\mu^+(v)}{\mu^+(u)+\mu^+(v)} = f_z \frac{\exp(-(u^2+v^2)/2\omega_0)}{\exp(-u^2/2\omega_0)+\exp(-v^2/2\omega_0)},$$

*where $f_z = \mu^+(0)$; see Johannesson and Thomas (2001) [96] for detailed presentation. For a stationary Gaussian load, $f_z = (2\pi)^{-1}\sqrt{\omega_2/\omega_0}$.*

### 5.3.3 THE NARROW-BAND APPROXIMATION

In this and the following subsection we assume that only the crossing spectrum $\mu^+(u)$ and the minimal upcrossing spectrum $k^+(u,v) = \min_{u \le z \le v}(z)$ (4.40) are available, and we give two possible estimates: The conservative narrow-band bound and the narrow-band Markov chain of turning points, MCTP.

Applications of the narrow-band bound for the rainflow damage intensity are discussed at several places in this book. It can be computed whenever the upcrossing frequencies $\mu^+(u)$ are available, since for any interval $[u,v]$, the oscillation intensity

$$\mu^{osc}(u,v) \leq k^+(u,v).$$

Then, by replacing $\mu^{osc}(u,v)$ by $k^+(u,v)$ in master-formula (4.47), the general narrow-band bound (4.42), $d^+ \geq d^{rfc}$, can be computed.

The bound can be very conservative, but an advantage is that one can use it also for asymmetric crossing spectra, common for non-Gaussian loads. In the special case, when $\mu^+(u)$ is a uni-modal function of $u$ with maximum at 0, say, it is a useful tool to estimate damage intensity; see Section 4.7.

A typical example, which we will see in Chapter 10, is when environmental loads are changing slowly. Sea conditions can be stationary for 30 minutes up to 3 hours, then the load consists of a couple of hundred waves up to few thousands. The severity of wave loads depends mostly on variability of experienced sea-states, described by the frequency of zero crossing waves $f_z = 1/\mu^+(0)$ and significant wave heights $h_s = 4\sigma$. During a sea state, a load can be approximated by a stationary Gaussian process with mean zero and standard deviation $\sqrt{\omega_0} = h_s/4$. Then, the expected damage is approximated by the narrow-band approximation (4.36),

$$d^{nb} = f_z \sigma^\beta \, 2^{3\beta/2}\Gamma(1+\beta/2) = f_z h_s^\beta \, 2^{-\beta/2}\Gamma(1+\beta/2). \tag{5.14}$$

In the formula, the zero-upcrossing frequency $f_z = \mu^+(0)$ is about 0.1 Hz in a "standard" sea state and, depending on mission, the significant wave height $h_s = 4\sigma$ can be even 10 meters. Consequently, it is the distribution of experienced $h_s$ during a mission that is the major factor that determines the risk for fatigue in marine applications. See also Section 10.4.2 where yearly variability of sea load damage is considered.

For values of $\beta \in [2,4]$, which are the most often used values in offshore applications, the approximation (5.14) can be further simplified by the fact that $0.469 < 2^{-\beta/2}\Gamma(1+\beta/2) \leq 0.5$, leading to the very simple approximation

$$d^{nb} \approx 0.5 \cdot f_z h_s^\beta. \tag{5.15}$$

This is an indication that significant amplitude seems to be the relevant parameter when studying Gaussian loads.

A hidden assumption for this approach is that during a sea-state, the probability that a wave crest is higher than $u$ is equal to

$$\mathsf{P}(H > u) = \mu^{osc}(0,u)/f_z \leq \mu^+(u)/\mu^+(0). \tag{5.16}$$

For the Gaussian sea often used in marine applications, one says that the narrow-band sea has Rayleigh distributed wave crests (amplitudes) and

$$\mu^{osc}(0,u)/f_z \leq \exp(-u^2/2\sigma^2) = \mathsf{P}(\sigma R > u), \quad \sigma = h_s/4,$$

where $R$ has a standard Rayleigh distribution. Similar narrow-band approximation of wave trough is minus Rayleigh.

In a more realistic setup, one is interested in stresses in some components, and hence the frequency of wave and significant wave-heights has to be re-scaled to zero crossing frequency of the stress function and its variance. This can be done by means of dedicated software; for an example of such analysis, see [143].

In tests, one would not use the sampled Gaussian load but rather the sequence of turning points. In the following example, we give a method to simulate a sequence of symmetric turning points which on average give the narrow-band bound which means that under stationary conditions, the oscillation intensity $\mu^{osc}(u,v) = k^+(u,v)$.

**Example 5.2** (Narrow-band symmetric turning points). *Consider a standardised Gaussian load having mean zero and variance one and frequency of zero upcrossings $f_z$. If the power spectrum of the load is unknown, the narrow-band approximation gives a bound for the damage intensity that only depends on $f_z$ (and the variance of the load here scaled to be one). For fatigue strength tests, one would like to simulate such a sequence of turning points.*

*The simplest model is to assume symmetry between crests and troughs. For such a load $\mu^+(u) = f_z \exp(-u^2/2)$ and*

$$\mu^{osc}(u,v) \leq k^+(u,v) = f_z \min(\exp(-u^2/2), \exp(-v^2/2)).$$

*The narrow-band approximation of minima and maxima postulates that the extremes are symmetric and Rayleigh distributed, viz. for $u < 0 < v$,*

$$\mathsf{P}(M_0 > v) = \mu^+(v)/\mu^+(0) = \exp(-v^2/2), v \geq 0,$$
$$\mathsf{P}(m_0 < u) = \exp(-u^2/2), u \leq 0.$$

*The Rayleigh variable can be simulated using uniformly, on $[0,1]$, distributed noise $U_i$, by $R_i = \sqrt{-2\ln U_i}$. Let us define the sequence of turning points $M_0, m_0, M_1, m_1, \ldots$ as $R_0, -R_0, R_1, -R_1, \ldots$. Since the intensity of min-Max cycles is equal to $k^+(x,y)$ and $\mu^+(u) = 0.5\exp(-u^2/2)$, then also $\mu^{osc}(u,v) = k^+(u,v)$. (This is because $\mu^{mM}(u,v) \leq \mu^{rfc}(u,v) = \mu^{osc}(u,v)$. Consequently, the distribution of rainflow range $H^{rfc}$ is $2R$ since its oscillation intensity is $\mu^{osc}(u,v) = k^+(u,v)$.*

The method could be useful to simulate wave loads in Gaussian slowly changing environment when only frequency of waves $f_z$ and significant wave heights $h_s = 4\sqrt{\omega_0}$ are known for a planned maritime operation. Often, one is assuming that the stationary sea condition rests for T hours, T is between 0.5 and 3 hours, and that during that period, there are on average $N = T f_z$ waves.

In the next section, we propose a Markov sequence of turning points having individual Rayleigh distributed crests and troughs but oscillation intensity $\mu^{osc}(u,v) < k^+(u,v)$ for $u < 0 < v$. Consequently, the damage intensity of the sequence is $d^{rfc} < d^+$ which may somewhat reduce the conservatism of the narrow-band bound.

### 5.3.4 NARROW-BAND MARKOV CHAIN OF TURNING POINTS

Consider a load having uni-modal upcrossing intensity $\mu^+(u)$ with maximum at zero. Define the two variables $M_0$ and $m_0$ having distributions

$$\begin{aligned} \mathsf{P}(M_0 > v) &= \mu^+(v)/\mu^+(0),\ v \geq 0, \\ \mathsf{P}(m_0 < u) &= \mu^+(u)/\mu^+(0),\ u \leq 0. \end{aligned} \tag{5.17}$$

Let $m_i$ and $M_i$ be two sequences of random variables having distributions (5.17). In addition, we assume that all maxima and minima are independent of each other. For a standardised narrow-band Gaussian load, the sequence of turning points consists of independent approximately Rayleigh distributed local extremes (maxima are positive and minima are negative).

The sequence is a narrow-band Markov chain of turning points, nbMCTP, and hence we can use (5.13) to find the probabilities $\mathsf{P}(B_0 \mid M_0 = x)$ for all pairs $(u,v)$, i.e. that the sequence $(m_0, M_1, m_1, \dots)$ down-crosses $u$ before it exceeds $v$. Note that except the case $u < 0 < v$, the equation has a trivial solution $\mathsf{P}(B_0 \mid M_0 = x) = \mathsf{P}(M_0 > v)$ if $u \geq 0$ and zero if $v \leq 0$. Now, for $u < 0 < v$, by independence of $R_i$, (5.13) writes

$$\mathsf{P}(B_0(u,v) \mid M_0 = x) = \mathsf{P}(B_0) = \mathsf{P}(m_0 < u) + \mathsf{P}(B_0)\int_u^v q_{[u,v]}(y|x)\,\mathrm{d}y,$$

and by (5.11)

$$\int_u^v q_{[u,v]}(y|x)\,\mathrm{d}y = \mathsf{P}(M_0 < v)\mathsf{P}(m_0 > u),$$

and hence

$$\mathsf{P}(B_0 \mid M_0 = x) = \frac{\mathsf{P}(m_0 < u)}{1 - \mathsf{P}(m_0 \geq u)\mathsf{P}(M_0 \leq v)}, \quad 0 \leq x \leq v.$$

Since $\mathsf{P}(B_0 \mid M_0 = x)$ does not depend on $x$, we have that

$$\mathsf{P}(H_0^+ > h \mid M_0 = v) = \mathsf{P}(B_0 \mid M_0 = v),$$

for $h = v - u$, and hence

$$\begin{aligned} \mu^{osc}(u,v) &= f_z \frac{\mathsf{P}(M_0 > v)\mathsf{P}(m_0 < u)}{1 - \mathsf{P}(m_0 \geq u)\mathsf{P}(M_0 \leq v)} \\ &= f_z \frac{\mathsf{P}(M_0 > v)\mathsf{P}(m_0 < u)}{\mathsf{P}(m_0 < u) + \mathsf{P}(M_0 > v) - \mathsf{P}(m_0 < u)\mathsf{P}(M_0 > v)}. \end{aligned} \tag{5.18}$$

Formally, the nbMCTP has frequency 0.5 of maxima (rainflow cycles) per time unit. However, when applying it to simulate Gaussian load, one needs to choose the number of $N$ of cycles generated by the MCTP in period $T$, viz. $N = f_z T$. That's why we replaced 0.5 by $f_z$ in (5.18).

Obviously, for $u > 0$, $\mathsf{P}(m_0 \geq u) = 0$, while for $v < 0$, we have $\mathsf{P}(M_0 \leq v) = 0$ and hence $\mu^{osc}(u,v) = k^+(u,v)$ if $u \geq 0$ or $v \leq 0$. Only for large intervals $[u,v]$, when $u < 0 < v$, the oscillation intensity $\mu^{osc}(u,v) < k^+(u,v)$. However, both models have the same upcrossing intensity $\mu^+(u)$. Consequently, for the damage $f(u,v) = (v-u)^\beta$, the damage intensity $d^{nbMCTP} = d$ is smaller than the narrow-band bound $d^{nb}$, while the rainflow damage intensity of narrow-band Markov chain of turning points $d^{nbMCTP} = d^{nb}$ if $\beta = 1$ and $d^{nbMCTP} < d^{nb}$ when $\beta > 1$.

Since the nbMCTP has crossing intensity $\mu^+(u)$, positive variation of this chain is, in average, the same as the positive variation of the load considered here; cf. Banach's theorem on total variation, Theorem 2.2. Now, since oscillation intensity $\mu^{nbMC}(u,v) < k^+(u,v)$, for $u < 0 < v$, consequently for $\beta > 1$ the sequence will have rainflow damage rate smaller than the narrow-band bound for the load. However, it will be only an approximation (not a bound) of the rainflow damage rate $d^{rfc}$ of the load.

In the following examples, we will evaluate the usefulness of the narrow-band MCTP in fatigue damage estimation for stationary Gaussian loads. In that test, we will simulate Gaussian loads with the given spectrum for a time span of length $T$. Then, the rainflow damage intensity (growth per time unite) $d^{rfc}$ is estimated. Next, the intensity $d^{nbMCTP}$ is estimated in the similar way, but the number of $N$ of local maxima is equal $f_z T$. The simplest test that statistical errors are small is to check $d^{rfc} \approx d^{nbMCTP} \approx d^{nb}$ for $\beta = 1$. For $\beta > 1$, we have $d^{nbMCTP}/d^{nb} < 1$ which shows the degree of reduction of the narrow-band bound of the true rate $d^{rfc}$

**Example 5.3** (The narrow-band MC of turning points for Gaussian loads). *Let $(U_{2i}, U_{2i+1})$ be a sequence of independent uniformly distributed pairs of variables and let $m_i = -\sqrt{-2\ln U_{2i}}$ and $M_i = \sqrt{-2\ln U_{2i+1}}$. It is easy to see that the sequence has $\mu^+(u) = 0.5\exp(-u^2/2)$ and hence $k^+(x,y) = 0.5\min(\exp(-x^2/2), \exp(-y^2/2))$, $x \le y$. The fraction $d^{nbMCTP}/d^{nb}$ is given in Table 5.1.*

**Table 5.1**
**Narrow-Band vs. nbMCTP Approximative Damage**

| $\beta$ | 1 | 2 | 3 | 4 | 5 |
|---|---|---|---|---|---|
| $d^{nbMCTP}/d^{nb}$ | 1 | 0.95 | 0.90 | 0.86 | 0.83 |

For Gaussian loads, the narrow-band MCTP simulation will give asymptotically the correct frequency of extremely large cycles, since for extreme levels $u, v$, the term $\exp(-(u^2+v^2)/2)$ in (5.18) is negligible and the narrow-band Markov chain approach gives the same result as the asymptotic Poisson limit given in (5.16). For this reason, a Markov chain of turning points can be a useful model to both make the narrow-band bound less conservative and give means to simulate random loads for variable amplitude loading both for Gaussian and some non-Gaussian loads when upcrossing intensity is uni-modal, e.g. Laplace loads presented in Chapter 9.

The narrow-band bound is a very useful way to estimate damage intensity for narrow-band processes. The question: How to test which load is narrow-band. For

Gaussian load, a popular irregularity measure is the irregularity factor, i.e. the fraction of frequencies of cycles and waves, i.e. local maxima and zero upcrossing frequencies, $1/\alpha_2 = \sqrt{\omega_4\omega_0}/\omega_2$. In the following example, we will use three Gaussian loads with different widths: A narrow JONSWAP spectrum, which is used for wind waves in the North Sea; the so-called low frequency white noise spectrum; and a wind gust speed normalised Davenport spectrum

$$S(\omega) = \frac{c\omega}{(1+c^2\omega^2)^{4/3}}, \qquad c = 1.5.$$

We let all three spectra be zero for $\omega > 10$. The irregularity factors $1/\alpha_2$ are 1.34, 1.85 and 7.34, respectively. In the following example, we will present the fractions $d^{nbMCTP}/d^{rfc}$. If the fraction is higher than one, then narrow-band MCTP overestimates the rainflow damage while if the ratio is below one, it underestimates it. Note that for $\beta = 1$, one always has $d^{rfc} = d^{nbMCTP} = d^{nb}$.

**Example 5.4** (Three Gaussian loads). *Next, we illustrate in Table 5.2 the efficiency of the narrow-band MCTP using three loads of increasing irregularity, illustrated in Figure 5.2; the narrow-band JONSWAP spectrum; the low frequency white noise and the Davenport wind gust spectrum, where very high frequencies have been removed. The sampling frequency is 20 Hz, and simulated signals have duration* $T = 28$ *hours. The time T is very long in order to limit statistical uncertainty in the estimated damages. The frequency of zero upcrossings* $f_z = 0.12, 0.18, 0.48$ *Hz and irregularity factors equal to 1.84, 1.34, 2.15, respectively. Consequently, the narrow-band MCTP should generate* $f_z \cdot T = 11990, 18340, 48150$ *cycles (local maxima), while the simulated Gaussian processes will have a factor 1.85, 1.34, 2.15 more cycles than the MCTP.*

*For the first two spectra, the MCTP model gives very good approximation, while for the Davenport spectrum, it gives about 20% overestimation of the damage intensity. It is still better than the narrow-band bound which for the Davenport spectrum gives over-estimation by factors 1.0, 1.30, 1.42, 1.45, 1.45.*

*The rainflow cycle plots in Figure 5.3 may give a hint to why the Davenport model stands out from the narrow-band Markov approximation – the accumulation of small range cycles with large positive or negative mean.*

**Table 5.2**

**Relative Damage Intensity from Narrow-Band MCTP Approximation Conpared with Simulated RFC-Damage for Three Types of Spectra**

| $\beta$ | 1 | 2 | 3 | 4 | 5 |
|---|---|---|---|---|---|
| $d^{nbMCTP}/d^{rfc}$ | 0.996 | 0.98 | 0.94 | 0.91 | 0.90 |
| $d^{nbMCTP}/d^{rfc}$ | 1.002 | 1.05 | 1.04 | 1.02 | 0.99 |
| $d^{nbMCTP}/d^{rfc}$ | 1.005 | 1.24 | 1.27 | 1.22 | 1.17 |

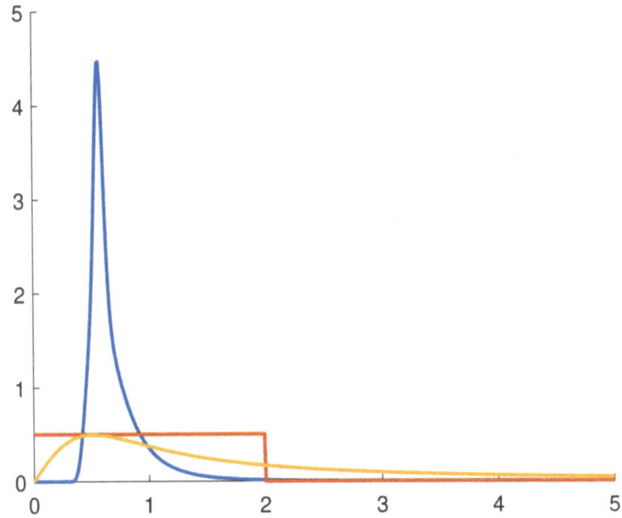


**FIGURE 5.2** Narrow JONSWAP spectrum, low frequency white noise spectrum and Davenport wind spectrum, used in Example 5.4.

In the case when the load has non-zero skewness and kurtosis exceeding 4, i.e. the signal is highly non-Gaussian, one can transform the non-Gaussian load so that the skewness and kurtosis get the values as Gaussian distribution. There are several approximations proposed in the literature, e.g., by Winterstein (1988) [212] or Ochi and Ahn (1994) [151].

The toolbox WAFO [207, 208] contains a non-parametric transformation by Rychlik et al. [182] as well as the Winterstein and Ochi/Ahn transformations which can be combined with the narrow-band MCTP model. However, such models do not give "transients" which may occur when a ship experiences slamming or a car drives through pothole. For such type of loads, we shall in Chapter 9 introduce the Laplace models which allow the variance of a signal to vary in a random manner.

## 5.4 GAUSSIAN AR(1) TIME SERIES AND THE ORNSTEIN-UHLENBECK PROCESS

Here, we will consider the Gaussian O-U process X(t) and its sampled version $X(i\,dt)$ which is Markov and an AR(1) time series. The O-U process is extremely irregular and has an infinite number of rainflow cycles in any time interval. However, the expected damage is finite when $\beta > 2$. We will use results from Section 5.2.2.2 and demonstrate that this approach "works" for any sampling step $dt$. i.e. the numerical method of the Fredholm equation (5.8) works. We will not use narrow-band bound $d^{nb}$ or construct the narrow-band MC of turning points. It is possible for any $dt > 0$. However, the bound tends to infinity as $dt$ approaches zero.

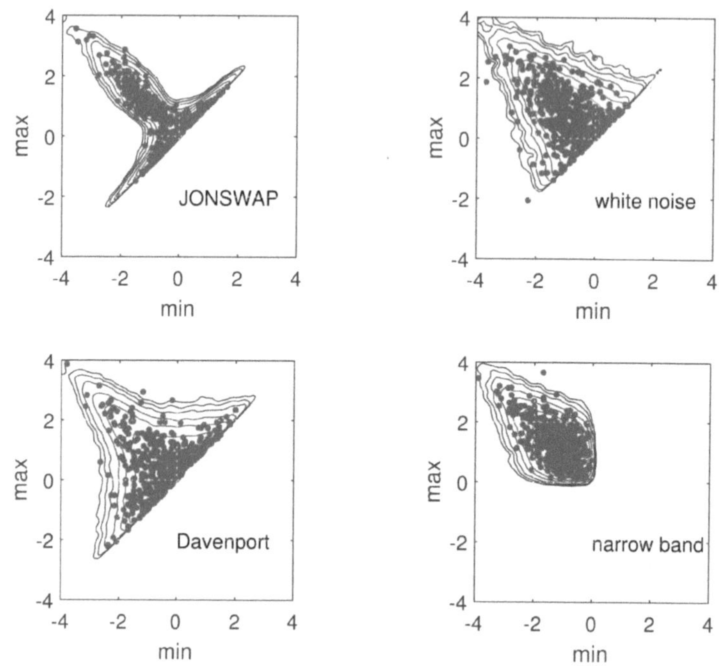


**FIGURE 5.3** Three rainflow cycle plots for Gaussian loads and one narrow-band Markov chain approximation for Gaussian load with Davenport spectrum in Example 5.4.

Here, we will investigate the accuracy of the numerical solution of the Fredholm equation (5.8) needed to compute the oscillation intensity for one-step Markov chain. For this, we shall use the Gaussian AR(1) time series $X_i$ which is a sampled O-U process $X(t)$, i.e. $X_i = X(i\,dt)$. As the sampling step $dt$ decreases towards zero, the oscillation intensity of the time series $X_i$ converges to the oscillation intensity of the O-U process, which is given by

$$\mu^{osc}(u,v) = \frac{c}{\sqrt{2\pi}} \left( \int_{u/\sigma}^{v/\sigma} \exp(z^2/2)\,\mathrm{d}z \right)^{-1}, \tag{5.19}$$

see Scheutzov (1994) [186] and Rychlik (1996) [175, 176] for mathematical details.

In Appendix 5.A, we will motivate formula (5.19) using a random walk approximation of the O-U process. Actually, one can derive the O-U process as a limit of random walk with unequal probability of jumping up and down. This interpretation of the O-U process was introduced by Kac 1947 [98]. After normalisation of formula (5.26), one obtains (5.19).

### 5.4.1 THE GAUSSIAN AR(1) TIME SERIES

Maybe the simplest random time series is the Gaussian AR(1) process defined as follows. Let $e_i$ be a sequence of independent zero-mean, variance one, Gaussian variables and let

$$X_{i+1} = \rho X_i + \sigma\sqrt{1-\rho^2}\cdot e_{i+1}, \tag{5.20}$$

$\rho \in [-1,1]$ and $X_0 \in N(0,\sigma^2)$. (We need to also assume that $e_{i+1}$ is independent of the series $X_0,\dots,X_i$.) Now, $X_i \in N(0,\sigma^2)$ is stationary with covariance $\mathsf{Cov}[X_i,X_j] = \sigma^2\rho^{|j-i|}$. Here, we discuss means to estimate the rainflow damage intensity for the time series and how it depends on the parameter $\rho$. Obviously, the case $\rho = 1$ gives a constant load and no damage occurs, while $\rho = -1$ is the constant amplitude load with

$$d^{rfc} = 0.5\mathsf{E}[(2|X_0|)^\beta] = \frac{0.5}{\sqrt{\pi}}2^{3\beta/2}\sigma^\beta\Gamma(0.5+\beta/2),$$

if $f(u,v) = (v-u)^\beta$. Simply, we know that this is a constant amplitude load with unknown amplitude, which has distribution $|X_0|$.

Finally, we construct a narrow-band bound for the Gaussian process $X^{pl}(t)$ which is a linear interpolation of the Gaussian time series $X_i$. Obviously, the frequency by which $X^{pl}(t)$ upcrosses the level $u$ is equal to $\mathsf{P}(X_0 < u < X_1)$. If $X_i$ is a sequence of independent random variables, then

$$\mathsf{P}(X_0 < u < X_1) = \mathsf{P}(X_0 < u)\mathsf{P}(X_0 > u).$$

For correlated Gaussian variables, e.g. for the zero-mean, variance one, AR(1) process, we have that

$$\begin{aligned}\mathsf{P}(X_0 < u < X_1) &= \frac{1}{2\pi}\int_{-\infty}^{u}\int_{u}^{\infty}\exp(-(y-x\rho)^2/2)\exp(-x^2/2)\,\mathrm{d}y\,\mathrm{d}x\\ &= \frac{1}{\sqrt{2\pi}}\int_{-\infty}^{u}\Phi(u-x\rho)\exp(-x^2/2)\,\mathrm{d}x.\end{aligned}$$

An explicit formula is known for $u = 0$, when it can be shown that for a stationary Gaussian time series $X_i$, with mean zero,

$$f_z = \mathsf{P}(X_0 < 0 < X_1) = \frac{1}{4} - \frac{1}{2\pi}\arcsin\rho, \tag{5.21}$$

where $f_z$ is the zero upcrossing frequency, an important parameter when studying damage accumulations in ship details. Note that Bendat's narrow-band approximation $d^{nb}$ for $X^{pl}(t)$ has to be computed numerically. Formula (5.21) is a convenient way to estimate the parameter $\rho$ in an AR(1) process.

The reason we present the AR(1) load model first is that it is a very simple Gaussian time series which is also a Markov chain, i.e. if one has observed the values of $X_0,\dots,X_i$, then the future value of $X_{i+1}$ depends on history only through the $X_i$ and $e_{i+1}$ ($e_{i+1}$ is assumed to be independent of $X_0,\dots,X_i$). The damage intensity of an AR(1) sequence will be studied first, and we shall check the accuracy of the proposed estimates by using the known explicit results, viz. the damage intensities for an iid sequence and of the Gaussian O-U process.

Since the kernel $f_{X_0,X_1}$ of the AR(1) time series is known, we can numerically estimate its oscillation intensity by means of Equations (5.5–5.8). In Section 5.4.3, we will present two types of checks of accuracy. In the first test, Example 5.5, we select a range of $dt$-values, from very large to quite small, leading from almost independent to very dependent Gaussian variables. Then, we solve numerically Equations (5.5–5.8) and use the master-formula to derive estimates of rainflow damage intensities.

In the second test, Example 5.6, we use $dt \approx 0$ and (5.3) to get the oscillation intensity and then evaluate (numerically) the master-formula to get the rainflow damage intensities. (Note that the estimation of the O-U damage intensity also involves numerical integration.)

### 5.4.2 RAINFLOW DAMAGE FOR O-U PROCESS

The O-U process $X(\tau)$, Definition 2.9, is a zero-mean stationary Gaussian process with covariance function

$$r(\tau) = \sigma^2 \exp(-c|\tau|),$$

and the power spectrum

$$S(\omega) = \frac{\sigma^2}{\pi} \frac{c}{\omega^2 + c^2}, \qquad \omega \in \mathbb{R}. \tag{5.22}$$

Since $\omega_2 = \infty$, the process is very irregular and its samples are not differentiable ($\mathsf{E}[N^+(u)] = \infty$). One can also show that it has infinitely many extremes (and hence rainflow cycles) in any time interval. But those cycles have also very small amplitudes, and as a consequence, the intensity of interval crossings $\mu^{osc}(u,v)$ is finite. The damage intensity however can be finite or infinite depending on the damage function.

Let the damage function be $f(u,v) = (v-u)^\beta$ and set $\sigma = 1$ and $c = 1$. We will show that for $\beta \leq 2$, the expected damage is infinite since small cycles (which are infinitely many) cannot be neglected. For $\beta > 2$, the expected damage of small cycles is finite but also large cycles cause finite damage, leading to finite damage intensity.

The damage intensity is given by master-formula (4.47) which now can be rewritten as follows:

$$d^{rfc} = \beta\,(\beta-1) \int_{-\infty}^{\infty} \int_0^{\infty} h^{\beta-2} \mu^{osc}(u, u+h)\, \mathrm{d}h\, \mathrm{d}u. \tag{5.23}$$

(Note that for $\beta > 1$, the second term in the master-formula is zero.)

As shown in [175, 176, 186], the interval crossing intensity for the O-U process is (5.19),

$$\mu^{osc}(u,v) = \frac{1}{\sqrt{2\pi}} \left( \int_u^v \exp(z^2/2)\, \mathrm{d}z \right)^{-1}. \tag{5.24}$$

Now, with $S(v) = \int_0^v \exp(z^2/2)\, \mathrm{d}z$ and $\phi(u)$ for the standard normal density, one has $\mu^{osc}(u,u+h) = (S(u+h) - S(u))^{-1}$ and hence $\mu^{osc}(u,u+h) \approx \phi(u)/h$ for small $h$. Consequently, $d^{rfc} = \infty$ for all $\beta \leq 2$. For $\beta > 2$, one can use that

$$\mu^{osc}(u,u+h) \leq h\,\phi(\min(|u|, |u+h|),$$

and the expected damage is finite even for large cycles.

**Table 5.3**
**Damage Fractions $\widehat{d}^{rfc}(\beta)/\bar{d}^{rfc}(\beta)$ for $\beta = 3,4,5$ for AR(1) Time Series in Example 5.5**

| $\beta$ | $dt =$ | 10 | 1.0 | 0.1 | 0.01 | 0.001 |
|---|---|---|---|---|---|---|
| 3 | | 0.99 | 1.00 | 1.00 | 0.99 | 1.02 |
| 4 | | 0.98 | 0.97 | 0.97 | 0.97 | 0.96 |
| 5 | | 0.95 | 0.94 | 0.94 | 0.94 | 0.90 |

### 5.4.3 THE ACCURACY OF APPROXIMATION (5.8)

The Gaussian AR(1) process and its continuous version, the O-U process, offer explicit expressions for the transitions probabilities in a load series. They are also easy to simulate, also in very long series. We will benefit from these properties and investigate how well simulated rainflow damage agrees with "exact" approximations from (5.5–5.8).

Consider a Gaussian O-U process with parameters $\sigma^2 = c = 1$ sampled at frequency $dt$. The resulting time series $X_i = X(idt)$ is an AR(1) process with $\mathsf{Cov}[X_0, X_i] = \exp(-|i|\,dt)$, i.e. it is generated as (5.20) with $\rho = \exp(-dt)$.

For $dt = 10, 1, 0.1, 0.01, 0.001$, the $\rho$-parameter in (5.20) is equal to $5 \cdot 10^{-5}$, 0.368, 0.905, 0.990, 0.999, respectively, leading from virtually independent to strongly dependent variables, with bivariate density

$$f_{X_0,X_1}(x,y) = \frac{1}{2\pi\sqrt{1-\rho^2}} \mathrm{e}^{-\frac{1}{2}\left(x^2+y^2-2\rho xy\right)}.$$

**Example 5.5** (Simulated damage and the Fredholm solution (5.8)). *Using the kernel $f_{X_0,X_1}$ of the AR(1) time series, we numerically estimate the oscillation intensity by means of Equations (5.5–5.8). Then, the master-formula is used to get rainflow damages from the oscillations intensities. Let us denote those numerically computed damage intensities by $\widehat{d}^{rfc}(\beta)$. The accuracy of this numerical estimates is evaluated by comparing $\widehat{d}^{rfc}(\beta)$ with damage intensities observed in very long simulations of AR(1) time series. The simulated damage intensities will be denoted by $\bar{d}^{rfc}(\beta)$.*

*In Table 5.3, we give the fractions $\widehat{d}^{rfc}(\beta)/\bar{d}^{rfc}(\beta)$ for $\beta = 3,4,5$. The presented results show that the numerical errors when using Equations (5.5–5.8) are small, except for very high values of $\beta$. These errors depend on the oscillation intensity $\mu^{osc}(u,v)$ for extreme levels $u,v$, which need to be better approximated. In this example, the approximation of oscillation intensity was obtained by first solving (5.8); we have used $n = 181$ equidistant grid points in the interval $[-3.5, 3.5]$ with step $du \approx 0.04$. We conclude that at least for the AR(1) Markov chain, the numerical solutions of* (5.8) *are sufficiently accurate.*

**Example 5.6** (Simulated damage and the master-formula). *Finally, consider the extremely small sampling step* $dt = 0.0001$ *of the O-U process* $X(t)$. *Then, the AR(1) process* $X_i = X(idt)$, *with correlation* $\rho \approx 1 - dt$ *and zero upcrossing frequency* $f_z = 0.0023$, *is a very accurate simulation of an O-U process. A long simulation of a Gaussian AR(1) process with* $dt = 0.0001$ *gave for* $\beta = 3, 4, 5$ *the rainflow damage intensities* $d^{rfc} = 17.4, 65.8, 310$.

*The simulated values could be compared with the exact* $d^{rfc}$ *computed using the oscillation intensity* (5.24) *of the O-U process and master-formula* (5.23). *The same grid as used for Table 5.3 was used. The resulted damage intensities are 17.8, 66.4, 312.0 for* $\beta = 3, 4, 5$. *We conclude that the use of the master-formula with moderately dense grid gives accurate estimates for the damage intensity for the O-U process.*

These examples illustrate that the accuracy of numerical derivations of the rainflow damage intensity for one-step Markov chain seems to be good. The second conclusion drawn from the speed by which the damage intensity increases as sample step $dt$ decreases is that one should pay some attention when choosing sampling frequency, particularly to very irregular loads. In the next section, we will look into this problem.

### 5.4.4 SAMPLING FREQUENCY FOR GAUSSIAN LOADS

Stationary Gaussian loads are very useful models for fatigue loads varying in a non-predicable manner. Many exact results are derived for Gaussian loads, and many approximations have their roots in a Gaussian model. One reason for the popularity is that the power spectral density $S(\omega)$ and its Fourier transform, the covariance function $r(\tau)$, together with the mean $m = \mathsf{E}[X(t)]$, completely define the Gaussian model; see Section 2.2.2.3. These functions are usually easy to estimate from observed load data. A very useful property of the Gaussian load model is that the output of linear systems with Gaussian input is also a Gaussian signal with spectral density $|H(\omega)|^2 S(\omega)$; see Section 2.2.3.3.

From the Gaussian model, one also derived the narrow-band bound and approximation for loads with Gaussian type crossing spectra, (4.36). With $f_z = \mu^+(0)$ denoting the apparent wave frequency, for the Gaussian model given by $f_z = \frac{1}{2\pi}\sqrt{\omega_2/\omega_0}$, and $\sigma^2$ for the process variance $\sigma^2$, and damage function $f(u,v) = (v-u)^\beta$, the narrow-band bound for rainflow damage intensity $d^{rfc}$ is

$$d^{rfc} \leq d^{nb}(\beta) = f_z \sigma^\beta \, 2^{3\beta/2} \Gamma(1+\beta/2).$$

For $\beta = 1$, the bound is simply,

$$d^{rfc} = d^{nb}(1) = f_z \sigma \sqrt{2\pi}.$$

Note that if $f_z = \infty$, then $d^{nb} = \infty$ but the rainflow damage intensity can be finite, e.g. for the O-U process that has $f_z = \infty$ but $d^{rfc}$ finite for $\beta > 2$. This is a somewhat artificial case but still rises a question how to proceed if one had measured relatively short noise signal with sampling step $dt$.

Obviously, good estimates of $f_z$ and $\sigma$ are essential if one wants to apply the narrow-band damage bound. One possibility is to estimate directly from an observed load signal. A more flexible, and often more reliable, variant is to chose some parametric families of spectra $S(\omega)$, typical for the particular application. Estimating spectral parameters from data, cut off high frequencies and compute spectral moments $\omega_0, \omega_2$, will give the needed $f_z$ and $\sigma$. However, one still faces the problem of modellings errors.

A complementary approach would be to rainflow-filter the signal, e.g. remove oscillations below a threshold $\varepsilon$, say. Such a signal will have finite crossing spectrum $\mu^+(u)$ and oscillation intensity $\mu^{osc}(u,v)$. The rainflow damage can be computed using the master-formula. However, one often wishes to make a fatigue test with randomly generated loads. How to obtain time signals from the oscillation matrix is called an inverse problem. It means that given an oscillation intensity matrix, one wishes to generate random loads, which in average give the oscillation intensities. Such inversion algorithms have been proposed, one combinatorial algorithm in [58], and another, which looks for Markov chain that would approximate the oscillations, in [177].

We finish this section with a discussion of possible oversampling of the measured signal in order to control whether the chosen sampling step $dt$ is adequate for the measured irregularity of the signal. Let us estimate the damage intensity in the sampled signal for the damage function $f(u,v) = v - u$. The rainflow damage is equal to the positive variation of the signal, which for long sequences is equal to half of the total variation

$$V = \sum_{i=1}^{N} |X(i\,dt) - X((i-1)dt)|.$$

Here, $N$ is the length of the measured signal. Consequently, the estimated damage intensity would be

$$d^{rfc} = V/2N\,dt \approx \mathsf{E}[|X(dt) - X(0)|]/2dt = \frac{\sqrt{2(r(0) - r(dt))}}{dt\sqrt{2\pi}}.$$

Obviously, if the covariance of the $X$-process is twice differentiable, then $\omega_2 = -r''(0)$ and hence an estimate of $d^{rfc}$ is approximately equal to the narrow-band bound. Here, we will use the observed damage computed for damage function with $\beta = 1$, when $d^{nb} = d^{rfc}$, viz.

$$d^{rfc} \approx d^{nb}(1) = f_z \sigma \sqrt{2\pi}.$$

i.e. for differentiable loads, $f_z$ is time scale, while $\sigma$ is load scale.

Having the sampled time series of the process, one has to decide if the load is a smooth, i.e. differentiable, function or not. We illustrate the problem to use the O-U process with variance $\sigma^2 = 1$ and covariance $r(t) = e^{-|t|}$, which is sampled with frequency 10 Hz ($dt = 0.1$). The measured signal is an AR(1) process; see Figure 5.4.

However, if $X$ were an irregular O-U process for which

$$\frac{1}{t}\mathsf{E}[\sum |x_{i+1} - x_i|] \approx \frac{1}{\sqrt{\pi\,dt}},$$

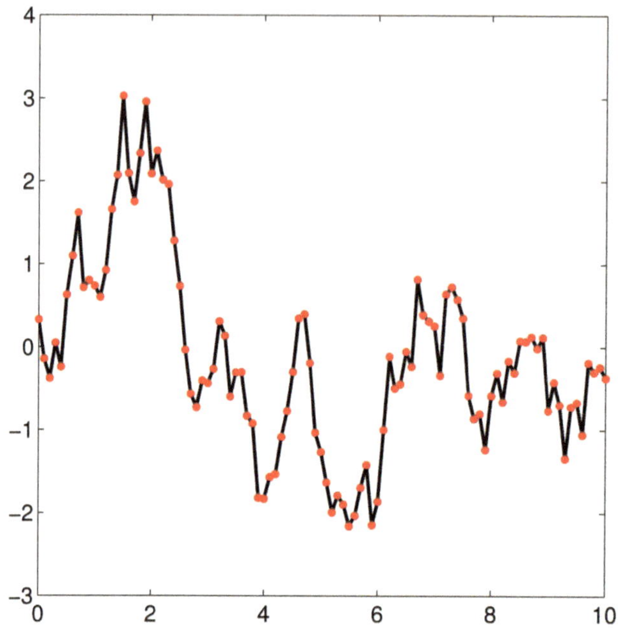


**FIGURE 5.4** The Ornstein-Uhlenbeck process $X$ sampled with 10 Hz frequency (red dots). The approximation $x^{pl}(\tau)$ the solid line.

the bound would grow to infinity as $dt$ tends to zero. This is not surprising since the expected rainflow damage in an O-U process is infinite for $\beta \le 2$. Consequently, the narrow-band $d^{nb}$ would increase as $1/\sqrt{dt}$.

Clearly, the behaviour of $\frac{1}{t}\mathsf{E}[\sum |x_{i+1} - x_i|]$ computed for different values of $dt$ would give us an idea which of the two cases we have.

# APPENDIX

## 5.A ORNSTEIN-UHLENBECK PROCESS AS THE LIMIT OF A RANDOM WALK

Marc Kac presented the Ornstein-Uhlenbeck process as a limit of random walk [98]. Consider an urn containing $N$ blue and yellow balls. At each step, a ball is chosen at random and replaced by a ball of the opposite colour. Let $X_n$ be the number of blue balls in the urn after $n$ steps. The time series $X_n$ is a Markov chain with transition probabilities $\mathsf{P}(X_1 = n+1 \mid X_0 = n) = n/N$, $\mathsf{P}(X_1 = n-1 \mid X_0 = n) = 1 - n/N$, for $0 < n < N$.

Let $X_0$ have a stationary distribution $\mathsf{P}(X_0 = n) = F_n$; see Chapter 8 on Markov chain models for details how to derive the distribution. As in Chapter 4, introduce $F_{n,n+1} = \mathsf{P}(X_0 = n, X_1 = n+1)$ and $F_{n+1,n} = \mathsf{P}(X_0 = n+1, X_1 = n)$. Obviously, all remaining $F_{ij} = 0$. Further, $F_{N,N+1} = F_{0,-1} = 0$, since $0, N$ are reflecting barriers.

Since $F_{ij} = F_{ji}$, we can use the electric circuit analogy, with resistances $R_i = 2/F_{i,i+1}$, connected in series to compute the interval crossing intensity $\mu^+(i,j)$. In order to find the resistances, one needs to estimate the stationary probabilities $F_i$, i.e. solve $N+1$ linear equations. This is a simple matter for small values of $N$; however, for large $N$ (e.g., $N \geq 500$), this can be done using numerical methods. The Matlab solver gave

$$F_{k,k+1} = F_i \cdot \frac{i+1}{N} \approx f_X(k)$$

where $X$ is normally distributed variable with mean $N/2$ and variance $(N+1)/4$. Consequently, the oscillation intensity is

$$\mu^{osc}(i,j) = \frac{1}{2\sum_{k=i}^{j+1} 1/f_X(k)},$$

since resistance $R_k = 2/f_x(k)$.

Let us standardise $X_k$ to have mean zero and variance one, viz. let

$$Y_k = \frac{k - N/2}{0.5\sqrt{N+1}}, \tag{5.25}$$

and let the load $Y_k$ vary between $-\sqrt{N}$ and $\sqrt{N}$ with step $dy = y_{y+1} - y_k = 2/\sqrt{N+1}$. The probability function of $Y_n$ is

$$\mathsf{P}(Y_0 = y_k) = \frac{dy}{\sqrt{2\pi}} e^{-y_k^2/2} = f_X(k)$$

and the oscillation intensity

$$\mu_X^{osc}(x_i, x_j) = \frac{2/(N+1)}{\sqrt{2\pi}} \frac{1}{\sum_{k=i}^{j+1} dy\, e^{y_k^2/2}}$$

leading to the asymptotic

$$\mu^{osc}(x,y) \approx \frac{c}{\sqrt{2\pi}} \frac{1}{\int_x^y e^{z^2/2}\,\mathrm{d}z}, \qquad c = 2/N. \tag{5.26}$$

We see that the standardised random walk process $Y_k$ (5.25) is approximating the O-U process with $c = 2/(N+1)$. Surprisingly, the oscillations tend to zero as $N$ increases without bound. Consequently, in order to get any sensible load model, we need to increase the frequency of rainflow cycles in $Y_k$. Here, we shall increase it by a factor $N+1$, i.e. it will be an approximation of O-U process with parameter $c = 2$.

We finish with estimation of the intensity of local maxima (rainflow cycles) in $Y_k$. The intensity $\mu^{rfc} = \mathsf{P}(X_{-1} < X_0, X_0 > X_1)$ is given by

$$\mu^{rfc} = \sum (i/N)(1 - i/N)\mathsf{P}(X_0 = i) \approx \mathsf{E}[X_0]/N - \mathsf{E}[X_0]^2/N^2 = 1/4,$$

since $\mathsf{Var}(X_0)/N^2$ is negligible.

## Notes

[1]That $\mathsf{P}(A_j) = \mathsf{P}(A_0)$ is given by (5.2) is obvious for iid variables; regardless of from what $j$ one starts observing the series, the probability of reaching above $v$ before below $u$ is divided in proportion.

[2]The rfc memory $H_0^+$ can also be used to give the oscillation intensity,

$$\mu^{osc}(u,v) = \mu_{Max} \int_v^\infty \mathsf{P}(H_0^+ > x - u \mid M_0 = x) f_{M_0}(x)\,\mathrm{d}x.$$

# 6 Gaussian Loads – Theory

## 6.1 INTRODUCTION

When the mechanical and ocean engineering communities in the 1950s accepted random loads and random waves as legitimate models, they inherited Steve Rice's stationary Gaussian process paradigm; [153]. They also inherited his work on level crossings in mechanical engineering, and cycles and extremes in ocean engineering. Both fields have inspired to much of the work reported in this book, and in this chapter, we dwell on some of the important parts in that development. We also try to throw some more light on how to interpret the statistical concepts we use throughout the book. The Slepian model for the behaviour of a stationary process in the neighbourhood of level crossings and local extremes is defined and explained.

The theory of statistical fatigue has developed in parallel with computational tools, mostly using Gaussian process theory. We give a brief introduction to such tools.

## 6.2 AMPLITUDE AND RANGE: A NON-TRIVIAL PROBLEM

Statistical extreme value theory for global extremes can offer quite general asymptotic results, under fairly simple conditions, as described in Section 3.4.1. Amplitude and range refer to local extremes, for which strong local dependence structure determines everything. The advent of fast computing power and advances in numerical linear algebra has made exact solutions possible for problems where only ad hoc or empirical solutions used to be available for these important variables.

### 6.2.1 THE INTRIGUING DISTRIBUTION OF M,m

The Gaßner proposal 1954 [67] of the crossing spectrum $N^+(u)$ as a measure of load amplitude distribution and Bendat's systematic use of the ratio $N^+(u)/N^+(0)$ for the Rayleigh distribution for load maxima [12] have been recurring themes earlier in this book. Actually, Bendat promoted Rice's work on random noise for general engineering public, in the book *Principles and Applications of Random Noise Theory* 1958 [13]. However, the Rayleigh distribution appears in this book only as the distribution of the envelope of a narrow-band Gaussian process.

From 1960, Julius Bendat got engaged as a consultant on statistical problems concerning "Flying vehicle vibration", and in 1963, he was contracted by NASA to write a report on probability for the prediction of random peaks, fatigue damage and catastrophic failures [12]. Here, he gave the recipe for how to use a standard Rayleigh variable $R$ to represent the peaks in narrow-band standard Gaussian loads.

In the meantime, the Swedish truck and bus manufacturer Scania-Vabis financed an ambitious research project on random fatigue, facing both theoretical and

DOI: 10.1201/9781003019947-6

experimental sides of load cycle distributions. The final report was published in the form of a PhD thesis by Sverker Sjöström, "On random load analysis", 1961, [189].

Sjöström's theoretical analysis was based on the representation (6.10) for the height of local maxima, and the analogue form for local minima, and his goal was to combine them to a distribution for the Max-min range $H = M^+ - M^-$, where $M^+$ and $M^-$ stand for a randomly selected pair of local maximum and minimum. Transforming the problem to the uncorrelated variables

$$H = M^+ - M^-,$$
$$M = (M^+ + M^-)/2,$$

he formulated the following conjecture for a Gaussian process with spectral width parameter $\alpha_2$.

**Conjecture 6.1.** *If $H$ and $M$ are statistically independent and the level crossing spectrum follows Rice's formula, then the half-range $H/2$ is Rayleigh distributed with mean $\mathsf{E}[H/2] = \alpha_2\sqrt{\pi/2}$, and the mean $M$ is normal with mean zero and variance $\omega_0(1-\alpha_2^2)$.*

Obviously, as $\alpha_2 \to 1$, the load becomes narrow-banded and Sjöström's result coincides with Bendat's.

Sjöström supported his independence assumption by many live experiments like the one in Figure 6.1. The innovative experiments were designed in conjunction with new statistical tools for quality development, a strategy that leads to the position as technical director of the Scania company.

The academic opponent at the PhD-defence 1961 was able to prove the statement in the conjecture – independence of $H$ and $M$ leads to the Rayleigh distribution – but he could not find any mathematical arguments for the hypothetical independence. After the "disputation", the opponent, Bertram Broberg, moved to the new Lund Institute of Technology as Vice-chancellor and professor in solid mechanics. He handed over the independence problem to the professor in mathematical statistics as a good PhD theme. The student could disprove the hypothesis – $H, M$ are always dependent in a Gaussian process; Lindgren 1970 [121]. Broberg's proof is now published as Lindgren and Broberg 2004 [127]. It is an open problem, i.e. how to construct a differentiable stationary non-Gaussian process that fulfils the requirements in the conjecture, if such a process exists.

New questions from Broberg's department led a new PhD student to investigate the mathematics and statistics of Endo's rainflow cycle count; the first result was the new local definition by Rychlik 1986 [169].

The toolbox WAFO [32, 207, 208] for crossings and fatigue analysis is another result of the Sjöström experience. It started as a Rice-series approximation by Lindgren 1972 [122, 128] and grew in parallel with research by Rychlik, Lindgren and students, e.g., by the introduction of the "Markov method" from Frendahl and Rychlik 1993 [62]. An important step in the development of WAFO to a reliable numerical tool for exact computation of the distribution of wave characteristics like amplitude/range, period, steepness, etc, was the improved handling of nearly singular

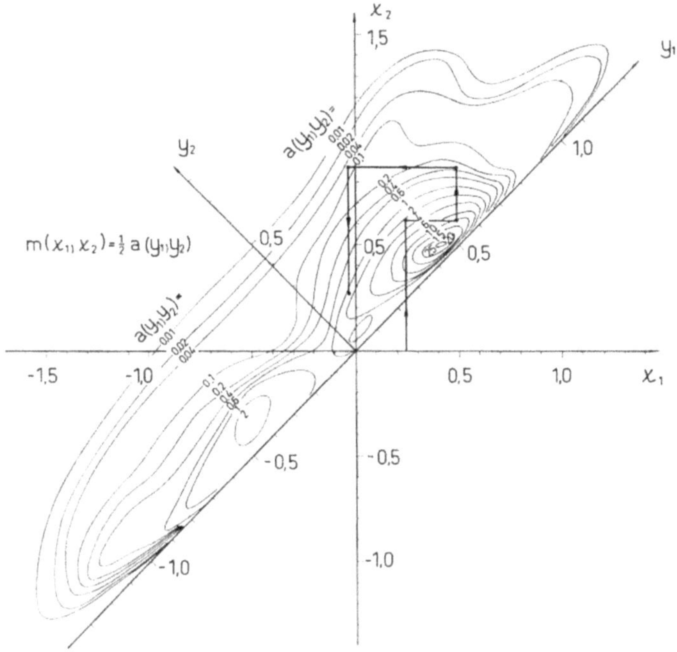


**FIGURE 6.1** Empirical distribution for $M^+, M^-$ for torque of half shafts on heavy truck. (Reproduced with permission from [189, Fig. 12]. Copyright © 1961 Scania CV AB.)

covariance matrices, Brodtkorb 2004, [30, 31, 126]. As an example of its capacity, we use it here to calculate the exact density function for the Max-min and rainflow range distributions for Gaussian loads with four very different spectra with very different narrowness.

**Example 6.1.** *We study Gaussian processes with the following spectra after truncation at* $\omega_c = 2.5$ *and normalised to* $\omega_0 = \omega_2 = 1$.

- *A lightly damped linear oscillator with relative damping 0.05, a very narrow-banded Gaussian process with* $\alpha_2 = 0.9426$.
- *Low frequency white noise with constant spectral density for* $\omega < \omega_c$ *with* $\alpha_2 = 0.7454$.
- *Combination of two JONSWAP spectra with peak period* $T_p^1 = 15\,s$, $T_p^2 = 5\,s$ *and total* $\alpha_2 = 0.6598$.
- *Standardised O-U process with* $\alpha_2 = 0.4254$.

*We computed, by* WAFO, *the exact density function* $f_{Mm}$ *for the four processes, and then, by the MCTP method, described in Section 5.3.2, we obtained also the rainflow cycle distribution. For comparison, we simulated, also by* WAFO, *more than 2 million Max-min and rfc pairs and their ranges. The result is shown in Figure 6.2*

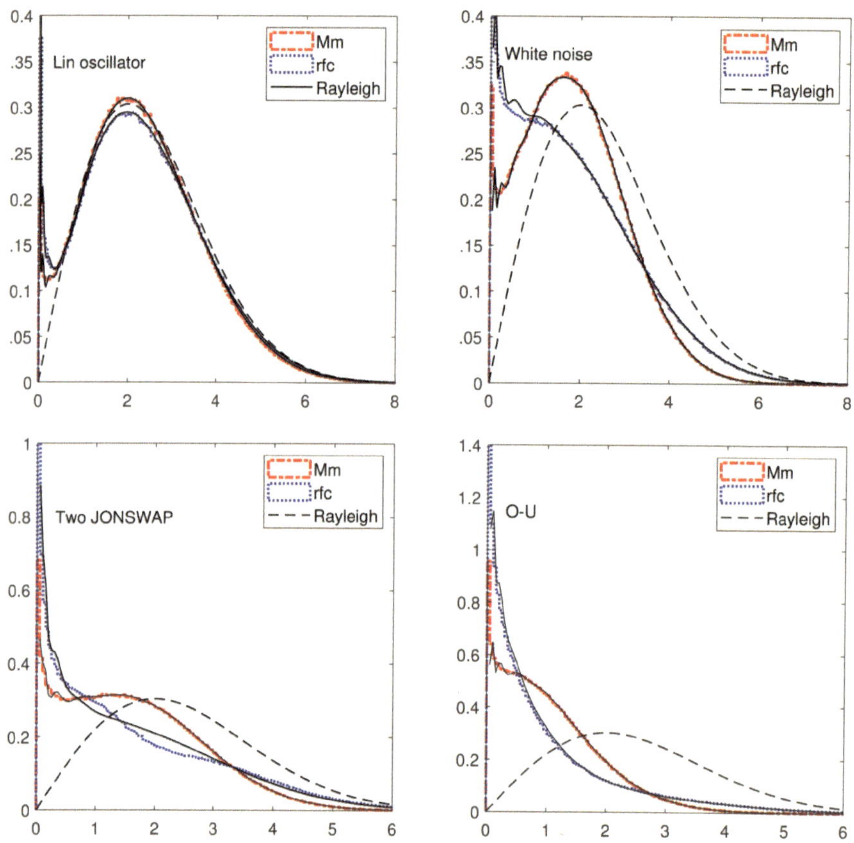


**FIGURE 6.2** Cycle range distributions: The four subfigures, one for each spectrum, show: Dash-dotted curve: Max-min range pdf; Dotted curve: Rfc range pdf; Histogram behind or near each pdf: Histogram for more than two million simulated ranges. The pdfs are computed by routines in WAFO [207]. Dashed curve: Rayleigh density for the narrow-band approximation.

*together with Bendat's Rayleigh density for range, 2R. The mean of the range distributions is found in the* $\beta=1$ *column in Table 6.1. It is equal to* $\alpha_2\sqrt{2\pi}$. *The constant* $f_{max} = \frac{1}{2\pi}\sqrt{\omega_4/\omega_2}$ *in the table is the expected number of maxima per time unit.*

*From the pdfs and from the simulated ranges, we computed* $\mathsf{E}[H^\beta]$ *for* $\beta$ = *1, 2, 3, 4, 5. The values are in Table 6.1 in the rows marked "pdf" and "emp", respectively. The narrow-band approximation, also an upper bound [174], is found in the last row in the table.*

*From the figure, one can draw several conclusions. First: For Max-min ranges, the agreement between the Monte Carlo simulations and the computed distributions from* WAFO *is excellent. Only for very small range-values, the computed distribution becomes somewhat irregular, depending on near singular*

**Table 6.1**
**Total Damage Intensity $d = f_{max}\mathsf{E}[H^\beta]$ for rfc and Max-Min Counts**

| | $\alpha_2$ | $f_{max}$ | $\mathsf{E}[H^\beta]$ | | | | |
|---|---|---|---|---|---|---|---|
| Osc 0.05, rfc pdf | 0.9427 | 0.1688 | 2.36 | 7.52 | 28.11 | 118.96 | 555.67 |
| emp | | | 2.36 | 7.42 | 27.53 | 115.57 | 534.65 |
| Osc 0.05, Mm pdf | | | 2.38 | 7.36 | 26.86 | 110.73 | 503.36 |
| emp | | | 2.36 | 7.27 | 26.35 | 108.04 | 489.41 |
| White noise, rfc pdf | 0.7454 | 0.2135 | 1.87 | 5.38 | 19.19 | 79.23 | 365.26 |
| emp | | | 1.87 | 5.36 | 19.04 | 77.83 | 353.29 |
| White noise, Mm pdf | | | 1.87 | 4.75 | 14.27 | 48.28 | 179.64 |
| emp | | | 1.87 | 4.75 | 14.27 | 48.22 | 179.48 |
| Two JONSWAP, rfc pdf | 0.6598 | 0.2412 | 1.65 | 4.69 | 16.72 | 69.17 | 319.46 |
| emp | | | 1.65 | 4.78 | 17.51 | 73.64 | 341.69 |
| Two JONSWAP, Mm pdf | | | 1.66 | 4.05 | 11.93 | 40.03 | 148.47 |
| emp | | | 1.65 | 4.03 | 11.87 | 39.75 | 147.16 |
| OU, rfc pdf | 0.4264 | 0.3742 | 1.07 | 2.49 | 8.39 | 34.75 | 164.20 |
| emp | | | 1.06 | 2.52 | 8.48 | 34.72 | 160.70 |
| OU, Mm pdf | | | 1.07 | 1.83 | 4.00 | 10.39 | 30.80 |
| emp | | | 1.07 | 1.82 | 3.98 | 10.34 | 30.54 |
| Rayleigh, nb | | | 2.51 | 7.99 | 30.02 | 127.6 | 598.5 |

*Note*: "pdf/emp": expectation from pdf computed by [207] and simulated with more than 2 million ranges, respectively. $\beta = 1, 2, 3, 4, 5$.

*covariance structures at close distance; see [31]. Also for rainflow ranges, there is very good agreement between simulation and computation, even if small deviations appear for the complicated 2-JONSWAP spectrum. More examples of the capacity of* WAFO *can be found in [126].*

*Second: Looking at the high range moments in the table, important for damage prediction, the agreement between simulations and numerical computations is very good for all the spectra.*

### 6.2.2 HOW TO INTERPRET THE DISTRIBUTION OF LOCAL MAXIMA

This section is mathematically intricate, but its intuitive interpretation is easy to grasp. The idea is to give a frequentistic interpretation of the distribution of the height $M$ of local maxima that explains the link between the empirical damage formula (6.14) and the theoretical statistical formula (6.15) later in this chapter.

Let us first rephrase the definition from Chapter 2, Definition 2.5,

$$N_T^{max} = \text{number of local maxima of } X(t) \text{ in } [0,T],$$
$$\mu_{max} = f_{max} = \text{expected number of local maxima per time unit, } \mathsf{E}[N_1^{max}].$$

To account for the height of the maxima, define

$$N_T^{max}(u) = \text{number of local maxima } \tau_i, 0 \le \tau_i \le T \text{ with } X(\tau_i) > u,$$
$$\mu_{max}(u) = \text{expected number of local maxima above } u \text{ per time unit, } \mathsf{E}[N_1^{max}(u)].$$

The *empirical probability* that a local maximum has height exceeding the threshold $u$ is thus

$$\mathsf{P}_T(M > u) = N_T^{max}(u)/N_T^{max},$$

and $\mathsf{P}_T(u) = 1 - \mathsf{P}_T(M > u)$ is the empirical distribution function of local maxima. The probability $\mathsf{P}_T$ measures how likely it is that a maximum chosen at random from the sequence of turning points $T_P(X)$ in $X(t)$, $t \in [0,T]$, has height above $u$. Clearly, $\mathsf{P}_T(\cdot)$ is a random quantity, an estimate of the "true" probability, dependent on the particular realisation of the process $X$. Our intuition demands that the longer the observation interval, larger $T$, the more reliable and stable should the estimate be.

**Theorem 6.1** ("Long-run distribution"). *If the process $X(t)$ is ergodic, the empirical probability $P_T(M > u)$ converges, as $T$ increases without bound, to a deterministic limit, called the ergodic or long-run distribution,*

$$\lim_{T\to\infty} \mathsf{P}_T(M > u) = \lim_{T\to\infty} \frac{N_T^{max}(u)}{N_T^{max}} = \frac{\mu_{max}(u)}{\mu_{max}}.$$

*Proof.* In Section 2.4.5, we described the class of stochastic processes for which time averages always converge towards the mathematical average, i.e. the expectation. For an ergodic load process also the sequence of local maxima is ergodic, which implies that, as stated,

$$\lim_{T\to\infty} \frac{N_T^{max}(u)}{T} = \mathsf{E}[N_1^{max}(u)] = \mu_{max}(u),$$
$$\lim_{T\to\infty} \frac{N_T^{max}}{T} = \mathsf{E}[N_1^{max}] = \mu_{max}. \qquad \square$$

Let us now consider the positive peak count and rewrite (4.24) in terms of $N_T^{max}$ and $N_T^{max}(u)$ to get

$$\begin{aligned} D^{pp}(T) &= \int_0^\infty 2\beta(2u)^{\beta-1} N_T^{max}(u)\,\mathrm{d}u \\ &= T\,\frac{N_T^{max}}{T} \int_0^\infty 2\beta(2u)^{\beta-1} \frac{N_T^{max}(u)}{N_T^{max}}\,\mathrm{d}u \\ &= T\,\frac{N_T^{max}}{T} \int_0^\infty 2\beta(2u)^{\beta-1} \mathsf{P}_T(M > u)\,\mathrm{d}u. \end{aligned} \tag{6.1}$$

By ergodicity, $N_T^{max}/T \to \mu_{max}$, the deterministic intensity of local maxima (cycles) and $\mathsf{P}_T(M > u) \to \mathsf{P}(M > u)$, and hence

$$\lim_{T\to\infty} \frac{1}{T} D^{pp}(T) = \lim_{T\to\infty} \frac{N_T^{max}}{T} \lim_{T\to\infty} \int_0^\infty 2\beta(2u)^{\beta-1}\mathsf{P}_T(M>u)\,\mathrm{d}u$$
$$= \mu_{max} \int_0^\infty 2\beta(2u)^{\beta-1}\mathsf{P}(M>u)\,\mathrm{d}u = \mu_{max}\,\mathsf{E}[(2M^+)^\beta], \tag{6.2}$$

and hence

$$d^{pp} = \mu_{max}\,\mathsf{E}[(2M^+)^\beta]. \tag{6.3}$$

In the last step in (6.2), we used a simple rule (2.5) for expectations: If $g(u)$ is a differentiable function of a random variable $X$ and $g(u) = 0$ for $u \le 0$, then

$$\mathsf{E}[g(X^+)] = \int_0^\infty \frac{dg(u)}{du} \mathsf{P}(X > u)\,\mathrm{d}u.$$

The interpretation of the limit of an empirical distribution as a probability distribution is the standard one used in this work in connection with level crossings and local extremes. In a later section, we shall give an explicit formula for the long-run probability $\mathsf{P}(M > u)$ and the intensity of cycles $\mu_{max}$ for a zero-mean Gaussian load $X(t)$ having power spectrum density $S(\omega)$ (such a process is ergodic). However, first we need to introduce the generalised Rice's formula for the intensity of marked crossings.

### 6.2.3 INTENSITY OF MARKED CROSSINGS

The height of local maxima, treated in the previous section, is an example of a marked crossing problem: A local maximum is a downcrossing zero of the process derivative and the height of the maximum is a "mark" attached to the crossing.

Let $z(t)$ be a smooth ergodic process, such that the joint density $f(\dot{z}, z)$ of $\dot{z}(t), z(t)$ exists, and say we are interested in the intensity $\mu_u$ of $\tau$ such that $z(\tau) = u$, when $z(t)$ is observed over a very long time interval. Suppose that, in addition to $z(t)$, one has a vector of processes $\mathbf{y}(t) = (y_1(t), \ldots, y_n(t))$ such that $z(t), \mathbf{y}(t)$ are jointly stationary and ergodic.[1]

Consider now statement $A$ about the outcome of $\mathbf{y}(t)$ for which we can say if it is true or not at any time $t$. For reasons that will become clear later, we also include $\dot{z}(t)$ in our interest and let $A$ be a statement about the joint behaviour of $\mathbf{y}(t)$ and $\dot{z}(t)$.

As before let

$$N_T(u) = \text{number of } t \in [0,T] \text{ such that } z(t) = u,$$

be the crossing count and let

$$N_T(A \mid u) = \text{number of } t \in [0,T] \text{ such that } z(t) = u \text{ and } A \text{ is true},$$

be the count of qualified crossings where $\mathbf{y}(t)$ satisfies condition $A$. Further define the intensities

$$\mu(z(t) = u) = \mathsf{E}[N_1(u)], \quad \mu(A \mid z(t) = u) = \mathsf{E}[N_1(A \mid u)].$$

Most often, we shall use the simplified notation $\mu(u) = \mu(z(t) = u)$ and $\mu(A \mid u) = \mu(A \mid z(t) = u)$. By the assumed ergodicity,

$$\mu(A \mid u) = \lim_{T\to\infty} \frac{N_T(A \mid u)}{T}$$

is the intensity of the solutions of the equation $z(t) = u$, such that $\mathbf{y}(t), \dot{z}(t)$ satisfy A. Obviously, $\mu(u) = \lim_{T\to\infty} \frac{N_T(u)}{T}$ is the intensity of the solutions of $z(t) = u$.

**Example 6.2.** *Let $z(t) = \dot{x}(t)$, $y(t) = x(t)$, and $A$ be the statement $A =$"$y(t) > u, \dot{z}(t) < 0$". Then, $N_T(A \mid 0)$ is the number of local maxima of $x$ observed in $[0,T]$ exceeding level $u$, while $\mu(A \mid 0)$ is the intensity of $\dot{x}(t)$-zeros such that the second derivative at $t$ is negative and $x(t) > u$.*

Now, the "long-run" probability of $A$, observed at points $t$ where $z(t) = u$, is defined as

$$\mathsf{P}(A \mid z(t) = u) = \lim_{T\to\infty} \frac{N_T(A \mid u)}{N_T(u)} = \lim_{T\to\infty} \frac{N_T(A \mid u)/T}{N_T(u)/T} = \frac{\mu(A \mid u)}{\mu(u)}. \tag{6.4}$$

Consider again statement $A$ about the joint behaviour of $\mathbf{y}(t)$ and $\dot{z}(t)$. Note that $\mathsf{P}(A) = \mathsf{P}((\mathbf{y}(0), \dot{z}(0)) \in A)$ (likelihood that $A$ is true at time zero) is usually not equal $\mathsf{P}(A \mid z(t) = u)$, as defined in Equation (6.4). The reason for this is that $\mathbf{y}(t), \dot{z}(t)$ at points where $z(t) = u$ can have a different distribution than $\mathbf{y}(t), \dot{z}(t)$ observed at $t = 1, 2, 3, \ldots$. This will be illustrated in Section 6.4 where the distribution of a derivative of a process taken at deterministic points is compared with the distribution of the derivative at points when the process takes the specific value $u$.

**Theorem 6.2.** *Under the general assumptions of this section, if the joint density $f(\dot{z}, z)$ of $\dot{z}(0), z(0)$ exists, then*

$$\mu(A \mid u) = \int \mathsf{P}(A \mid u, \dot{z})|\dot{z}| f(\dot{z}, u)\, \mathrm{d}\dot{z}, \tag{6.5}$$

*for almost all $u$, where $\mathsf{P}(A \mid u, \dot{z})$ is the conditional probability*

$$\mathsf{P}(A \mid u, \dot{z}) = \mathsf{P}\Big((\mathbf{y}(0), \dot{z}(0)) \in A \mid z(0) = u, \dot{z}(0) = \dot{z}\Big).$$

*In addition, if also the density $f(\mathbf{y}, \dot{z}, z)$ of $\mathbf{y}(0), \dot{z}(0), z(0)$ exists, then*

$$\mu(A \mid u) = \int_A |\dot{z}| f(\mathbf{y}, \dot{z}, u)\, \mathrm{d}\mathbf{y}\, \mathrm{d}\dot{z}. \tag{6.6}$$

**Remark 6.1** (Important interpretation)**.** *Formula* (6.6) *will be our standard reference for long-run intensities and conditional intensities of cycle counts. It will also be the basis for expected damage and cycle-related distributions.*

The proof that (6.6) is true for all levels $u$ for the Gaussian $z(t), \mathbf{y}(t)$ can be found in Leadbetter et al. [115, Ch. 10]. Note that $\mu(A \mid u)$ is often entering the formulas for expected damage as an integrand and hence the almost everywhere validity of (6.5) is all that is needed. This is important when the expected damage is computed for non-Gaussian processes.

### 6.2.4 LOCAL MAXIMA AND MINIMA IN GAUSSIAN LOADS

#### 6.2.4.1 Representation of Local Maxima

In this section, we have collected some useful facts concerning the distribution of local maxima and minima in Gaussian loads and how they are related to the power spectral distribution and the spectral moments; see also Section 2.2.2.3.

Let us consider a stationary Gaussian process $X(t)$ with mean zero and spectral distribution having finite fourth spectral moment, i.e. $\omega_0, \omega_2, \omega_4 < \infty$. This implies that the second-order derivative $\ddot{X}(t)$ exists and $\ddot{X}(t) \in N(0, \omega_4)$, while $\dot{X}(t) \in N(0, \omega_2)$, $X(t) \in N(0, \omega_0)$.

The distribution we seek will contain two parameters, the standard deviation $\sqrt{\omega_0}$ of the process and the spectral width parameter

$$\alpha_2 = \frac{\text{intensity of upcrossings of the mean } \mathsf{E}[x(0)]}{\text{intensity of local maxima}}. \tag{6.7}$$

For a stationary Gaussian process, we have (by Rice's formula) that the intensity of upcrossings of the mean $\mathsf{E}[X(0)]$ is given by $f_z = \frac{1}{2\pi}\sqrt{\omega_2/\omega_0}$, while the intensity of local maxima (intensity of zero down crossings of the derivative $\dot{X}(t)$) is $\mu_{max} = \frac{1}{2\pi}\sqrt{\omega_4/\omega_2}$ and hence the parameter $\alpha_2$ is equal to

$$\alpha_2 = \frac{\omega_2}{\sqrt{\omega_0 \omega_4}}. \tag{6.8}$$

We turn first to the long-run distribution of the height of local maxima, i.e. the distribution of $X(\tau)$ for $\tau$ such that $\dot{X}(\tau) = 0$ and $\ddot{X}(\tau) < 0$, given by

$$\mathsf{P}(M \le x) = \frac{\mathsf{P}(X(t) \le x; \ddot{X}(t) < 0 \mid \dot{X}(t) = 0)}{\mathsf{P}(\ddot{X}(t) < 0 \mid \dot{X}(t) = 0)}.$$

**Theorem 6.3.** *For a zero-mean stationary Gaussian process $x(\tau)$, the long-run distribution of the height of local maxima $M$ is given by*

$$\mathsf{P}(M \le x) = \mathsf{P}\left(\sqrt{\omega_0}\left(\alpha_2 R + \sqrt{1-\alpha_2^2}\,U\right) \le x\right),$$

*where $R$ has a standard Rayleigh distribution, $f_R(r) = r\exp(-r^2/2), r \ge 0$, while $U$ is standard normal and independent of $R$.*

*Proof.* We begin by computing $\mathsf{P}(X(t) \le x,\ \ddot{X}(t) < 0 \mid \dot{X}(t) = 0)$ by means of Lemma 6.2 in Appendix 6.4. In order to use the lemma, define $Z(t) = \dot{X}(t)$ and $Y(t) = X(t)$ and set $u = 0$ in the condition $Z(t) = u$. We have $m_Y = 0 = m_Z = 0$, and the covariance matrix of $(Z(t), \dot{Z}(t), Y(t))$ is

$$\Sigma = \begin{bmatrix} \omega_2 & 0 & 0 \\ 0 & \omega_4 & -\omega_2 \\ 0 & -\omega_2 & \omega_0 \end{bmatrix},$$

with $\sigma_Z^2 = \omega_2$, $\sigma_{\dot Z}^2 = \omega_4$, $\rho_{ZY} = 0$, $\rho_{\dot Z Y} = -\alpha_2$, $\sigma_Y^2 = \omega_0$. Lemma 6.2 gives the representation

$$\dot Z_0 = \sqrt{\omega_4} R,$$
$$Y_0 = \sqrt{\omega_0}\left(-\alpha_2 R + \sqrt{1-\alpha_2^2}\,U\right), \tag{6.9}$$

where $R$ is double Rayleigh distributed, i.e. has pdf $f(r) = \frac{|r|}{2} e^{-r^2/2}$, while $U$ is standard normal distributed and independent of $R$. The two-sided $R$-variable represents the second-order derivative at points where the derivative is zero, i.e. a local minimum or maximum. Finally, $\mathsf{P}(M \le x)$ can be derived as follows:

$$\mathsf{P}(M \le x) = \mathsf{P}(Y_0 \le x \mid \dot Z_0 < 0) = \mathsf{P}(Y_0 \le x \mid R < 0).$$

This finishes the proof. □

The theorem is a *representation theorem*, meaning that it allows us to use two simple "synthetic variables" $R, U$ to describe the statistical variation of the complicated long-run behaviour of local maxima. Instead of making a long simulation of the process $x(t)$, taking the trouble to identify the local maxima to register their heights and get the empirical distribution, one can simulate independent pairs of $(R,U)$ and generate $M = Y_0$ according to (6.9). The justification for the procedure, given by Rice 1945 [168], is that the long-run distribution of the height of local maxima is the same as the distribution of the variable $Y_0$,

$$M \overset{\mathscr{L}}{=} \sqrt{\omega_0}\left(\alpha_2 R + \sqrt{1-\alpha_2^2}\,U\right). \tag{6.10}$$

The probability $\mathsf{P}(M \le x)$ can be computed exactly by means of formula (6.30). In Section 6.5, we will present a more complicated and very useful representation model.

## 6.3 EXPECTED DAMAGE FOR POSITIVE PEAK COUNT

### 6.3.1 THE EXPECTED DAMAGE

For a smooth stress function $x(t)$, $t \in [0,T]$, having finite number $N$ of local extremes in sequence $m_1, M_1, m_2, \ldots, M_N$, a permutation $k(i)$ defines a cycle count and then, by means of the damage function $f(u,v)$, it gives the total damage

$$D(T) = \sum_{i=1}^{N} f(m_{k(i)}, M_i). \tag{6.11}$$

Although we are mainly interested in the expected rainflow damage, we shall first consider the *positive peak* count damage $D^{pp}$, which can be computed using any cycle count, e.g. min-Max or rainflow, and gives $f(u,v) = (2v^+)^\beta$, viz.

$$D^{pp}(T) = \sum_{i=1}^{N} (2M_i^+)^\beta, \tag{6.12}$$

where $M_i^+ = \max(0, M_i)$. The reason for considering $D^{pp}(T)$ is that it is simple to analyse and can serve as an illustration for problems in defining the distribution of cycle ranges $S_i = M_i - m_{k(i)}$.

If $N$ were independent of the sequence of turning points and the maxima $M_i$ equally distributed with probability density function $p_{M_1}(x)$, then

$$\mathsf{E}[D^{pp}(T)] = \mathsf{E}[N]\mathsf{E}[(2M_1^+)^\beta] = \mathsf{E}[N]\int_0^\infty (2x)^\beta p_{M_1}(x)\,\mathrm{d}x. \tag{6.13}$$

However, for stationary loads, usually $N$ is not independent of the sequence of turning points $T_P(x)$. (The number of cycles should be negatively correlated with the sizes of ranges.) Further $M_1$ does not have the same distribution as $M_2$, the fact that we start to count cycles at time zero influences their distributions. We conclude that finding the distribution of the local maxima is a subtle matter that needs careful discussion.

Let us return to the definition of positive peak damage (4.24),

$$D^{pp}(T) = \sum_{i=1}^{N}(2M_i^+)^\beta = \int_0^\infty 2\beta(2u)^{\beta-1}N^{pp}(u)\,\mathrm{d}u, \tag{6.14}$$

where $N^{pp}(u)$ is the number of maxima exceeding level $u$. (Note that $N^{pp}(u) \geq N^+(u)$.) For a random (Gaussian) load, the counts $N$ and $N^{pp}(u)$ are random variables, and hence the expected pp damage accumulated in $[0,T]$ is given by

$$\begin{aligned}\mathsf{E}[D^{pp}(T)] &= \int_0^\infty 2\beta(2u)^{\beta-1}\mathsf{E}[N^{pp}(u)]\,\mathrm{d}u \\ &= \mathsf{E}[N^{pp}(0)]\int_0^\infty 2\beta(2u)^{\beta-1}\mathsf{E}[N^{pp}(u)]/\mathsf{E}[N^{pp}(0)]\,\mathrm{d}u. \end{aligned} \tag{6.15}$$

Obviously, $\mathsf{E}[N^{pp}(0)] = \mathsf{E}[N]$, while $\mathsf{E}[N^{pp}(u)]/\mathsf{E}[N^{pp}(0)] = \mathsf{P}(M > u)$. The question is weather $\mathsf{P}(M > u) = \int_u^\infty p_{M_1}(x)\,\mathrm{d}x$, the probability density of local maximum used in (6.13). This will be discussed in the following subsection.

### 6.3.2 AN APPLICATION OF (6.10) TO POSITIVE PEAK DAMAGE

We now use the representation (6.10) to get an expression for the pp-damage rate $d^{pp}$. Since $\mathsf{E}[D^{pp}(T)] = T \cdot d^{pp}$, the pp-damage rate is given by (6.3) as

$$d^{pp} = \mu_{max}\mathsf{E}[(2M^+)^\beta] = \frac{1}{2\pi}\sqrt{\frac{\omega_4}{\omega_2}}\mathsf{E}\left[\left(2\sqrt{\omega_0}\left(\alpha_2 R + \sqrt{1-\alpha_2^2}U\right)^+\right)^\beta\right],$$

which we rewrite as

$$\begin{aligned} d^{pp} &= f_z h_s^\beta \cdot 2^{-\beta}\alpha_2^{-1} \cdot \mathsf{E}\left[\left(\left(\alpha_2 R + \sqrt{1-\alpha_2^2}U\right)^+\right)^\beta\right] \\ &= c(\beta, \alpha_2) \cdot d^{nb}. \end{aligned} \tag{6.16}$$

Here, the factor $c(\beta,\alpha_2) = d^{pp}/d^{nb} \geq 1$ is given by

$$c(\beta,\alpha_2) = \frac{\mathsf{E}\left[\left(\left(\alpha_2 R + \sqrt{1-\alpha_2^2}\,U\right)^+\right)^\beta\right]}{\alpha_2 \mathsf{E}[R^\beta]}, \tag{6.17}$$

and hence $\alpha_2 c(\beta,\alpha_2) \leq 1$.

It is generally assumed that the positive peak count gives damages far exceeding the rainflow damage. This is motivated by a fact that for small $\alpha_2$, there are many cycles with positive maxima and very small ranges. For those cycles, the damage $(2v^+)^\beta >> (v-u)^\beta$. In Figure 6.3, the correction factor $c(\beta,\alpha_2)$ is presented (dashed line) for $\beta = 4$.

It can be shown, see Remark 6.2, that $d^{pp} \geq d^{nb}$, and hence $d^{pp}$ is of limited interest to bound the expected rainflow damage. However, in Remark 6.3, a possible application of $d^{pp}$ is discussed.

**Remark 6.2.** *Using the same reasoning as when proving (6.3), and hence also (6.16), one can show that for zero-mean Gaussian loads*

$$\mathsf{E}[D^{HM}(T)] = T \cdot \mu_{max}\left(\mathsf{E}[(2M^+)^\beta] - \mathsf{E}[(2M^-)^\beta]\right) = T \cdot d^{nb}, \tag{6.18}$$

*where* $a^- = \max(-a, 0)$. *Now, it is clear that*

$$\mathsf{E}[(2M^+)^\beta] - \mathsf{E}[(2M^-)^\beta] \leq \mathsf{E}[(2M^+)^\beta],$$

*and hence*

$$\mathsf{E}[D^{HM}(T)] \leq \mathsf{E}[D^{pp}(T)], \tag{6.19}$$

*for zero-mean symmetric loads.*

Finally, note that when $\alpha_2 \to 0$ then $\mu_{max} \to +\infty$, while

$$\left(\mathsf{E}[(2M^+)^\beta] - \mathsf{E}[(2M^-)^\beta]\right) \to 0.$$

This is one more example that when average properties are of interest, then it is more convenient to work with the derivatives $\dot{D}(t)$ (marked crossings of $x(\tau)$) than with damage increase per cycle $f(m_{k(i)}, M_i)$ (marked crossings of $\dot{x}(\tau)$).

**Remark 6.3** (Lower bound for $\mathsf{E}[D^{HM}(T)]$). *The main problem with the pp-damage is that it may change drastically if one adds small, for fatigue irrelevant, wiggles to the load. A natural solution would be to filter the signal to remove the cycles with small ranges. Here, we limit ourselves to zero-mean Gaussian loads which are symmetric. Such processes are often used to model the sea surface movement. A rough method, sometimes used by offshore engineers to evaluate fatigue damage caused by wave loads, is to define stress cycle range to be proportional to the wave* *range, which is the difference between the largest and lowest sea level between two consecutive zero upcrossings, respectively. Inspired by this approach, we propose to keep only the crests and troughs of "apparent waves" in the sequence of turning points $T_P(x)$.*

*More precisely, for a zero-mean Gaussian load $X(t)$ with $\omega_4 < \infty$, i.e. finite intensity of cycles, remove from the sequence of turning points $T_P(x)$ all maxima and minima except the largest local maxima and the smallest minima between the two consecutive zero upcrossings. Denote by $m_i^{tr}, M_i^{cr}$, $i = 1, \dots, n$, the new sequence of turning points. Obviously for any damage function and cycle count, the damage computed from the filtered signal will be smaller than from the original one.*

*Now, it is easy to see that the crossing spectrum $\widetilde{\mu}(u)$, say, computed from the filtered turning points has the following property; $\widetilde{\mu}(u) = f_z \mathsf{P}(M^{cr} > u)$ if $u > 0$ while $\widetilde{\mu}(u) = f_z \mathsf{P}(m^{tr} < u)$ if $u < 0$, which is symmetric around zero. Consequently, $\widetilde{\mu}(u) < \mu(u)$ and hence*

$$T d^{nb} > T \widetilde{d}^{nb} = T \cdot f_z \mathsf{E}[(2M^{cr})^{\beta}],$$

*computed for the filtered signal, is smaller than $T \cdot d^{nb}$ and hence*

$$T \cdot d^{nb} > T \cdot f_z \mathsf{E}[(2M^{cr})^{\beta}].$$

*Finally, since $M^{cr} \geq M^+$,*

$$T \cdot d^{nb} > T \cdot f_z \mathsf{E}[(2M^+)^{\beta}] = T \cdot \alpha_2 d^{pp} = T \cdot \alpha_2 c(\beta, \alpha_2) d^{nb}. \tag{6.20}$$

In Figure 6.3, the correction factor $\alpha_2 c(\beta, \alpha_2)$, for $\beta = 4$, is shown as a solid line. We can see that for low values of $\alpha_2$, it is much smaller then 1. We suspect that

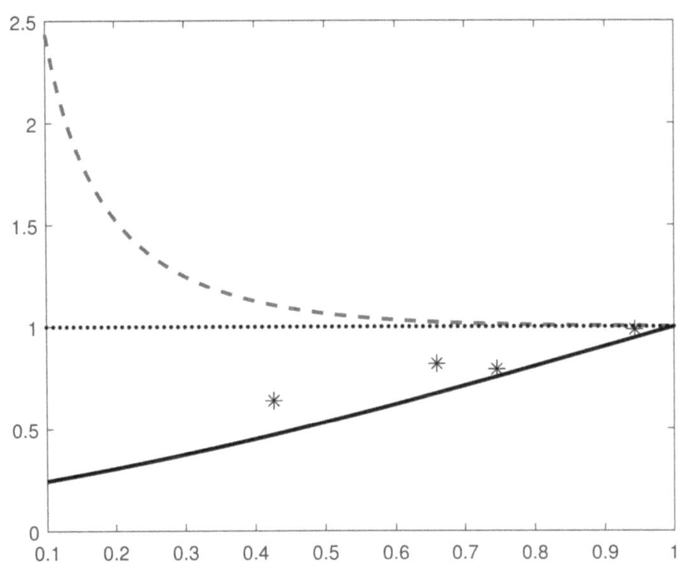


**FIGURE 6.3** The factor $c(\beta, \alpha_2)$ for $\beta = 4$ as function of $\alpha_2$, defined by (6.17), the dashed line. The solid line is the factor $\alpha_2 c(\beta, \alpha_2)$. Stars represent $D^{rfc}(T)/(T \cdot d^+)$ computed from data in Table 6.1 for the four spectra in Example 6.1.

approximately the rainflow damage intensity $d^{rfc} \leq d^{nb}$ is actually bounded from below by $\alpha_2\, c(\beta, \alpha_2)\, d^{nb}$. We have tested several Gaussian loads with different spectra and estimated the damage intensity $d^{rfc} \approx D^{rfc}(T)/T$. The results are shown in Figure 6.3 as stars. We can see that the rainflow damage intensities have values anywhere between the two bounds. This implies that $\alpha_2\, d^{pp}$ could serve as approximate lower bound for the rainflow damage intensity in zero-mean Gaussian loads in some cases.

## 6.4 CROSSING DERIVATIVES WITH SOME CONSEQUENCES

The derivative of a stationary Gaussian process is also a stationary Gaussian process. Sampling the derivative process at fixed intervals results in a sample of normal variables. It is easy to convince oneself that sampling the derivative only at times when the process upcrosses a fixed level $u$ does not produce a normal sample, not even the positive part of a normal distribution, but in fact a Rayleigh distribution. This "crossing bias" comes from the fact that level upcrossings are more likely to occur during time-periods when the derivative is large than when the process lies almost flat.

As a consequence of the crossing bias, we need to incorporate the biased derivative distribution in the fatigue cycle analysis as soon as the cycle count includes level crossings or local extremes, (zero crossings of the derivative).

In this section, we first analyse the crossing bias on the derivative and then present a useful tool, called the *Slepian model*, where the biased distribution is used to improve reconstruction of the process, conditioned of crossings.

### 6.4.1 DERIVATIVES AT CROSSINGS, CONDITIONING

#### 6.4.1.1 Derivative at Crossings

Consider a stationary ergodic Gaussian process $Z(t)$. Suppose that $\mathsf{E}[Z(t)] = m$ and that the spectral measure has the first two spectral moments finite, i.e. $\omega_0, \omega_2 < \infty$, so the derivative process $\dot{Z}(t)$ exists and is also Gaussian. Let us denote by $A$ the statement that the observed derivative exceeds the level $v$. As we have mentioned before, the derivative of a stationary Gaussian process is a zero-mean stationary Gaussian process with variance equal to the second-order spectral moment $\omega_2$, i.e. $\dot{Z}(t) \in N(0, \omega_2)$. This means that the fraction of derivatives exceeding $v$, observed at times $t = 0, 1, 2, 3, \ldots$ is equal to

$$\mathsf{P}(\dot{Z}(0) > v) = 1 - \Phi(v/\sqrt{\omega_2}).$$

Now, we shall demonstrate that the fraction of derivatives observed only at points $s_i$, such that $Z(s_i) = u$, will have a different value, i.e.

$$\mathsf{P}(\dot{Z}(0) > v) \neq \mathsf{P}(\dot{Z}(s_i) > v \mid Z(s_i) = u).$$

**Lemma 6.1.** *Under the general assumptions of this section, the distribution* $F^u(v) = \mathsf{P}(\dot{Z}(s) \leq v \mid Z(s) = u)$ *has density*

$$f^u(v) = \frac{|v|}{2\omega_2} e^{-\frac{v^2}{2\omega_2}}. \tag{6.21}$$

*Proof.* For $A =$"$\dot{Z}(s) > v$", using (6.5), we have that the intensity $\mu(A \mid u)$ is given by

$$\mu(A \mid u) = \int_{-\infty}^{\infty} \mathsf{P}(\dot{Z}(0) > v \mid Z(0) = u, \dot{Z}(0) = \dot{z})|\dot{z}|f(u,\dot{z})\, \mathrm{d}\dot{z}$$

$$= \int_{v}^{\infty} |\dot{z}|f(u,\dot{z})\, \mathrm{d}\dot{z},$$

where $f(u,\dot{z})$ is the density of $(Z(0),\dot{Z}(0))$. By our assumptions, $Z(0) \in N(m,\omega_0)$ is independent of $\dot{Z}(0) \in N(0,\omega_2)$. Consequently, $f(u,\dot{z}) = f(u)f(\dot{z})$ and hence

$$\mu(A \mid u) = f(u)\int_{v}^{\infty} |\dot{z}|f(\dot{z})\, \mathrm{d}\dot{z} = 2f(u)\sqrt{\frac{\omega_2}{2\pi}}\int_{v}^{\infty} \frac{|\dot{z}|}{2\omega_2}e^{-\frac{\dot{z}^2}{2\omega_2}}\, \mathrm{d}\dot{z}.$$

Clearly, $\mu(u) = \mu(A \mid u)$, for $v = -\infty$, and hence

$$\mu(u) = 2f(u)\sqrt{\frac{\omega_2}{2\pi}}\int_{-\infty}^{\infty} \frac{|\dot{z}|}{2\omega_2}e^{-\frac{\dot{z}^2}{2\omega_2}}\, \mathrm{d}\dot{z} = 2f(u)\sqrt{\frac{\omega_2}{2\pi}}.$$

Consequently

$$\mathsf{P}(\dot{Z}(s) > v \mid Z(s) = u) = \frac{\mu(A \mid u)}{\mu(u)} = \int_{v}^{\infty} \frac{|\dot{z}|}{2\omega_2}e^{-\frac{\dot{z}^2}{2\omega_2}}\, \mathrm{d}\dot{z},$$

which after differentiation leads to (6.21). □

#### 6.4.1.2 General Approach to Crossing Conditioning

Before we turn to some more complicated examples taken from naval engineering, we shall sketch a general approach to the computation of

$$\mathsf{P}((\mathbf{Y}(s),\dot{Z}(s)) \in A \mid Z(s) = u)$$

for Gaussian processes $\mathbf{Y}(s), Z(s)$.

(a) First, write down a formula for the long-run distribution of the vector $\mathbf{Y}(s), \dot{Z}(s)$ observed at $s$ such that $Z(s) = u$.
(b) Then, find the density $f^u(\mathbf{y},\dot{z})$. (Such density always exists if $\mathbf{Y}(0), \dot{Z}(0), Z(0)$ has a non-degenerated Gaussian density.)
(c) Finally, compute

$$\mathsf{P}((\mathbf{Y}(s),\dot{Z}(s)) \in A \mid Z(s) = u) = \int_A f^u(\mathbf{y},\dot{z})\, \mathrm{d}\mathbf{y}\, \mathrm{d}\dot{z}. \tag{6.22}$$

In order to simplify notation, one is often introducing variables $\mathbf{Y}_u, \dot{Z}_u$, which have the density $f^u(\mathbf{y},\dot{z})$, i.e. we can formally write

$$\mathsf{P}((\mathbf{Y}_u,\dot{Z}_u) \in A) = \mathsf{P}((\mathbf{Y},\dot{Z}) \in A \mid Z(s) = u) = \int_A f^u(\mathbf{y},\dot{z})\, \mathrm{d}\mathbf{y}\, \mathrm{d}\dot{z}. \tag{6.23}$$

The variables $\mathbf{Y}_u, \dot{Z}_u$ are special cases of the so-called Slepian model process, see Lindgren and Rychlik (1991) [132] for detailed presentation.

### 6.4.1.3 Conditioning on Crossing Derivative

Now, one can ask the question about the long-run probability that the derivative observed at upcrossings of the level $u$ exceeds level $v$, i.e. we wish to know the proportion of $s$ such that $Z(s)=u$ and $\dot{Z}(s)>0$ which also satisfies the condition $\dot{Z}(s)>v$, $v\geq 0$. After a short reflection, one can see that it is just a small variation of Lemma 6.1,

$$\mathsf{P}(\dot{Z}_u > v \mid \dot{Z}_u > 0) = \frac{\mathsf{P}(\dot{Z}(s) > v, \dot{Z}(s) > 0 \mid Z(s) = u)}{\mathsf{P}(\dot{Z}(s) > 0 \mid Z(s) = u)} = e^{-\frac{v^2}{2\omega_2}}.$$

We conclude that the derivative at upcrossing is Rayleigh distributed with parameter $\omega_2$ independently of the level $u$.

The following lemma gives an explicit formula for $Y_u, \dot{Z}_u$ for the one-dimensional $Y(s)$, i.e. when $\mathbf{Y}(s) = Y(s)$.

**Lemma 6.2.** *Suppose $Z(t), \dot{Z}(t), Y(t)$ is a stationary Gaussian vector valued process. Let $m_Z = \mathsf{E}[Z(t)]$, $m_Y = \mathsf{E}[Y(t)]$ and let the covariance matrix $\Sigma$ of the vector $Z(t), \dot{Z}(t), Y(t)$ be*

$$\Sigma = \begin{bmatrix} \sigma_Z^2 & 0 & \sigma_Z\sigma_Y\rho_{ZY} \\ 0 & \sigma_{\dot{Z}}^2 & \sigma_{\dot{Z}}\sigma_Y\rho_{\dot{Z}Y} \\ \sigma_Z\sigma_Y\rho_{ZY} & \sigma_{\dot{Z}}\sigma_Y\rho_{\dot{Z}Y} & \sigma_Y^2 \end{bmatrix}. \tag{6.24}$$

*The variables $Y_u, \dot{Z}_u$, with distribution defined by (6.23), have the following representation:*

$$\begin{aligned} \dot{Z}_u &= \sigma_{\dot{Z}} R, \\ Y_u &= m_u + \sigma_Y\left(\rho_{\dot{Z}Y} R + \sqrt{1-\rho_{ZY}^2-\rho_{\dot{Z}Y}^2}\, U\right), \end{aligned} \tag{6.25}$$

*where $R$ has a double Rayleigh distribution with density $f(r) = \frac{|r|}{2} e^{-r^2/2}$, independent of $U \in N(0,1)$. The non-random part of (6.25) is*

$$m_u = m_Y + \frac{\sigma_Y}{\sigma_Z}\rho_{ZY}(u - m_Z), \tag{6.26}$$

*and $\rho_{ZY}$, $\rho_{\dot{Z}Y}$ are correlations between $Y(0)$ and $Z(0)$, $\dot{Z}(0)$, respectively.*

*Obviously, the long-run distribution $F^u(y,v)$ of $Y(t), \dot{Z}(t)$ at points that satisfy $Z(s) = u$, i.e. $\mathsf{P}(\dot{Z}(s) \leq v, Y(s)) \leq y \mid Z(s) = u)$ is given by*

$$\mathsf{P}(Y_u \leq y, \dot{Z}_u \leq v) = \mathsf{P}\left(\rho_{\dot{Z}Y} R + \sqrt{1-\rho_{ZY}^2-\rho_{\dot{Z}Y}^2}\, U \leq \frac{y-m_u}{\sigma_Y},\ R \leq \frac{v}{\sigma_{\dot{Z}}}\right).$$

*Proof.* We need to demonstrate that

$$\mathsf{P}(Y(s) \leq y, \dot{Z}(s) \leq v \mid Z(s) = u) = \frac{\mu(A \mid u)}{\mu(u)},$$

where $A =$ "$Y(t) \leq y$ and $\dot{Z}(t) \leq v$" is equal to

$$\mathsf{P}\left(\rho_{\dot{Z}Y}R + \sqrt{1-\rho_{ZY}^2-\rho_{\dot{Z}Y}^2}\,U \leq \frac{y-m_u}{\sigma_Y},\ R \leq \frac{v}{\sigma_{\dot{Z}}}\right).$$

We begin by computing the intensity $\mu(A \mid u)$, i.e. the intensity of points $s$, such that $Z(s) = u$ and $A$ is true. From formula (6.5), we know that

$$\mu(A \mid u) = \int_{-\infty}^{\infty} \mathsf{P}(Y(0) \leq y, \dot{Z}(0) \leq v \mid Z(0) = u, \dot{Z}(0) = \dot{z})|\dot{z}| f(u,\dot{z})\,\mathrm{d}\dot{z},$$

where

$$\mathsf{P}(Y(0) \leq y, \dot{Z}(0) \leq v \mid u, \dot{z}) = \begin{cases} \mathsf{P}(Y(0) \leq y \mid Z(0) = u, \dot{Z}(0) = \dot{z}) & \text{if } \dot{z} \leq v, \\ 0 & \text{if } \dot{z} > v. \end{cases}$$

Now we know, from (2.37) and (2.38) in Section 2.1.5.2, that the conditional density of $Y(0)$, given $Z(0) = z, \dot{Z}(0) = \dot{z}$, is Gaussian with mean

$$\begin{aligned} m(z,\dot{z}) &= m_Y + \sigma_Y[\sigma_Z\rho_{ZY}\ \sigma_{\dot{Z}}\rho_{\dot{Z}Y}]\begin{bmatrix} \sigma_Z^2 & 0 \\ 0 & \sigma_{\dot{Z}}^2 \end{bmatrix}^{-1}\begin{bmatrix} z-m_Z \\ \dot{z} \end{bmatrix} \\ &= m_Y + \frac{\sigma_Y}{\sigma_Z}\rho_{ZY}(z-m_Z) + \frac{\sigma_Y}{\sigma_{\dot{Z}}}\rho_{\dot{Z}Y}\dot{z}, \end{aligned} \tag{6.27}$$

and variance

$$\begin{aligned} \sigma^2 &= \sigma_Y^2 - \sigma_Y^2[\sigma_Z\rho_{ZY}\ \sigma_{\dot{Z}}\rho_{\dot{Z}Y}]\begin{bmatrix} \sigma_Z^2 & 0 \\ 0 & \sigma_{\dot{Z}}^2 \end{bmatrix}^{-1}\begin{bmatrix} \sigma_Z\rho_{ZY} \\ \sigma_{\dot{Z}}\rho_{\dot{Z}Y} \end{bmatrix} \\ &= \sigma_Y^2(1-\rho_{ZY}^2-\rho_{\dot{Z}Y}^2). \end{aligned} \tag{6.28}$$

Consequently, we have

$$\begin{aligned} \mu(A \mid u) &= \frac{1}{\sqrt{2\pi}\sigma_Z}e^{-\frac{(u-m_Z)^2}{2\sigma_Z^2}} \int_{-\infty}^{v} \Phi\left(\frac{y-m(u,\dot{z})}{\sigma}\right)\frac{|\dot{z}|}{\sqrt{2\pi}\sigma_{\dot{Z}}}e^{-(\dot{z}^2/2\sigma_{\dot{Z}}^2)}\,\mathrm{d}\dot{z} \\ &= \frac{1}{\pi}\frac{\sigma_{\dot{Z}}}{\sigma_Z}e^{-\frac{(u-m_Z)^2}{2\sigma_Z^2}} \int_{-\infty}^{v} \Phi\left(\frac{y-m(u,\dot{z})}{\sigma}\right)\frac{|\dot{z}|}{2\sigma_{\dot{Z}}^2}e^{-(\dot{z}^2/2\sigma_{\dot{Z}}^2)}\,\mathrm{d}\dot{z} \\ &= \frac{1}{\pi}\frac{\sigma_{\dot{Z}}}{\sigma_Z}e^{-\frac{(u-m_Z)^2}{2\sigma_Z^2}} \int_{-\infty}^{v} \mathsf{P}\left(m_u + \frac{\sigma_Y}{\sigma_{\dot{Z}}}\rho_{\dot{Z}Y}\dot{z} + \sigma U < y\right)\frac{|\dot{z}|}{2\sigma_{\dot{Z}}^2}e^{-(\dot{z}^2/2\sigma_{\dot{Z}}^2)}\,\mathrm{d}\dot{z} \\ &= \frac{1}{\pi}\frac{\sigma_{\dot{Z}}}{\sigma_Z}e^{-\frac{(u-m_Z)^2}{2\sigma_Z^2}} \int_{-\infty}^{v/\sigma_{\dot{Z}}} \mathsf{P}\left(m_u + \sigma_Y\rho_{\dot{Z}Y}r + \sigma U < y\right)\frac{|r|}{2}e^{-r^2/2}\,\mathrm{d}r. \end{aligned} \tag{6.29}$$

Since $\mu(u) = \frac{1}{\pi}\frac{\sigma_{\dot{Z}}}{\sigma_Z}e^{-\frac{(u-m_Z)^2}{2\sigma_Z^2}}$, the lemma is proved. □

Note that formula (6.29) can be evaluated using integration by parts, see the following lemma for a related result.

**Lemma 6.3.** *Let $U,R$ be independent, standard Gaussian and Rayleigh distributed variables. For $a > 0$ and any $x$,*

$$\mathsf{P}(aU + bR > x) = \Phi\left(-\frac{x}{a}\right) + e^{-\frac{x^2}{2\sigma^2}} \cdot \frac{b}{\sigma}\Phi\left(\frac{x/a}{\sigma/b}\right), \tag{6.30}$$

*where $\sigma = \sqrt{a^2+b^2}$.*

The last two lemmas give an explicit way to compute the distribution of $Y_u, \dot{Z}_u$, i.e. $F^u(y,\dot{z}) = \mathsf{P}(Y_u \le y, \dot{Z}_u \le \dot{z})$. The representation, given in (6.25), helps to write the probabilities $\mathsf{P}((Y(s),\dot{Z}(s)) \in A \mid Z(s) = u)$ in a more transparent way, see the example given in the following subsection. Similar results can be derived for vector valued process $\mathbf{Y}(t)$ and even for $\mathbf{Y}(t)$ containing infinitely many processes $Y_i(t)$, see Lindgren and Rychlik (1991) [132].

**Remark 6.4.** *Suppose $(Y(0),\dot{Z}(0))$ is Gaussian. Then*

$$\mathsf{P}(Y(0) \le y, \dot{Z}(0) \le v) = \mathsf{P}\left(\rho_{\dot{Z}Y}R + \sqrt{1-\rho^2_{\dot{Z}Y}}\,U \le \frac{y-m_Y}{\sigma_Y},\ R \le \frac{v}{\sigma_{\dot{Z}}}\right),$$

*where $U,R$ are independent standard Gaussian, i.e. $U,R \in N(0,1)$. Consequently, if the process $Y(\cdot)$ is independent of $Z(\cdot)$, then the values of $Y(s)$ observed at points $s$ where $Z(s) = u$, $Y_u$ have the same distribution as $Y(0)$.*

## 6.5 THE SLEPIAN MODEL PROCESS APPROACH

A Slepian process model is a multivariate version of the representation of $Y_u$ in Lemma 6.2. The structure is similar to that of $Y_u$, with one deterministic part depending on the (known) crossed level and one stochastic part which contains the (unknown) derivative $R$ at the crossing and the remaining variability, represented by a non-stationary Gaussian residual process instead of the single normal variable $U$ present in (6.25).

### 6.5.1 THE PROCESS NEAR u-UPROSSINGS

In this section, we present the most fundamental example of a Slepian model process, namely a model that represents the variability of a stationary Gaussian process when it is observed in the neighbourhood of upcrossings of a level $u$. The term Slepian model was established 1977 in [123], after David Slepian, who introduced the idea 1962 [191] for crossings of the mean level. The Slepian model near-level crossings is a multidimensional version of Lemma 6.2.

Let $Z(t), t \in \mathbb{R}$, be a differentiable stationary Gaussian process with mean zero, covariance function $r(t)$ and spectral density $S(\omega)$ with spectral moments $\omega_0 = r(0) = \mathsf{Var}[Z(t)]$, $\omega_2 = -r''(0) = \mathsf{Var}[\dot{Z}(t)]$.

As the mark, we take the values of the $Z(t)$-process at $n$ different points located around $t$,

$$\mathbf{Y}(t) = Z(t+\mathbf{s}) = \big(Z(t+s_1), Z(t+s_2), \ldots, Z(t+s_n)\big).$$

Then, $\big(Z(t), \dot{Z}(t), \mathbf{Y}(t)\big)$ is an $(n+2)$-dimensional stationary Gaussian vector process. Write $\mathbf{Y}(t) \in A$ if $\mathbf{Y}(t)$ satisfies the statement $A$.

Using (6.5) in Theorem 6.2 and that $\dot{Z}(t) > 0$ at upcrossings, we get that the intensity of $u$-level upcrossings $\tau$ where $\mathbf{Y}(\tau)$ satisfies $A$ is

$$\mu_u^+(A) = \int_{\dot{z}=0}^{\infty} \mathsf{P}(\mathbf{Y}(0) \in A \mid Z(0) = u, \dot{Z}(0) = z)\, \dot{z} f_{Z(0),\dot{Z}(0)}(u, \dot{z})\, \mathrm{d}\dot{z}.$$

Dividing by $\mu_u^+ = \int_0^\infty \dot{z} f_{Z(0),\dot{Z}(0)}(u,\dot{z})\, \mathrm{d}\dot{z}$, the intensity of $u$-upcrossings, we arrive at our basic formula for marks at $u$-upcrossings,

$$\mathsf{P}(\mathbf{Y}(\tau) \in A \mid Z(\tau) = u, \text{upcrossing}) = \frac{\mu_u^+(A)}{\mu_u^+}. \tag{6.31}$$

### 6.5.2 THE SLEPIAN PROCESS IN THE GAUSSIAN CASE

In the Gaussian case, $Z(t)$ and $\dot{Z}(t)$ are independent, $f_{Z(0),\dot{Z}(0)}(u,z) = f_{Z(0)}(u) f_{\dot{Z}(0)}(z)$, which considerably simplifies (6.31). Cancelling $f_{Z(0)}(u)$ in the right-hand side quotient and using $f_{\dot{Z}(0)}(z) = 1/\sqrt{2\pi\omega_2}\exp(-z^2/2\omega_2)$,[2] we get the final form of the distribution,

$$\begin{aligned} &\mathsf{P}(\mathbf{Y}(\tau) \in A \mid Z(\tau) = u, \text{upcrossing}) \\ &\quad = \int_{\dot{z}=0}^{\infty} \mathsf{P}(\mathbf{Y}(0) \in A \mid Z(0) = u, \dot{Z}(0) = z)\, \frac{z}{\omega_2} \exp\left(-\frac{z^2}{2\omega_2}\right) \mathrm{d}z. \end{aligned} \tag{6.32}$$

We first consider the conditional probability in (6.32).

**Lemma 6.4.** *The conditional distribution of*

$$\mathbf{Y}(0) = Z(\mathbf{s}) = \big(Z(s_1), Z(s_2), \ldots, Z(s_n)\big)$$

*given* $Z(0) = u, \dot{Z}(0) = z$ *is n-dimensional normal with mean and covariance*

$$\mathsf{E}[Z(s_k) \mid u, z] = \frac{u r(s_k)}{\omega_0} - \frac{z r'(s_k)}{\omega_2}, \tag{6.33}$$

$$\mathsf{Cov}[Z(s_j), Z(s_k) \mid u, \dot{z}] = r(s_j - s_k) - \frac{r(s_j) r(s_k)}{\omega_0} - \frac{r'(s_j) r'(s_k)}{\omega_2}. \tag{6.34}$$

*Proof.* The conditional distribution of $\mathbf{Y}(0) = Z(\mathbf{s})$ is n-variate normal and completely determined by mean values and pairwise covariance; see Section 2.1.5.2.

Therefore, it suffices to consider the case $n = 2$ and use formulas (2.37) and (2.38) applied to the covariance matrix for $(Z(s_1), Z(s_2), Z(0), \dot{Z}(0))$,

$$\Sigma = \begin{bmatrix} \omega_0 & r(s_1 - s_2) & r(s_1) & -r'(s_1) \\ r(s_2 - s_1) & \omega_0 & r(s_2) & -r'(s_2) \\ r(s_1) & r(s_2) & \omega_0 & 0 \\ -r'(s_1) & -r'(s_2) & 0 & \omega_2 \end{bmatrix}.$$

The calculations are straightforward. □

The lemma takes us halfway to the distribution of the mark $\mathbf{Y}(\tau)$ at the $u$-upcrossings points $\tau$. The crux is that we do not have access to the derivatives $\dot{Z}(\tau)$, only to their distribution, which is Rayleigh.

The solution is to replace $z$ in (6.33) with a Rayleigh distributed variable $R$ with the density

$$f_R(r) = (r/\omega_2)\exp(-r^2/(2\omega_2)), \quad r > 0, \tag{6.35}$$

from (6.32). This gives the *regression part* of the Slepian representation,

$$Z_u^{reg}(s) = \frac{ur(s)}{\omega_0} - \frac{Rr'(s)}{\omega_2}. \tag{6.36}$$

Figure 6.4 (a) and (b) shows five realisations of these two terms in a Slepian process.

To represent the normal residual variation around the regression part, we will define a zero-mean, non-stationary Gaussian process $\Delta(s), s \in \mathbb{R}$, independent of the $R$-variable, and with covariance function

$$r_\Delta(s,t) = \mathrm{Cov}[\Delta(s), \Delta(t)] = r(s-t) - \frac{r(s)r(t)}{\omega_0} - \frac{r'(s)r'(t)}{\omega_2}. \tag{6.37}$$

Note that the $\Delta$-process does not involve the crossed level $u$ or the derivative $R$. Figure 6.4(c) shows five realisations of a residual process.

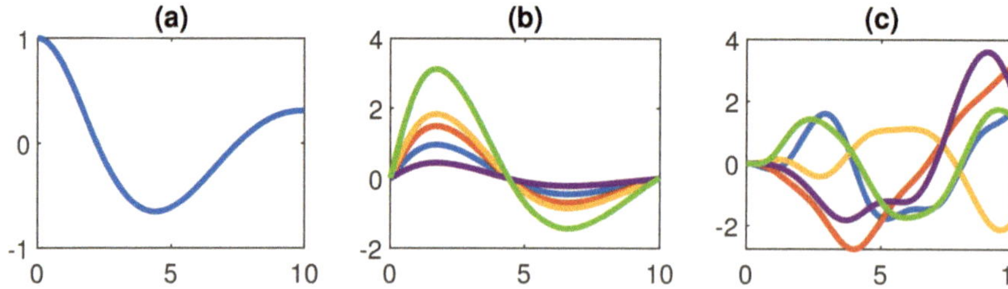


**FIGURE 6.4** Five realisations of the three components in a Slepian model. (a) Regression term on the crossed level. (b) Regression term on the random derivative at crossing. (c) Non-stationary Gaussian residual process.

We have now arrived at the main Slepian theorem.

**Theorem 6.4** (Slepian model after $u$-upcrossing)**.** *The Slepian model around $u$-upcrossings for a zero-mean, differentiable, stationary Gaussian process $Z(t), t \in \mathbb{R}$, with covariance function $r(t)$ has the form*

$$Z_u(s) = \frac{ur(s)}{\omega_0} - \frac{Rr'(s)}{\omega_2} + \Delta(s). \tag{6.38}$$

*Here $R$ is a Rayleigh variable with density* (6.35) *and $\Delta(s), s \in \mathbb{R}$, a non-stationary, zero-mean Gaussian process, independent of $R$, and with covariance function* (6.37).

*The distribution of $Z_u(\mathbf{s})$ is equal to the long-run empirical distribution of $Z(\tau_k + \mathbf{s})$ when $\tau_k$ runs through all $u$-upcrossings of $Z(t)$.*

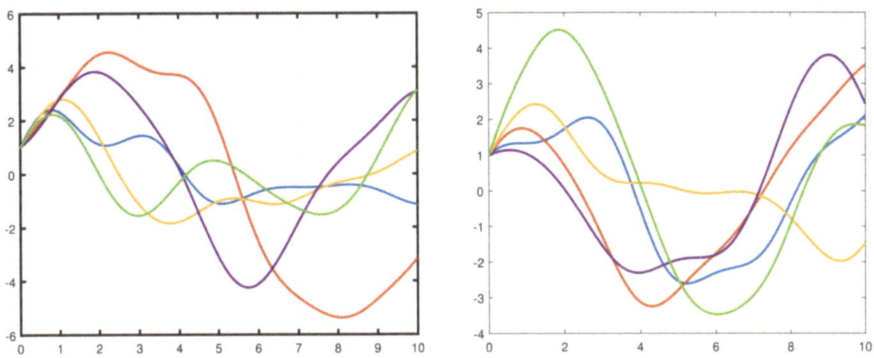


**FIGURE 6.5** Left: Five independent samples of Gaussian process after a randomly chosen upcrossing of level $u = 1$. Right: Five independent realisations of Slepian process after upcrossing of level $u = 1$ with components (a–c) as in Figure 6.4.

Figure 6.5 shows (left) five randomly selected upcrossings of level $u = 1$ together with a 10-second long view of the process. To the right, there are five realisations of the Slepian model. The $Z$-process is a typical sea-wave process with Pierson-Moskowitz spectrum with $H_s$ =7 m and $T_p$ =11 s.

**Remark 6.5.** *Slepian models can be built for more complicated crossing problems in Gaussian and some non-Gaussian processes, [158]. An early reference is Lindgren 1970 [121] on local maxima in a Gaussian load process. The article was inspired by the work by Sjöström 1961 [189] on random fatigue.*

**Example 6.3** (Slepian model at local maximum)**.** *The Slepian model at a local maximum in a Gaussian process can be used to approximate the distribution of the Max-min cycle, $f_{M,m}(x,y)$. To condition on a local maximum in $x(t)$ is equivalent to condition on a zero downcrossing of the derivative $\dot{x}(t)$. This leads to the Slepian model,*

*[125, Theorem 8.11],*

$$X_{max}(t) = -R\frac{r''(t)}{\sqrt{\omega_4}} + \Delta^{max}(t), \tag{6.39}$$

*where $R$ is a standard Rayleigh variable and the residual $\Delta^{max}(t)$ is a zero-mean non-stationary Gaussian process, independent of $R$ and with covariance function*

$$r_{\Delta^{max}}(s,t) = r(s-t) - \frac{r'(s)r'(t)}{\omega_2} - \frac{r''(s)r''(t)}{\omega_4}. \tag{6.40}$$

### 6.5.3 HOW TO USE A REPRESENTATION THEOREM

The Rice representation (6.10) of the height of a local maximum in a Gaussian process is simple to use for simulation and computing. The Slepian models (6.38) for level crossings and (6.39) for local maxima give more details but are still simple to generate and analyse. We shall give three examples of the use of these models.

#### 6.5.3.1 The Kowalewski Hybrid Model for the $(m, M)$ Density

Kowalewski (1964) [106, 107] proposed to approximate the probability density (pdf) $f_{mM}(u,v)$ of the min-Max pair $(m,M)$ in Gaussian loads by

$$f(u,v) = \frac{v-u}{4\alpha_2^2\omega_0\sqrt{2(1-\alpha_2^2)}} \exp\left(-\frac{(v-u)^2 + 4vu\alpha_2^2}{8\alpha_2^2(1-\alpha_2^2)\omega_0}\right), \quad \text{for } u \le v. \tag{6.41}$$

The Kowalewski approximation has the correct marginal densities for minimum and maximum, as given by (6.10), i.e. $\int_{-\infty}^{+\infty} f(x,y)\,\mathrm{d}x = f_M(y)$ and $\int_{-\infty}^{+\infty} f(x,y)\,\mathrm{d}y = f_m(x)$. However, the joint density is an accurate approximation of $f_{mM}(x,y)$ only for narrow band spectrum when $\alpha_2 \approx 1$.

The logic behind the Kowalewski approximation is the representation (6.10) of maximum and the analogue for the minimum,

$$M = \frac{h_s}{4}\left(\alpha_2 R + \sqrt{1-\alpha_2^2}\,U\right), \quad m = \frac{h_s}{4}\left(-\alpha_2 R + \sqrt{1-\alpha_2^2}\,U\right),$$

where $R, U$ are independent Rayleigh and Gaussian distributed random variables, respectively. The question of dependence between the Rayleigh and normal variables for the maximum and for the minimum is left unanswered.

Bishop and Sherratt (1990) [17] proposed to use a Markov chain of turning points having the Kowalewski kernel (6.41) to approximate the sequence of turning points in stationary Gaussian loads.

#### 6.5.3.2 A Slepian Approximation for the $(M, m)$ Distribution

Consider the Slepian model (6.39)

$$X_{max}(t) = -R\frac{r''(t)}{\sqrt{\omega_4}} + \Delta^{max}(t),$$

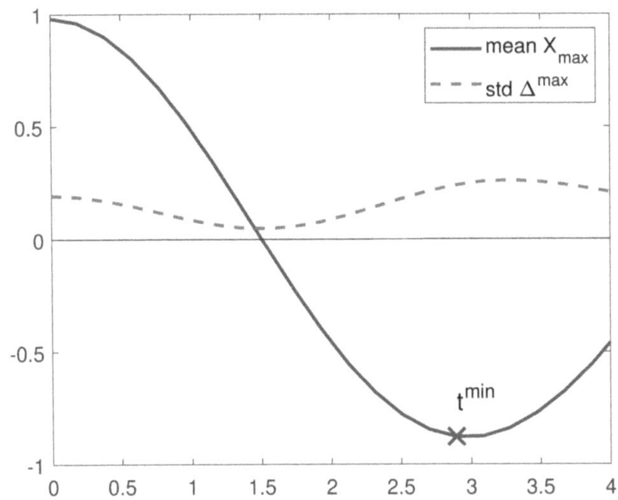


**FIGURE 6.6** Mean of model $X_{max}(t)$ and standard deviation of Gaussian residual $\Delta^{max}(t)$.

where $R$ is standard Rayleigh and the Gaussian process $\Delta^{max}(t)$ has covariance function (6.40). It is easy to verify that

$$X_{max}(0) = -Rr''(0)/\sqrt{\omega_4} + \Delta^{max}(0) = \sqrt{\omega_0}\left(\alpha_2 R + \sqrt{1-\alpha_2^2}U\right)$$

agreeing with (6.10), the distribution of a local maximum $M$.

The corresponding local minimum is

$$m = X_{max}(t^{min}), \quad t^{min} = \text{first minimum of } X_{max}(t), t > 0. \tag{6.42}$$

Looking at Figure 6.6, we see that it is the Rayleigh term that dominates the model near the maximum, and the standard deviation is small. A reasonable approximation of the minimum time $t^{min}$ in (6.42) is therefore to take the minimum of the mean function $\mathsf{E}[X_{max}(t)]$, marked $t^{min}$ in Figure 6.6.

**The Slepian approximation of the Max-min distribution:** With

$$t^{min} = \min\{t > 0; \mathsf{E}[\dot{X}_{max}(t)] = 0\},$$

the Slepian approximation of the Max-min pair $(M,m)$ is

$$\begin{bmatrix} M \\ m \end{bmatrix} \stackrel{\mathscr{L}}{=} R \begin{bmatrix} \sqrt{\omega_0}\alpha_2 \\ -r''(t^{min})/\sqrt{\omega_4} \end{bmatrix} + \begin{bmatrix} \Delta^{max}(0) \\ \Delta^{max}(t^{min}) \end{bmatrix}. \tag{6.43}$$

The distribution (6.43) is a bi-variate version of Rice's Max-representation (6.10). It differs from the Kowalewski approximation in two ways: The Rayleigh-term for the minimum has an adjusted coefficient and the Gaussian residuals are correlated in the correct way. The exact density function can be found in [129, Fact 2.].

In a simulation experiment, we simulated about 170 000 pairs of $Max, min$-values with the linear oscillator in Example 6.1 and compared the sampled distribution with the Kowalewski approximation (pdf) and the exact $Mm$-distribution computed in WAFO. The results are shown in Figure 6.7. It appears that the Kowalewski approximation is closer to the true $Mm$-distribution than the simple Slepian model for this very narrow spectrum, at least in the outer parts of the distribution. For the Pierson-Moskowitz spectrum used in Example 8.5, the agreement is not as good, as seen in Figure 8.4.

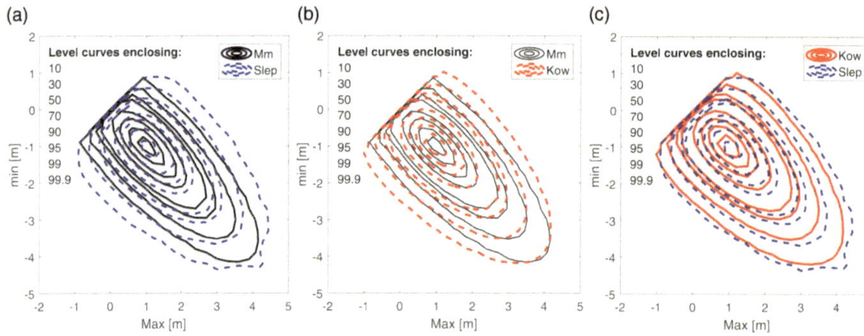


**FIGURE 6.7** (a) Slepian (dashed) (b) Kowalewski (dashed) approximations of the Max-min distribution compared with the exact pdf (solid). (c) Comparison of Kowalewski (solid) to Slepian (dashed).

#### 6.5.3.3 Simulation of Oscillation Intensity

We turn now to a method to estimate the matrix $\mu^{rfc} = \mathsf{E}[N_1^{rfc}(u_j, u_i)]$. Let $x(\tau)$ be a stationary Gaussian process with spectral density function $S(\omega)$, covariance function $r(s)$ and spectral moments $\omega_0 = r(0)$ and $\omega_2 = -\ddot{r}(0)$. Since $x(0)$ and $\dot{x}(0)$ are uncorrelated, they are also independent.

Now let us introduce a non-stationary Gaussian process $\Delta(\tau)$ which is independent of $x(0), \dot{x}(0)$, viz.

$$\Delta(\tau) = x(\tau) - x(0)\frac{r(\tau)}{r(0)} - \dot{x}(0)\frac{\dot{r}(\tau)}{\ddot{r}(0)}. \tag{6.44}$$

(Actually, $\Delta(\tau)$ is the residual in the linear regression of $x(\tau)$ on $(x(0), \dot{x}(0))$, and by simple covariance computations, using (2.38), it is easy to demonstrate that the variables are uncorrelated.)

Define a Gaussian process indexed by $v, z$ as follows:

$$\begin{aligned} x_{v,z}(\tau) &= v\frac{r(\tau)}{r(0)} + z\frac{\dot{r}(\tau)}{\ddot{r}(0)} + \Delta(\tau) \\ &= x(\tau) + (v - x(0))\frac{r(\tau)}{\omega_0} - (z - \dot{x}(0))\frac{\dot{r}(\tau)}{\omega_2}. \end{aligned} \tag{6.45}$$

It follows from (2.37) and (2.38) that the distribution of $x_{v,z}(\tau)$ is equal to the conditional distribution of $x(\tau)$ given that $x(0) = v, \dot{x}(0) = z$. Consequently,

$$P(x^-(0) < u \mid x(0) = v, \dot{x}(0) = z))$$

can be estimated by computing the distribution of $x^-(0)$ defined by (4.18) for the process $x_{v,z}(\tau)$. This can be done according to the following algorithm:

1. For the spectrum $S(\omega)$, find $\omega_0, \omega_2$ and the covariances $r(s)$, $\dot{r}(s)$ ($s \in [0,t]$ and $r(s) = \int e^{is\omega} S(\omega)\, \mathrm{d}\omega$.).
2. Simulate $K$ samples of the process $x(s)$ ($s \in [0,t]$), $x_m(s)$, $m = 1, \dots, K$, say.
3. Choose a suitable quadrature of $z f_{\dot{x}(0)}(z)$, $z_l, h_l$, $l = 1, \dots, L$, and compute $\delta_l = z_l f_{\dot{x}(0)}(z_l) h_l$, i.e. $\int_0^\infty z f_{\dot{x}(0)}(z)\, \mathrm{d}z \approx \sum_{l=1}^L \delta_l$.
4. Choose discretisation of the levels $u < v$, $u_1 > u_2 > \dots > u_J$ with step $\delta u = u_i - u_{i+1}$.
5. For $m = 1, \dots, K$: compute, for $\tau \in [-t, 0]$,

$$y(\tau) = x_m(-\tau) + (u_J - x_m(0)) r(-\tau)/\omega_0 + \dot{x}_m(0) \dot{r}(-\tau)/\omega_2.$$

6. For $i = J-1, J-2, \dots, 1$: modify $y(\tau) := y(\tau) + \delta u\, r(-\tau)/\omega_0$.
7. For $l = 1, \dots, L$: compute $x_{u_i, z_l}(\tau)\, \mathrm{d}z = y(\tau) - z_l \dot{r}(-\tau)/\omega_2$ and find $x^-(0)$.
8. For $j = i+1, \dots, J$: if $x^-(0) < u_j$, $N_{ij}^{rfc} := N_{ij}^{rfc} + \delta_l$.
9. Finally, set $\mu_{ij}^{rfc} \approx N_{ij}^{rfc} f_{x(0)}(u_i)/K$.

**Remark 6.6.** *For zero-mean Gaussian x, we have that $N^{rfc}(u,v) = N^{rfc}(-v,-u)$ and hence only $u_i > 0$, and $u_i$ such that $|u_j| \le u_i$ need to be considered in steps (6.) and (8.).*

**Remark 6.7.** *If K is not very large, then for $u_i$ high and $u_j$ low (j close to J while i close to 1), there can be some values of $\mu_{ij}^{rfc}$ that are zero since no $x^-(0)$ were below $u_j$. For such extreme levels, the Poisson limit theorem can be used to estimate $\mu_{ij}^{rfc}$, viz.*

$$\mu_{ij}^{rfc} = \mathsf{E}[N_1^{rfc}(u_j, u_i)] \approx f_z \cdot \frac{\exp(-(u_i^2 + u_j^2)/2\omega_0)}{\exp(-u_i^2/2\omega_0) + \exp(-u_j^2/2\omega_0)},$$

*see Johannesson and Thomas (2001) [96] for detailed presentation.*

**Remark 6.8.** *Why "compute" $\mu_{ij}^{rfc}$ by means of algorithm (1–9) instead of estimating it by $\widehat{\mu}_{ij}^{rfc}$? Both methods have their advantages. Estimate $\widehat{\mu}_{ij}^{rfc}$ combined with some kernel smoothing would be an efficient way to compute $N_{ij}^{rfc}$ for $(u_j, u_i)$ close to the mean. For more extreme values $u_j, u_i$, it is better to use the algorithm. Note also that the algorithm can be changed and the Monte Carlo simulation of $\mathsf{P}(x^-(0) < u_j \mid x(0) = u_i, \dot{x}(0) = z_l)$ replaced by some numerical integration algorithm.*

**Remark 6.9.** *A more sophisticated numerical method to compute the rainflow intensity* $\mu^{rfc}(u,v)$ *is to use the interval oscillation intensity introduced in Definition 4.8,*

$$\mu^{rfc}(u,v) = \mu^{osc}(u,v) = \frac{1}{T}\mathsf{E}[N_T^{osc}(u,v)],$$

*where* $N_t^{osc}(u,v)$ *is the number of oscillations across* $[u,v]$ *starting at* $0 < s < T$*, defined as*

$$\#\{s; x(s) = u, x(t) = v, \textit{ and } u < x(\tau) < v \textit{ for all } \tau, s < \tau < t\}.$$

*The expected number of oscillations can be calculated by a Slepian model.*

**Remark 6.10.** *Note that the expected Holm-de Maré and positive peak damages can be estimated from many relatively short simulations of x by taking the average of the simulated damages. This is not the case for the rainflow damage because* $\mathsf{E}[D^{rfc}(t)] \le t \cdot d^{rfc}$ *where* $d^{rfc} = \lim_{t\to\infty} \mathsf{E}[D^{rfc}(t)]/t$*. The reason for this bias is that ranges for some cycles are systematically underestimated (*$\tau^-$ *cannot be smaller than zero).*

## APPENDIX

## 6.A PROPERTIES OF SMOOTH RANDOM PROCESSES

In this appendix, we discuss some special properties of smooth random processes that are important for rainflow cycle analysis. For the more basic properties, we refer to Section 2.2.4 in Chapter 2.

### 6.A.1 BULINSKAYA'S LEMMA

We consider only stationary smooth random processes, more specifically, we assume that the process has as many derivatives as we need for a particular application and that the process and the derivatives have probability density functions (pdf) $f_{x(0)}(x)$, $f_{\dot{x}(0)}(x)$, $f_{\ddot{x}(0)}(x)$ that are bounded. (These assumptions are satisfied for smooth stationary Gaussian loads.) We begin with a very important result, Bulinskaya's lemma, [33], given here for completeness.

**Lemma 6.5.** *Let* $x(\tau)$, $0 \le \tau \le t$, *be a continuously differentiable process. If, for each* $\tau$, *the random variable* $x(\tau)$ *has the probability density* $f_{x(\tau)}(x)$ *which is bounded by* $c < \infty$ *(the same c for all* $\tau \in [0,t]$*), then, for any u,*

$$\mathsf{P}(\#\{\tau \in [0,t] : x(\tau) = u, \dot{x}(\tau) = 0\} > 0) = 0.$$

From the lemma, it follows that for a fixed level $u$, all solutions of $x(\tau) = u$ are either strict upcrossings or downcrossings, e.g.,

$$N_t^+(u) = \#\{\tau \in [0,t] : x(\tau) = u, \dot{x}(\tau) > 0\}. \tag{6.46}$$

Since we also assumed that the probability density of the derivative is bounded, then by Bulinskaya's lemma, the probability of finding $\tau \in [0,t]$ such that $\dot{x}(\tau) = \ddot{x}(\tau) = 0$ is zero. This implies that all solutions of $\dot{x}(\tau) = 0$, $\tau \in [0,t]$, are locations of turning points (local maxima or minima), and there are no terrace points in the load.

### 6.A.2 RICE'S FORMULA AND GENERALISATIONS

In Section 2.2.4.2, we presented and proved Rice's formula for the expected number of level crossings in a stationary process. Here, we elaborate on some consequences for random load processes and fatigue analysis. We assume the load satisfies the conditions for Bulinskaya's lemma, so we can assume that all level crossings are either up- or downcrossings and that it never happens that simultaneously $x(\tau) = u, \dot{x}(\tau) = 0$, and that all points where $\dot{x}(\tau) = 0$ is either a strict local maximum or minimum.

We start with some details about Rice's formula for the expected number of solutions to the equation $x(\tau) = u$ per time unit,

$$\begin{aligned} \mathsf{E}[N(u)] &= \mathsf{E}[\#\{\tau, 0 \le \tau < 1, x(\tau) = u\}] \\ &= 2\mathsf{E}[\#\{\tau, 0 \le \tau < 1, x(\tau) = u, \dot{x}(\tau) > 0\}] = 2\mathsf{E}[N^+(u)]. \end{aligned}$$

In Section 2.3.2, we proved Rice's formula

$$\mathsf{E}[N(u)] = \int_{-\infty}^{\infty} |z| \, f_{\dot{x}(0),x(0)}(z,u) \, \mathrm{d}z, \tag{6.47}$$

under somewhat vague assumptions, and in fact we reached only the conclusion that the formula is true for "almost all" levels $u$. In fact, for a differentiable Gaussian load, the formula is true for all levels $u$, see Leadbetter et al. (1983) [115] for a proof.

For stationary loads, the derivative $\dot{x}(0)$ is uncorrelated with $x(0)$ and for Gaussian loads, they are independent. Denoting $\mathsf{Var}[x(\tau)] = \sigma_x^2 = \omega_0$ and $\mathsf{Var}[\dot{x}(\tau)] = \sigma_{\dot{x}}^2 = \omega_2$ the explicit form follows easily,

$$\begin{aligned} \mathsf{E}[N(u)] &= \int_{-\infty}^{\infty} |z| \, f_{\dot{x}(0)}(z) f_{x(0)}(u) \, \mathrm{d}z \\ &= \int_{-\infty}^{\infty} |z| \frac{1}{\sigma_{\dot{x}}\sqrt{2\pi}} e^{-z^2/2\sigma_{\dot{x}}^2} \, \mathrm{d}z \times \frac{1}{\sigma_x \sqrt{2\pi}} e^{-u^2/2\sigma_x^2}, \\ &= \frac{1}{\pi} \frac{\sigma_{\dot{x}}}{\sigma_x} e^{-u^2/2\sigma_x^2} = \frac{1}{\pi} \sqrt{\frac{\omega_2}{\omega_0}} e^{-u^2/2\omega_0}. \end{aligned} \tag{6.48}$$

Expressed in terms of significant load-range $h_s = 4\sqrt{\omega_0}$ and average period $t_z = 2\pi\sqrt{\omega_0/\omega_2} = 1/f_z$ Rice's formula (6.48) becomes

$$\mathsf{E}[N(u)] = 2 f_z e^{-8(u/h_s)^2}, \quad \mathsf{E}[N^+(u)] = f_z e^{-8(u/h_s)^2}. \tag{6.49}$$

## Notes

[1]Here, we break the rule for notation and use lower case letters, $z(t), \mathbf{y}(t)$ for random processes. The reason is that we make empirical statements about each sample function.

[2]$\int_0^\infty x\exp(-x^2/2a)\,\mathrm{d}x = a$.

# 7 Gaussian Loads – Applications

## 7.1 INTRODUCTION

In Chapter 5, we described how the rainflow cycle counting algorithm stood up against different simple time series load models and sampling schemes.

There are many reasons to use the Gaussian process class as a standard model for random fatigue loads. First of all, many environment loads can be represented by a stationary Gaussian process after judicious choice of power spectrum. Second, statistics important for fatigue prediction, like cycle counts, can be extracted from the Gaussian distribution of the process, analytically or by dedicated numerical tools. In this chapter, we will present some of these methods and tools based on the Gaussian model, and consider both approximations valid for different spectrum types, and exact cycle properties in detail from the full spectrum.

The following two chapters deal with two types of statistical load models that can generate realistic load series, the Markov models and the Laplace model, a flexible variation of the Gaussian model. Markov models are most useful in the design of fatigue testing studies, based on observed load cycle data from customer's use statistics or planned studies.

For both Gaussian and Markov models, there are explicit methods to find the rain flow damage, which can be used to construct dedicated types of load variations. Each model has its advantages and disadvantages when it comes to fatigue damage prediction under changing conditions,

In this chapter, we focus on simple measures to characterise fatigue damage, the expected fatigue damage, $\mathsf{E}[D(T)]$, and the time-dependent expected fatigue increase per time unit, $d(t) = \mathsf{E}[\dot{D}(t)]$. Obviously $\mathsf{E}[D(T)] = \int_0^T d(t)\,\mathrm{d}t$.

## 7.2 GAUSSIAN LOADS, DAMAGE MODELS

The present chapter is focused on expected damage under the broad assumption of a Gaussian load process, and we will consider a sequence of gradually more complex damage models and approximations. The statistical properties of a damage measure will of course depend on the statistics of the load process, and we choose to investigate their properties for Gaussian loads with various spectral properties. In our presentation, we will introduce some very useful tools, characteristic for Gaussian processes.

The considered damage models are as follows:

**min-max count:** Exact damage intensity $d^{mM}$ for stationary twice differentiable Gaussian process; Lindgren 1972 [122, 128].

DOI: 10.1201/9781003019947-7

**rfc count:** Exact damage intensity $d^{rfc}$ for stationary differentiable Gaussian process; Rychlik 1988 [171].

**Narrow-band:** Damage intensity $d^{nb}$ for Gaussian process with narrow spectrum and $f_z = \mathsf{E}[N_1^+(0)]$; Bendat 1964, [12].

**Holm-de Maré:** Damage intensity for Holm-de Maré model 1988, [84].

**cc count:** Damage intensity $d^{cc}$ for the crossing-cycle count, defined in Section 4.2.2.

**Positive-peak count:** Damage intensity $d^{pp}$ for cycles with range $2M^+$ when $M^+$ is the random height of positive local maxima in a Gaussian process. Since the positive peak damage has been discussed in the previous chapter, we will not study it here.

The link between load process and fatigue increase will be level crossings and cycle counts as defined in Chapter 4, and some form of damage function as in Section 4.3. Special focus will be on the statistical distribution of the rainflow cycles for a stationary Gaussian process, one of the few load models which allow the exact computation of the cumulative rainflow cycle count; see Section 4.6. The methods can also be used for functions of Gaussian vector valued loads and for computation of exact distributions for Markov loads.

Surprisingly, all damage accumulation models are (for smooth load $x(t)$, e.g. stationary Gaussian load with finite fourth spectral moment $\omega_4$) absolutely continuous, which means that the damage rate $\dot{D}(t)$ exists and $D(T) = \int_0^T \dot{D}(t)\,\mathrm{d}t$. For simplicity only, we consider the damage functions $f(u,v) = (v-u)^\beta$ and $f(u,v) = G(v)$.

Having a derivative is an important property for a damage model, i.e. counting method and damage function, because for stationary smooth Gaussian loads $x(t)$, we can compute damage intensity at time $t = 0$ or any other fixed time. It is an alternative to finding the distribution of cycle ranges $S_i = M_i - m_{k(i)}$ occurring at random times $t_i$. However, both approaches have their merits, it can be easiest seen in the case of positive peak count.

## 7.3 THE DERIVATIVE OF MIN-MAX AND RFC DAMAGE

The derivative of the total damage is the rate of damage growth at time $t$. It can be used for stationary loads to estimate damage intensity, i.e. expected damage in some time interval as well as the damage variance. Since it is causal, i.e. it depends only on load variability prior to time $t$, it can be used to estimate damage rate for loads whose properties change with time, e.g., due to seasonal effects. Examples will be given in Chapter 11.

In the case of stationary load, the damage derivatives are also stationary processes, and it is meaningful to talk about the min-Max and rainflow damage intensities

$$d^{mM} = \mathsf{E}[\dot{D}^{mM}(0)], \quad d^{rfc} = \mathsf{E}[\dot{D}^{rfc}(0)],$$

such that the expected damage over a finite interval $[0,T]$ is $\mathsf{E}[D^{mM}(T)] = T \cdot d^{mM}$, $\mathsf{E}[D^{rfc}(T)] = T \cdot d^{rfc}$, say.

For the damage function $f(u,v) = (v-u)^\beta$, the min-Max and rainflow derivatives have the following general form, with $t^- = \arg\min_{t^{mM/rfc} < \tau < t} x(\tau)$,

$$\dot{D}(t) = \beta(x(t) - x(t^-))^{\beta-1}\dot{x}(t)^+, \quad t \geq 0, \tag{7.1}$$

with different notations for mM and rfc counts. More precisely, define for each $t$, for mM count $t^{mM} = t$ if $\dot{x}(t) < 0$ and otherwise

$$t^{mM} = \sup\{s \in (-\infty, t) : \dot{x}(s) < 0\}. \tag{7.2}$$

Similarly, for the rfc count,

$$t^{rfc} = \sup\{s \in (-\infty, t) : x(s) > x(t)\}. \tag{7.3}$$

Note that for Gaussian load with non-degenerated spectrum, the probability is zero that for fixed $t$, $\dot{x}(t) = 0$. Furthermore

$$-\infty < t^{rfc} \leq t^{mM} < t,$$

and hence $\dot{D}^{mM}(t) \leq \dot{D}^{rfc}(t)$.

### 7.3.1 APPROXIMATION OF MIN-MAX AND RFC DAMAGE RATES

We consider a zero-mean stationary Gaussian process with covariance function $r(\tau)$. We let $t = 0$ and use the same symbol $d$ for the min-Max and rfc damages, depending on time $t^{rfc}$, $t^{mM}$. Then

$$\begin{aligned} d &= \int_{\infty}^{\infty}\int_0^{\infty} \mathsf{E}[\dot{D}(0) \mid X(0) = v, \dot{X}(0) = r]\, f(v,r)\,\mathrm{d}r\,\mathrm{d}v \\ &= \int_{\infty}^{\infty}\int_0^{\infty} \mathsf{E}[\beta H^{\beta-1} \mid X(0) = v, \dot{X}(0) = r]\, r f(v,r)\,\mathrm{d}r\,\mathrm{d}v, \end{aligned} \tag{7.4}$$

where $H = X(0) - X(\tau^-)$, $\tau^- = t^{mM}$ or $t^{rfc}$, depending on which of the two counting methods were used.

The representation (7.4) is an exact recipe for how to compute the damage $d$, and the Slepian approach from Section 6.5 gives the means to compute exact as well as approximate solutions. For example, the model in Example 6.3 is easy to simulate, since one does not need long time histories of the process to get an accurate estimate of the Max-min distribution.

### 7.3.2 THEORETICAL METHOD – GENERAL PRINCIPLES

We have demonstrated, in Remark 4.6, the relation to the left semi-ranges introduced in (4.3),

$$N_T^{rfc}(u,v) = \#\{\tau : x(\tau) = v, \dot{x}(\tau) > 0, m_\tau^- < u\}, \tag{7.5}$$

where $m_\tau^-$ is the minimum of $x(t), \tau^- \leq t \leq \tau$ for $\tau^- = \sup\{s < \tau; x(s) = x(\tau)\}$.

This formula is very similar to the one derived for the min-Max damage. It is just the location of the minimum that changes. It means that $m_\tau = x(t^{mM})$ is replaced by $m_\tau^- = x(t^{rfc}) \leq x(t^{mM})$. Consequently

$$N_T^{mM}(u,v) = \#\{t \in [0,T] : x(t) = v, \dot{x}(t) > 0, x(t^{mM}) < u\}, \tag{7.6}$$

$$N_T^{rfc}(u,v) = \#\{t \in [0,T] : x(t) = v, \dot{x}(t) > 0, x(t^{rfc}) < u\}, \tag{7.7}$$

We can see that $N_T^{rfc}(u,v) \geq N_T^{Mm}(u,v)$.

The Markov chain (MC) approximation of the rainflow count is defined as the rainflow count for Markov chain of turning points (MCTP) defined by the min-Max count (7.6) treated in Section 8.3.2.

From here on, we need some simple adaptations in that formula $\mathsf{E}[N^+(u)]$ as given by (6.48) is replaced by

$$\mathsf{E}[N_1^{rfc}(u,v)] = \mu(A \mid x(s) = v),$$

that can be computed by means of (6.5).

However, since the event $''m_0^- < u''$ depends on infinitely many random variables, the computation of $\mathsf{P}(m_0^- < u \mid x(0) = v, \dot{x}(0) = z)$ has to be done by some approximation method.

Suppose that we managed to find $\mathsf{E}[N_1^{rfc}(u,v)] = \mu(h^-(0) < u \mid x(s) = v)$. Then, we still need to compute the integrals in the master-formula. It is often done by replacing all integrands by matrices and the integrals by double sums. This means that the levels $u, v$ should be discretised to an equidistant partition of the cycle values with step $\delta u$, $u_1 > u_2 > \ldots > u_J$. Let us introduce the $(J,J)$ counting matrix $N^{rfc}$ with element $(i,j)$ defined by

$$N_{ij}^{rfc} = N_t^{rfc}(u_j, u_i). \tag{7.8}$$

The counting matrix $N$ can be defined for any cycle count, e.g. $N^{mM}$, $N^{cc}$ for min-max or cc count, respectively. Having the cycle matrix $X$ the component

$$N_{ij} = \sum_{m<i,n>j} X_{mn}, \quad X_{ij} = N_{ij} + N_{(i+1)(j-1)} - N_{i(j-1)} - N_{(i+1)j}. \tag{7.9}$$

We can see that there is a one-to-one relation between the cycle matrix $X$ and cumulative counting matrix $N$.

**Example 7.1.** *Consider the sequence of turning points in Figure 7.1 and the resulting mM-cycle plot. The rainflow cycle matrix and the corresponding counting matrix are*

$$X^{rfc} = \begin{bmatrix} 0 & 0 & 0 & 0 & 0 & 1 & 1 & 0 \\ 0 & 0 & 0 & 0 & 1 & 1 & 0 & 1 \\ 0 & 0 & 0 & 0 & 1 & 0 & 0 & 0 \\ 0 & 0 & 0 & 0 & 0 & 0 & 0 & 0 \\ 0 & 0 & 0 & 0 & 0 & 0 & 0 & 0 \\ 0 & 0 & 0 & 0 & 0 & 0 & 1 & 0 \\ 0 & 0 & 0 & 0 & 0 & 0 & 0 & 0 \\ 0 & 0 & 0 & 0 & 0 & 0 & 0 & 0 \end{bmatrix}, \quad N^{rfc} = \begin{bmatrix} 0 & 0 & 0 & 0 & 0 & 0 & 0 & 0 \\ 2 & 2 & 2 & 2 & 2 & 1 & 0 & 0 \\ 5 & 5 & 5 & 5 & 4 & 2 & 1 & 0 \\ 6 & 6 & 6 & 6 & 4 & 2 & 1 & 0 \\ 6 & 6 & 6 & 6 & 4 & 2 & 1 & 0 \\ 6 & 6 & 6 & 6 & 4 & 2 & 1 & 0 \\ 7 & 7 & 7 & 7 & 5 & 3 & 1 & 0 \\ 0 & 0 & 0 & 0 & 0 & 0 & 0 & 0 \end{bmatrix}.$$

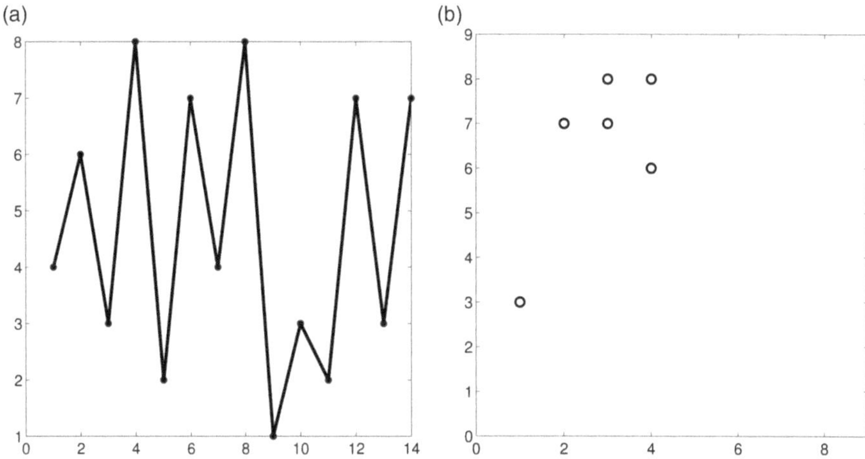


**FIGURE 7.1** (a) A sequence of turning points. (b) 2d-plot of *mM* cycles.

*For $f(u,v) = (v-u)^2$, the damage is*

$$D^{rfc} = 25 + 36 + 9 + 16 + 36 + 4 + 1 = 127.$$

We also compute the oscillation matrix $\mu^{rfc}$ containing the intensities of crossing of intervals defined using the levels $u_j < u_i$, viz.

$$\mu_{ij}^{rfc} = \frac{1}{T}\mathsf{E}[N_{ij}^{rfc}]\,(= \mathsf{E}[N_1^{rfc}(u_j, u_i)]). \tag{7.10}$$

Now with $f_{ij} = -\frac{\partial^2 f}{\partial u \partial v}(u_j, u_i)$ the rainflow damage intensity is

$$d^{rfc} = \int_{-\infty}^{\infty}\int_{-\infty}^{v} \frac{\partial^2 f(u,v)}{\partial u \partial v}\mathsf{E}[N_1^{rfc}(u,v)]\,\mathrm{d}u\,\mathrm{d}v \approx \sum_{i>j} \mu_{ij}^{rfc} \cdot f_{ij} \cdot (\delta u)^2.$$

## 7.4 THE NARROW-BAND APPROXIMATION

We recall the idea behind Bendat's (1964) [12] narrow-band approximation of fatigue damage, Section 4.7.1, and give some alternative formulations for Gaussian loads.

Assuming that the load amplitude $H/2$ has a Rayleigh distribution, Bendat proposed the following expressions for the expected total damage during time $T$,

$$\mathsf{E}[D^{nb}(T)] = T\,f_z\,\mathsf{E}[(2H)^\beta] = \int_0^\infty 2(2h)^{\beta-1}\mathsf{E}[N^+(h)]\,\mathrm{d}h, \tag{7.11}$$

since $P(H > h) = E[N^+(h)]/T\,f_z$.

Inspired by formula (7.11), we define the observed narrow band damage by

$$D^{nb}(T) = \int_0^\infty 2(2h)^{\beta-1}N^+(h)\,\mathrm{d}h. \tag{7.12}$$

The damage is absolutely continuous, and its derivative is given by

$$\dot{D}^{nb}(t) = 2\beta\,(2x(t)^{+})^{\beta-1}\dot{x}(t)^{+}, \tag{7.13}$$

where $x^{+} = \max(0,x)$. Consequently, employing that $x$ is a stationary Gaussian process, $x(t)$ and $\dot{x}(t)$ are independent and the damage rate can be computed as

$$d^{nb} = \mathsf{E}[\dot{D}^{nb}(0)] = \int_0^\infty 2(2x)^{\beta-1} f_{x(0)}(x)\,\mathrm{d}x \int_0^\infty z f_{\dot{x}(0)}(z)\,\mathrm{d}z.$$

The integrals are easily evaluated, and the explicit result is given in (7.14).

Expression (7.13) follows from Banach's theorem for positive variation, Theorem 2.2. An elementary proof of the result can be given by some modification of Theorem 2.3 in Holm and de Maré 1988 [84].

For easy reference, we give two alternative formulations of the narrow-band approximation from (4.36) expressed in terms of significant wave amplitude and apparent frequency.

**Theorem 7.1.** *For a zero-mean stationary Gaussian load $X(t)$ with significant amplitude $h_s = 4\sigma_X$ and apparent frequency $f_z = \mu^{+}(0) = \frac{1}{2\pi}\sigma_{\dot{X}}/\sigma_X$, the narrow-band approximation of the damage intensity with damage function $(v-u)^\beta$ is*

$$d^{nb} = f_z h_s^\beta\, 2^{-\beta/2}\Gamma(1+\beta/2) \tag{7.14}$$

$$= \frac{1}{\pi}\sigma_{\dot{X}}\sigma_X^{\beta-1} 2^{3\beta/2-1}\Gamma(1+\beta/2), \tag{7.15}$$

*and the expected accumulated damage is approximated as $\mathsf{E}[D^{nb}(T)] = T d^{nb}$.*

For values of parameter $\beta \in [2,4]$, which are the most often used values in offshore applications, $0.469 < 2^{-\beta/2}\Gamma(1+\beta/2) \le 0.5$ and hence

$$d^{nb} \approx 0.5 \cdot f_z h_s^\beta. \tag{7.16}$$

This is an indication that significant load-range $h_s$ seems to be the relevant parameter when studying Gaussian loads.

## 7.5 EXPECTATION OF HOLM-de MARE DAMAGE

The Holm-de Maré damage (H-M damage) accumulation model is based on the concept of an *exhaustion function* proposed by Holm and de Maré 1988 [84]. This method is derived using the following rules:

1) The Palmgren-Miner linear damage accumulation rule. This rule states that a component fails when the total damage exceeds a prescribed value. The total damage is the sum of the damages caused by each load, independent of the order in which the different load cycles are applied to the component.

2) A time invariance assumption, which states that the damage caused by a load depends only on the sequence of extremes of the load and not on the particular form of the load function between its extreme values.

Assume that the damage is accumulated during the increase of stress or elongation process. Let the damage accumulated when the process grows from $y$ to $x$ be given by $G(x) - G(y)$, where $G$ is called an exhaustion function. Similarly, one can define damage increase when the stress is decreasing; however, it can have a different exhaustion function. Here, we do not consider this case.

Consider a load $x(t)$, $t \in [0.T]$, and the sequence of turning points $T_P(x) = \{m_i, M_i\}, i = 1, \ldots, n$, with $x(t_i^-) = m_i$ , $x(t_i^+) = M_i$. The H-M damage [84] is defined as the positive variation of the exhaustion function $G$,

$$D^{HM}(T) = \sum_i^{n_t} G(M_i) - G(m_i).$$

Obviously, this is the min-Max damage with the damage function $f(u,v) = G(v) - G(u)$. Consequently, $D^{HM}(t)$ is differentiable and by (7.1), the H-M damage rate is

$$\dot{D}^{HM}(t) = \dot{D}^{mM}(t) = \dot{G}(x(t))\dot{x}(t)^+. \tag{7.17}$$

(Note that the H-M damage intensity is not limited to Gaussian loads.) The following theorem gives two useful expressions for the H-M damage for a differentiable exhaustion function.

**Theorem 7.2** (Holm-de Maré damage). *For a differentiable load $x(t)$,*

$$D^{HM}(T) = \sum_i^{n_T} G(M_i) - G(m_i) = \int_0^T \dot{G}(x(t))\dot{x}^+(t)\,\mathrm{d}t. \tag{7.18}$$

*Expressed in terms of $N_T^+(z)$, the number of upcrossings of level $z$ by $x$ in $[0,T]$, the damage is*

$$D^{HM}(T) = \sum_i^{n_T} G(M_i) - G(m_i) = \int_{-\infty}^{\infty} \dot{G}(z) N_T^+(z)\,\mathrm{d}z. \tag{7.19}$$

*Proof.* For a differentiable $G$, a change of variables $z = x(t)$, gives

$$G(M_i) - G(m_i) = \int_{m_i}^{M_i} \dot{G}(z)\,\mathrm{d}z = \int_{t_i^-}^{t_i^+} \dot{G}(x(t))\dot{x}^+(t)\,\mathrm{d}t,$$

which shows (7.18). Expression (7.19) is Banach's theorem for positive variation, Theorem 2.2. □

Note that the double integral in master-formula (4.31) is zero, and hence the H-M damage depends only on the crossing spectrum, i.e. it is independent of the cycle count method. Consequently, the H-M damage function will give the same damage for any cycle count under the assumption that $f(u,v) = G(v) - G(u)$. We will see that for the damage computed with $f(u,v) = (v-u)^\beta$, the H-M damage is the bound for damages based on cycle counts, see Lemma 7.2.

The two representations of $D^{HM}(T)$ are equivalent but have different merits, with focus on local (in time) damage and total damage, respectively. We will illustrate the use on the exhaustion function

$$G(z) = 0.5\,(2z)^\beta \text{ if } z \geq 0, \text{ and } G(z) = -0.5\,(-2z)^\beta \text{ if } z < 0, \tag{7.20}$$

with derivative

$$\dot{G}(z) = \beta\,|2z|^{\beta-1}.$$

**Total damage:** For this $G$-function, and stationary, zero-mean load, the total damage is

$$D^{HM}(T) = \int_{-\infty}^{\infty} \beta(|2z|^{\beta-1})\,N_T^+(z)\,\mathrm{d}z,$$

and the expected damage is

$$\mathsf{E}[D^{HM}(T)] = T\int_{-\infty}^{\infty} \beta(|2z|^{\beta-1})\,\mathsf{E}[N^+(z)]\,\mathrm{d}z, \tag{7.21}$$

by stationarity. For the Gaussian load, the last formula can be evaluated using Rice's formula for $\mathsf{E}[N_t^+(z)]$. The result is identical to the narrow-band approximation,

$$\mathsf{E}[D^{HM}(T)] = \mathsf{E}[D^{nb}(T)].$$

Obviously, for stationary loads with symmetric crossing spectra, the H-M and narrow-band damage approximations will coincide, and hence, in some sense, the H-M expected damage is a generalisation of Bendat's narrow-band approximation to loads with asymmetric crossing spectra.

**Damage intensity:** The representation (7.18) has focus on the development of damage in time and is particularly useful for non-stationary loads, treated in Chapters 10 and 11. For the $G$-function (7.20),

$$D^{HM}(T) = \int_0^T \dot{G}(x(t))\dot{x}^+(t)\,\mathrm{d}t = \int_0^T \beta|2x(t)|^{\beta-1}\dot{x}^+(t)\,\mathrm{d}t. \tag{7.22}$$

Thus, the damage increase at time $t$ is

$$\dot{D}^{HM}(t) = \beta|2x(t)|^{\beta-1}\dot{x}^+(t), \tag{7.23}$$

with expectation, i.e. damage intensity,

$$d^{HM}(t) = \mathsf{E}[\dot{D}^{HM}(t)] = \beta\mathsf{E}[|2x(t)|^{\beta-1}\dot{x}^+(t)].$$

Again, since for a Gaussian load, $x(t)$ and $\dot{x}(t)$ are independent, the expectation is easy to evaluate with the expected equivalent results

$$d^{HM}(t) = \begin{cases} f_z h_s^\beta\, 2^{-\beta/2}\Gamma(1+\beta/2), \\ \frac{1}{\pi}\sigma_{\dot{X}}\sigma_X^{\beta-1}\, 2^{3\beta/2-1}\Gamma(1+\beta/2). \end{cases} \tag{7.24}$$

We finish with some remarks on narrow band and H-M damage:

- The two damage formulas give the same expected damage for stationary zero-mean Gaussian loads with the same $f_z$ and $h_s$.
- If the load has positive skewness, the narrow-band damage should have higher expected damage than the H-M damage, and if it has negative skewness, the narrow-band damage should be smaller than the H-M damage in average.

## 7.6 THE EXPECTED $D^{CC}(T)$ DAMAGE

The crossing-cycle count "cc", defined in Section 4.2.2, is the cycle count that gives the highest damage. It is bound to produce exaggerated cycle ranges since it pairs extreme minima to extreme maxima without respect to location; see Figure 4.3. As a consequence, they give the largest damage measured by the damage function $f(u,v) = (v-u)^\beta$, for $\beta > 1$. The analysis is valid for all loads with finite crossing spectrum and is not limited to Gaussian loads.

The cumulative crossing-cycle count is defined as

$$N_T^{cc}(u,v) = \min_{u \le z \le v} N_T^+(z).$$

The master-formula (4.31) is applicable and gives the result, for $\beta > 1$,

$$D^{cc}(T) = \int_{-\infty}^{\infty} \int_{-\infty}^{v} \beta(\beta-1)(v-u)^{\beta-2} N_T^{cc}(u,v)\, \mathrm{d}u\, \mathrm{d}v,$$

and hence in order to compute $\mathsf{E}[D^{cc}(T)]$, one needs to find $\mathsf{E}[N_T^{cc}(u,v)]$. This is not an easy problem. However, the following bound with the minimal crossing spectrum $k^+(u,v)$ could be used,

$$\mathsf{E}[N_T^{cc}(u,v)] = \mathsf{E}[\min_{u \le z \le v} N_T^+(z)] \le T \cdot \min_{u \le z \le v} \mathsf{E}[N^+(z)] = T \cdot k^+(u,v), \tag{7.25}$$

by (4.40), to bound $\mathsf{E}[D^{cc}(T)]$. Now, we can obtain the upper intensity bound $d^+$,

$$\mathsf{E}[D^{cc}(T)] \le T \int_{-\infty}^{\infty} \int_{-\infty}^{v} \beta(\beta-1)(v-u)^{\beta-2} k^+(u,v)\, \mathrm{d}u\, \mathrm{d}v = T \cdot d^+.$$

**Lemma 7.1.** *For stationary zero-mean Gaussian loads and damage function* $f(u,v) = (v-u)^\beta$,

$$d^+ = d^{nb} = f_z h_s^\beta\, 2^{-\beta/2} \Gamma(1+\beta/2). \tag{7.26}$$

*Proof.* Some lines of calculus are needed to demonstrate the lemma, see, e.g. Rychlik (1993) [174] or Rychlik (2003) [179]. □

The following lemma combines the results for zero-mean Gaussian loads.

**Lemma 7.2.** *For a stationary zero-mean Gaussian load with non-degenerated frequency spectrum, the damage function* $f(u,v) = (v-u)^\beta$ *and exhaustion function*

$$G(z) = 0.5\,(2z)^\beta \text{ if } z \geq 0, \text{ and } G(z) = -0.5\,(-2z)^\beta \text{ if } z < 0,$$

*one has that*

$$\mathsf{E}[D^{rfc}(T)] \leq \mathsf{E}[D^{cc}(T)] < T \cdot d^+ = \mathsf{E}[D^{HM}(T)] = \mathsf{E}[D^{nb}(T)] = T \cdot d^{nb}. \quad (7.27)$$

*Proof.* Only the inequality has to be demonstrated. As it was shown before, for any cycle $(u,v)$, $(v-u)^\beta < G(v) - G(u)$ if $u \neq -v$. Consequently, for Gaussian processes having non-degenerated to one frequency spectrum and $G$ as above $D^{cc}(T) < D^{HM}(T)$, and hence even

$$\mathsf{E}[D^{cc}(T)] < \mathsf{E}[D^{HM}(T)],$$

i.e. for non-degenerated Gaussian loads $\mathsf{E}[N_T^{cc}(u,v)] < T \cdot k^+(u,v)$. □

Obviously, $d^+$ and $\mathsf{E}[D^{HM}(T)]$ are defined for any stationary load and can be computed if the crossing spectrum $\mathsf{E}[N_T^+(u)]$ is available. We may ask weather always $\mathsf{E}[D^{HM}(T)] = T \cdot d^+$? The answer is negative: $E[D^{HM}(T)] \geq T \cdot d^+$. In order to prove this claim, we need an alternative way to compute $d^+$.

**Theorem 7.3.** *For a stationary load with asymmetric continuous crossing spectrum* $\mathsf{E}[N^+(u)]$, *the damage function* $f(u,v) = (v-u)^\beta$ *and exhaustion function*

$$G(z) = 0.5\,(2z)^\beta \text{ if } z \geq 0, \text{ and } G(z) = -0.5\,(-2z)^\beta \text{ if } z < 0,$$

*one has that*

$$\mathsf{E}[D^{rfc}(T)] \leq \mathsf{E}[D^{cc}(T)] < T \cdot d^+ < \mathsf{E}[D^{HM}(T)]. \quad (7.28)$$

The proof is found in Appendix 7.A.

For multi-modal crossing spectra, the H-M damage is a very conservative bound for the rainflow damage. By using another definition of the $G$-function, one can decrease this conservatism.

## 7.7 OTHER SUGGESTED APPROXIMATIONS OF $D^{rfc}$

In previous sections, we have shown that:

- For a stationary load the rainflow damage intensity $d^{rfc} \leq d^+$, the general damage intensity bound (4.7).
- For zero-mean Gaussian load the Bendat's narrow-band damage intensity $d^{nb} = d^+$.
- The intensity of positive peak damage $d^{pp} = c(\beta, \alpha_2) \cdot d^{nb} \geq d^{nb}$, depending on the spectral width parameter $\alpha_2$ (6.7–6.8); $c(\beta,1) = 1$ and $c(\beta,0) = \infty$.
- Further $\alpha_2 d^{pp} \leq d^{nb}$ and we argue that approximately $\alpha_2 d^{pp} \leq d^{rfc}$.

In the following section, some approximations, proposed in the literature, for the rainflow damage intensity $d^{rfc}$ will be presented for stationary zero-mean Gaussian loads. The loads are defined by the spectral power density having a finite cut-off frequency. Such loads are defined by the spectral moments $\omega_i = \int_0^\infty \omega^i S(\omega)\,d\omega$. Consequently, even the damage intensity $d^{rfc}$ is a function of the spectral moments. Loads in some applications can have a typical spectrum, and hence one tries to find the approximations (functions of spectral moments) which works for those spectra. This explains why there are many approximations proposed in the literature. In the following, we will present some of these.

Somewhat different notation listed in Table 7.1, will be used, and we hope that it will not cause confusion. Now, $t_z$ will be denoted by $T_2$, while $d^{nb}$ by $d_2(t)$. Furthermore, only damage function $f(u,v) = (v-u)^\beta$ will be considered. The example is taken from Bengtsson and Rychlik (2009) [15].

**Table 7.1**
**Notation Used in Section 7.7.1**

| Variable | Notation | Alt. Notation from the Literature | Definition |
|---|---|---|---|
| apparent frequency | $f_z$ | - | $\frac{1}{2\pi}\sqrt{\omega_2/\omega_0}$ |
| zero-crossing period | $t_z = 1/f_z$ | $T_2$ | $2\pi\sqrt{\omega_0/\omega_2}$ |
| peak frequency | $f_p$ | - | $\arg\max S(\omega)/2\pi$ |
| peak period | $t_p = 1/f_p$ | - | - |
| generalised period | - | $T_z$ | $2\pi(\omega_0/\omega_z)^{1/z}$ |
| maximi-frequency | $f_{max}$ | - | $\frac{1}{2\pi}\sqrt{\omega_4/\omega_2}$ |
| maximi-period | $t_{max} = 1/f_{max}$ | $T_4$ | $2\pi\sqrt{\omega_2/\omega_4}$ |

### 7.7.1 SPECTRAL MOMENTS-BASED APPROXIMATIONS

A stationary zero-mean load is considered with power spectral density $S(\omega)$ and spectral moments of general order $z$,

$$\omega_z = \int_0^\infty \omega^z S(\omega)\,d\omega, \quad z \geq 0. \tag{7.29}$$

Let the generalised average period be

$$T_z = 2\pi(\omega_0/\omega_z)^{1/z}, \tag{7.30}$$

where $T_2 = t_z$ is the average distance between upcrossings of the mean level. By Hölder's inequality, $(\omega_{z_1}/\omega_0)^{1/z_1} \leq (\omega_{z_2}/\omega_0)^{1/z_2}$ for $z_1 \leq z_2$, and hence we have

$$T_{z_1} \geq T_{z_2} \text{ if } z_1 \leq z_2. \tag{7.31}$$

A general type of approximation for the rainflow damage intensity $d^{rfc}$ also includes the *damage exponent* $\beta$ from the damage function $(v-u)^\beta$,

$$d_z(\beta) = \frac{\omega_0^{\beta/2}}{T_z} 2^{3\beta/2}\Gamma(1+\beta/2) = \frac{T_2}{T_z} d^{nb}. \tag{7.32}$$

For a fixed $\beta$, $d_z(\beta)$ is an increasing function of $z$. Obviously, $d_z(\beta) \geq d^{rfc}(\beta)$, for $z \geq 2$, and since $d_2(1) = d^{rfc}(1)$, it is the sharpest bound among $d_z$.

**The single moment approximation:** A simple approximation was derived by Lutes and Larsen (1990) [134], where the authors let $z$ vary with $\beta$, viz. letting $z = 2/\beta$ gives the so-called "single moment" approximation,

$$d^{LL}(\beta) = \frac{w_0^{\beta/2}}{T_{2/\beta}} 2^{3\beta/2}\Gamma(1+\beta/2) = \frac{w_{2/\beta}^{\beta/2}}{2\pi} 2^{3\beta/2}\Gamma(1+\beta/2). \tag{7.33}$$

Obviously, for $\beta \geq 2$, one has that $d^{LL}(\beta) = d_{2/\beta}(\beta) \leq d_1(\beta)$. Here, we shall compare $d_1(\beta)$, $d_2(\beta)$, and $d^{LL}(\beta)$ with other approximations.

**Some other approximation methods:** For a very broad-banded spectrum and high values of $\beta$, the bound $d_2(\beta)$ tends to be very conservative; see the values for the Ornstein-Uhlenbeck process in Table 6.1. A simple way to reduce the conservatism is to cut off high frequencies by splitting the spectrum into two,

$$S(\omega) = S_1(\omega) + S_2(\omega).$$

Principally, this means that the load is represented as a smoother part with added noise. Let $T_2(i)$, $\omega_0(i)$, $i = 1,2$, denote periods $T_2$ and energy $\omega_0$ computed for spectrum $S_i(\omega)$, respectively, and then use the approximation

$$d^{SOm}(\beta) = \left( \frac{(\omega_0(1)+\omega_0(2))^{\beta/2}}{T_2(1)} + \frac{m_0(2)^{\beta/2}}{T_2(2)} \right) 2^{3\beta/2}\Gamma(1+\beta/2). \tag{7.34}$$

This approximation is closely related to the approximation proposed by Sakai and Okamura (1995) [185],

$$d^{SO}(\beta) = \left( \frac{\omega_0(1)^{\beta/2}}{T_2(1)} + \frac{\omega_0(2)^{\beta/2}}{T_2(2)} \right) 2^{3\beta/2}\Gamma(1+\beta/2). \tag{7.35}$$

As reported by Benasciutti and Tovo (2007) [11], the Sakai-Okamura method may give non-conservative predictions when the high-frequency component has too much energy. This happens because approximation (7.35) does not take into account that the largest cycles have ranges that depend on both processes. The interaction is included in approximation (7.34). In the special case, when spectrum is bimodal, frequencies around the lower peak represent a slowly varying load part while the frequencies around the higher mode represent oscillations superimposed around the slowly varying component. If the energy in the high frequencies is not large, one can use this particular structure of $S(\omega)$ to reduce conservatism of $d_2(\beta)$. In such a case, there is a natural division of the spectrum into two components $S_1$ and $S_2$, which implies that $x(\tau) = x_1(\tau) + x_2(\tau)$ consists of two independent Gaussian processes with spectra $S_1$ and $S_2$, respectively.

Another approach is the method by Jiao and Moan (1990) [87], where the envelop process of the fast component is used to add the interaction term between $x_1$ and $x_2$ to give the approximation

$$d^{JM}(\beta) = \frac{c}{\sqrt{\omega_0\omega_2}}\left(\left(\frac{\omega_0(1)}{\omega_0}\right)^{2+\beta/2}\left(1-\frac{\sqrt{\omega_0(2)}}{\sqrt{\omega_0(1)}}\right)+\frac{\beta\Gamma((\beta+1)/2)\sqrt{\pi\omega_0(1)\omega_0(2)}}{\omega_0\Gamma(1+\beta/2)}\right) + \frac{\sqrt{\omega_2(2)\omega_0(2)}}{\sqrt{\omega_0\omega_2}}, \tag{7.36}$$

where $c = \sqrt{\omega_0(1)\omega_2(1) + \omega_0(2)\omega_2(2) - (\omega_1(2))^2}$. Without going into details, such as how to split the spectrum into two components, it can be noted that sometimes using the envelop to model damage due to interaction between $x_1$ and $x_2$ can be too conservative.

### 7.7.2 APPROXIMATIONS BASED ON SPECTRAL WIDTH

A special type of approximations use the spectral bandwidth parameters

$$\alpha_z = \frac{\omega_z}{\sqrt{\omega_0\omega_{2z}}},$$

introduced in Remark 2.2. Using the fact that $\alpha_z \le 1$ one can make the approximation smaller than the upper bound $d_2$ – recall that $d_2 = d^{nb}$, the narrow band approximation – e.g. by approximating $d^{rfc}$ by $\alpha_z d_2(\beta)$. We already had an example that $d_1(\beta) = \alpha_1 d_2(\beta)$, and another one can be found in Benasciutti (2004) [9], where it was proposed to use $\alpha_{0.75}^2 d_2(\beta) \le d_2(\beta)$ as an approximation of $d^{rfc}$.

Note that our approximate lower bound for the rainflow damage is $\alpha_2 d^{pp} = \alpha_2 c(\beta,\alpha_2) d_2(\beta)$, see (6.20), and differs from the approximations discussed here. It was also designed to make $d^{nb}$ less conservative, potentially a lower bound for the $d^{rfc}$.

Tovo (2002) [204] proposed to use

$$d^{BT}(\beta) = (p + (1-p)\alpha_2^{\beta-1}) d_2(\beta), \tag{7.37}$$

with $p = \min((\alpha_1 - \alpha_2)/(1-\alpha_1), 1)$. Often, $p = 1$ which gives $d^{BT}(\beta) = d_2(\beta)$. Benasciutti (2004) [9] presented an improvement for the constant $p$,

$$p = (\alpha_1 - \alpha_2)\frac{1.112e^{2.11\alpha_2}(1-\alpha_1-\alpha_2+\alpha_1\alpha_2) + (\alpha_1-\alpha_2)}{(\alpha_2-1)^2}. \tag{7.38}$$

Wirsching and Light (1980) [214] proposed the following approximation

$$d^{WL}(\beta) = \left( a(\beta) + (1 - a(\beta)) \left( 1 - \sqrt{1 - \alpha_2^2} \right)^{b(\beta)} \right) d_2(\beta), \tag{7.39}$$

where $a(\beta) = 0.926 - 0.033\beta$ and $b(\beta) = 1.587\beta - 2.323$.

### 7.7.3 APPROXIMATIONS OF RAINFLOW RANGE DISTRIBUTION

If the long-run distribution of the rainflow range $S^{rfc}$ were known, one could have used $d^{rfc} = \frac{1}{T_4}\mathsf{E}[(S^{rfc})^\beta]$. As was demonstrated in Example 6.1 and Figure 6.2, the distribution can be computed by WAFO for stationary Gaussian processes.

We turn now to presentation of the approximations of the $H = S^{rfc}$-distribution proposed in the literature.

Many authors, see e.g. Bishop and Sherratt (1990) [17], define the narrow-band approximation by assuming that the rainflow range has a $2\sqrt{\omega_0}R$ distribution, where $R$ is Rayleigh distributed, which is the case when the spectral bandwidth parameter $\alpha_2$ converges to one. Some simple algebra gives us this approximation,

$$\frac{1}{T_4}\mathsf{E}[(2\sqrt{m_0}R)^\beta] = d_4(\beta). \tag{7.40}$$

As we mentioned before, for fixed $\beta$, the approximation $d_z(\beta)$ is an increasing function of $\beta$, and hence (7.40) is also a bound. A successful way to reduce the conservatism of this approach was suggested by Dirlik (1985) [53]. Based on a large simulation study and least-square fitting, Dirlik proposed a correction to the narrow-band approximation by a mixture of an exponential and two Rayleigh distributions. The method leads to the following approximation

$$d^{Dirlik}(\beta) = \frac{1}{T_4}\left( D_1 \left( Q/\sqrt{2} \right)^\beta \Gamma(1+\beta) + \left( D_2 (R/2)^{\beta/2} + D_3 \right) \Gamma(1+\beta/2) \right), \tag{7.41}$$

where $Q, R, D_1, D_2, D_3$ are functions of $\omega_i$, $i < 4$, see Benasciutti (2004) [9] for discussion and some simple properties. Some authors claim that Dirlik's method is superior to other existing approximate methods. In some limited tests involving Gaussian loads with spectra used in offshore engineering, we shall demonstrate that although the formula is useful, it is not always the best approximation.

A similar approach to Dirlik's was proposed by Zhao and Baker (1992) [219]. The authors fitted Weibull and Rayleigh distributions to the observed rainflow ranges in simulated loads. Then, a regression type of approach was used to find the relation between values of fitted parameters and some moments in used spectra. The proposed relation is as follows: with $b = 1.1$ (if $\alpha_2 > 0.9$ then $b = 1.1 + 9(\alpha_2 - 0.9)$) and with

$$p = \frac{1 - \alpha_2}{1 - (8 - 7\alpha_2)^{-1/b}\Gamma(1 + 1/b)\sqrt{2/\pi}},$$

the approximation of $d^{rfc}$ is

$$d^{ZB}(\beta) = (1-p)d_2(\beta) + p\frac{\omega_0^{\beta/2}}{T_4}(8-7\alpha_2)^{-\beta/b}2^{\beta}\Gamma(1+\beta/b). \tag{7.42}$$

A slightly modified version of the approximation is obtained by replacing the factor $8-7\alpha_2$ by a function of $\alpha_{0.75}$. This leads to somewhat more complicated computations, which can be found in [219]. As reported in Benasciutti (2004) [9], the modification improves the accuracy of the Zhao-Baker approximation.

Yet another type of approach to derive approximation of $d^{rfc}$ damage intensity is to approximate the sequence of turning points $T_P(x)$ by means of a Markov chain (MC) sequence. This approach will be discussed in detail in a following chapter. Here, we shall give numerical values of the two approximations of this type, proposed by Frendahl and Rychlik (1993) [62] and Bishop and Sherratt (1990) [17].

**Example 7.2.** *In this example, we shall compare the accuracy of the different damage intensity approximations given in this section using Gaussian loads with different spectra. Details on the spectra are found in Appendix 7.B and also the results for each approximation and each individual spectrum. Three single peak and six double peak spectra have been used.*

*The rainflow damage intensities are estimated by means of simulations of 5000-second-long stationary Gaussian loads using the normalised spectra having* $H_s = 1$ *and* $T_2 = 1$*. For the spectra, the narrow band bound* $d_2(\beta) = 2^{-\beta/2}\Gamma(1+\beta/2)$.

*This normalisation will also be used in Chapter 11, where the coefficient of variation of the accumulated damage in sea loads with Pierson-Mokowitz spectra is analysed for the case when the spectral parameters change slowly during a year (so-called seasonal effects).*

*Table 7.3 in the appendix gives the result for each approximation method and each spectrum together with the rfc damage intensity, the bound* $d_2(\beta)$*, the alternative* $d_1(\beta)$ *and eight other approximations proposed in the literature.*

*In Table 7.2, the results are summarised by showing the average over the nine spectra. The upper half of the table shows the relative error, expressed as the difference between the approximation and the correct value divided by the correct value. The lower half shows the relative absolute difference. The correct values* $d^{rfc}(\beta)$ *are obtained by means of simulation.*

*We first analyse the results for the three spectral moment-type approximations of rainflow damage intensity* $d^{rfc}$*;* $d_2 (= d^{nb})$*,* $d_1$ *and single moments approximation (7.33) denoted* $d^{LL}$.

*From the results presented in the table for the three approximations, we draw several conclusions:*

*1) Since the average relative errors and average absolute errors of the narrow band bound* $d_2(\beta)$ *differ only by sign, it seems that the statistical errors in estimation of* $d^{rfc}$ *are small.*

*2) For damage exponent* $\beta = 1$*, all damage rates employing cycle counting, e.g. rfc, mM, H-M, we have* $d_2(1) = d^{rfc} = d^{mM} = d^{HM}$ *(by Banach's theorem about total* *variation). Further, by (7.32), we have that* $d_1(\beta) \le d_2(\beta)$ *and since the average values of the relative errors and the absolute values of the relative errors for* $d_1(1)$, *which are* $-0.161$ *and 0.161, are large and differ only in sign, we conclude that for the considered spectra* $d_1(1) < d^{rfc}(1)$ *for* $1 \le \beta < 2$. *So, it seems that* $d_1(\beta)$ *is a non-conservative estimate of* $d^{rfc}$. *(Note that exponent* $\beta < 2$ *is seldom used in practice.)*

*3) The single moment approximation (7.33) is another attempt to lower the conservatism of the narrow-band bound. For* $\beta \ge 2$, *one has that* $d^{LL}(\beta) = d_{2/\beta}(\beta) \le d_1(\beta) \le d^{nb}$. *The result of the comparisons for the nine spectra in Table 7.2 shows that the* $d^{LL}$ *has improved the accuracy of* $d_1(\beta)$ *for* $\beta < 2$, *however to the cost of some small non-conservatism for* $\beta > 2$. *Generally* $d_1$ *and* $d^{LL}$ *have successfully made the conservatism of the narrow band bound smaller.*

*4) Finally, from the table, it can be seen that the approximations by Lutes-Larsen, Wirsching-Light, Benasciutti-Tovo, Zhao-Baker and Frendahl-Rychlik give more accurate results than the approximations by Dirlik, Jiao-Moan and Bishop-Sherratt. (The methods of Bishop-Sherratt and Frendahl-Rychlik will be presented in the next chapter.)*

**Remark 7.1.** *The relative errors in Table 7.2 are averages over the nine spectra in Figure 7.2. A recent extension of the Benasciutti-Tovo method, [197], uses an artificial neural network to adapt the p-parameter* (7.38) *not only to spectral moments but also to the damage exponent* $\beta$.

**Table 7.2**

**Upper Table: Average Relative Error,** $(d^{approx.} - d^{rfc})/d^{rfc}$**, for the Different Damage Intensity Approximation Methods and Spectra Given in Appendix***
**Lower Table: Average Absolute Relative Error** $\left(| d^{approx.} - d^{rfc} | /d^{rfc}\right)$

| | $\beta$ | $d_2(\beta)$ | $d_1(\beta)$ | LL | WL | Dir | BT | ZB | JM | FR | BS |
|---|---|---|---|---|---|---|---|---|---|---|---|
| $\frac{d^{app.} - d^{rfc}}{d^{rfc}}$ | 1 | 0.007 | −0.161 | 0.007 | 0.302 | 0.200 | 0.007 | 0.007 | −0.083 | 0.057 | 0.009 |
| | 2 | 0.179 | −0.040 | −0.040 | 0.055 | 0.205 | 0.015 | 0.002 | 0.139 | 0.026 | −0.147 |
| | 3 | 0.258 | 0.014 | −0.059 | 0.047 | 0.191 | −0.018 | 0.015 | 0.272 | 0.007 | −0.217 |
| | 4 | 0.293 | 0.038 | −0.072 | 0.028 | 0.182 | −0.046 | 0.024 | 0.359 | 0.001 | −0.234 |
| | 5 | 0.309 | 0.052 | −0.078 | −0.004 | 0.182 | −0.060 | 0.031 | 0.426 | 0.003 | −0.227 |
| $\frac{\lvert d^{app.} - d^{rfc}\rvert}{d^{rfc}}$ | 1 | 0.009 | 0.161 | 0.009 | 0.302 | 0.200 | 0.010 | 0.010 | 0.083 | 0.057 | 0.011 |
| | 2 | 0.179 | 0.040 | 0.040 | 0.115 | 0.205 | 0.022 | 0.086 | 0.139 | 0.043 | 0.153 |
| | 3 | 0.258 | 0.044 | 0.059 | 0.175 | 0.191 | 0.022 | 0.134 | 0.272 | 0.045 | 0.231 |
| | 4 | 0.293 | 0.053 | 0.072 | 0.187 | 0.182 | 0.046 | 0.147 | 0.359 | 0.050 | 0.256 |
| | 5 | 0.309 | 0.065 | 0.080 | 0.184 | 0.183 | 0.064 | 0.140 | 0.426 | 0.060 | 0.253 |

*Note*: Rainflow damage intensities, $d^{rfc}$, are obtained from simulations.

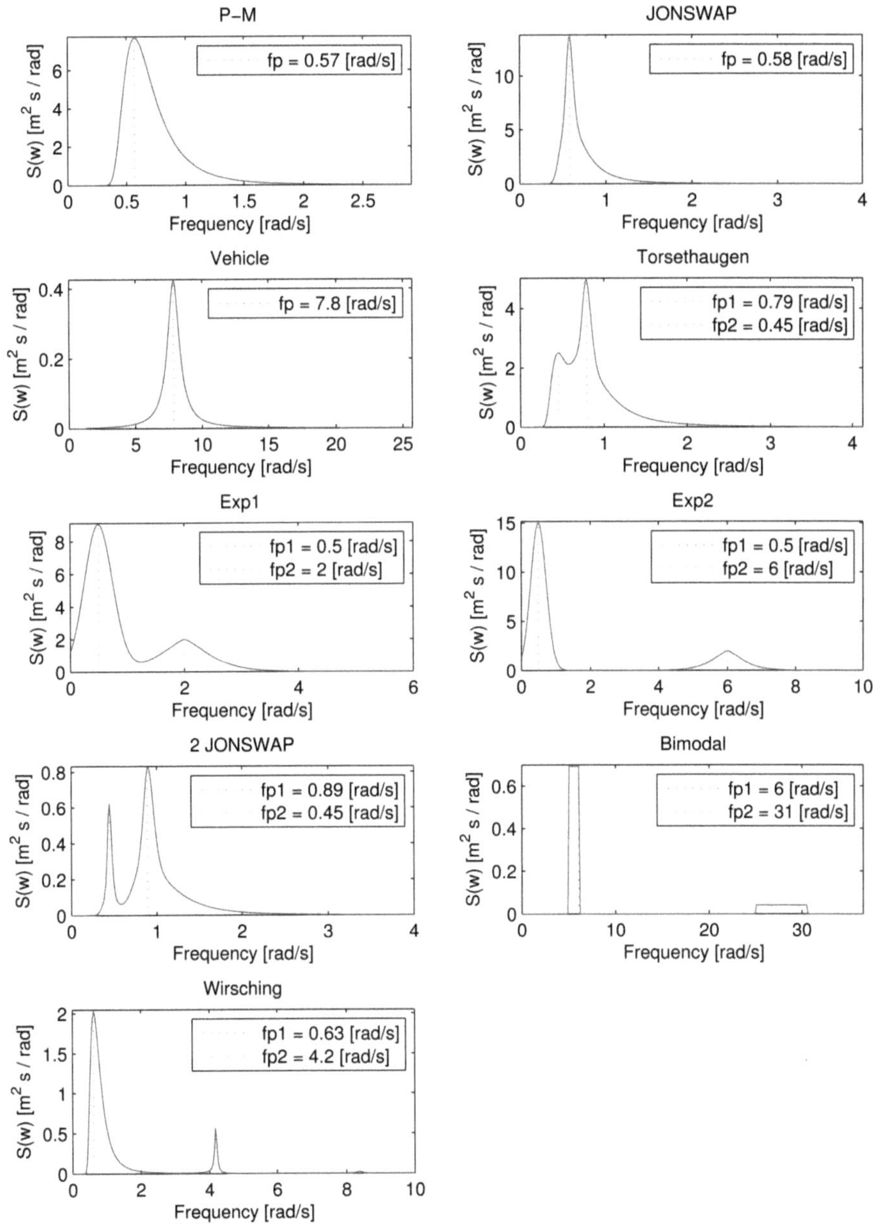


**FIGURE 7.2** Spectra used to produce Table 7.3. (Reproduced with permission from [15]. Copyright © 2009 Elsevier.)

# APPENDICES

## 7.A PROOF OF THEOREM 7.3

In the proof, we shall employ the simple fact that for any positive real valued function $g(\mathbf{x})$ defined on $R^n$,

$$g(\mathbf{x}) = \int_0^\infty 1_{\{\mathbf{u}:g(\mathbf{u})>z\}}(\mathbf{x})\,\mathrm{d}z. \tag{7.43}$$

We shall use (7.43) with $g(u,v) = k^+(u,v)$.

Let us assume that $\mathsf{E}[N^+(u)]$ is a continuous function of $u$ and define

$$\mathscr{C}_u = \{(x_i,y_i) : \mathsf{E}[N^+(x_i)] = \mathsf{E}[N^+(y_i)] = u \text{ and } \mathsf{E}[N^+(z)] > u \text{ for all } z \in (x_i,y_i)\}.$$

Obviously, $\cup_{i=-\infty}^{\infty}(x_i,y_i) = \{z \in \mathbb{R} : \mathsf{E}[N^+(z)] > u\}$. Since $x_i < y_i$, $\mathscr{C}_u$ is a cycle count, and let $\mathscr{D}_u$, denote the damage caused by all pairs $(x_i,y_i)$, viz.

$$\mathscr{D}_u = \sum_{i=-\infty}^{\infty} f(x_i,y_i).$$

This damage can also be computed by means of (4.31) if the counting distribution $N(x,y)$ of the cycle count is known. Here, it is easy to find $N(x,y)$ since

$$N(x,y) = \sum_{i=-\infty}^{\infty} 1_{(x_i,y_i)}(x)1_{(x_i,y_i)}(y) = 1_{\{(r,s):k^+(r,s)>u\}}(x,y).$$

Consider a damage function $f(u,v) = (v-u)^\beta$, $\beta > 1$, (the general case can be treated in the similar way). Then, by means of (4.31),

$$\mathscr{D}_u = \int_{-\infty}^{\infty}\int_{-\infty}^{v} 1_{\{(r,s):k^+(r,s)>u\}}(x,v)\beta(\beta-1)(v-x)^{\beta-2}\,\mathrm{d}x\,\mathrm{d}v,$$

and hence, employing Fubini's theorem and using (7.43),

$$\begin{aligned}\int_0^\infty \mathscr{D}_u\,\mathrm{d}u &= \int_{-\infty}^{\infty}\int_{-\infty}^{v}\int_0^\infty 1_{\{(r,s):k^+(r,s)>u\}}(x,v)\,\beta(\beta-1)(v-x)^{\beta-2}\,\mathrm{d}u\,\mathrm{d}x\,\mathrm{d}v\\ &= \int_{-\infty}^{\infty}\int_{-\infty}^{v} k^+(x,v)\,\beta(\beta-1)(v-x)^{\beta-2}\,\mathrm{d}x\,\mathrm{d}v = d^+.\end{aligned}$$

The expected H-M damage can also be computed in an alternative way, viz.

$$\begin{aligned}\mathsf{E}[D^{HM}(t)] &= t\int_{-\infty}^{\infty}\beta(|2z|^{\beta-1})\mathsf{E}[N^+(z)]\,\mathrm{d}z\\ &= t\int_0^\infty\int_{-\infty}^{\infty}\beta(|2z|^{\beta-1})1_{\{x:\mathsf{E}[N^+(x)]>u\}}(z)\,\mathrm{d}z\,\mathrm{d}u\\ &= t\int_0^\infty\sum_{i=-\infty}^{\infty}\int_{x_i}^{y_i}\beta(|2z|^{\beta-1})\,\mathrm{d}z\,\mathrm{d}u\\ &= t\int_0^\infty\sum_{i=-\infty}^{\infty} G(y_i)-G(x_i)\,\mathrm{d}u,\end{aligned} \tag{7.44}$$

where $(x_i, y_i)$ are cycles in $\mathscr{C}_u$ and clearly depend on $u$. The equality (7.44) implies that the expected H-M damage can be written as $\int_0^\infty \mathscr{D}_u \, du$ where the $\mathscr{D}$ is computed using the exhaustion function $G$, while in computation of $d^+$, $\mathscr{D}$ was defined using the damage function $f(u,v) = (v-u)^\beta$, $\beta > 1$. Consequently, as it was shown before that, if $\mathscr{C}_u$ contains a single asymmetric cycle $(x,y)$, say, then $G(y) - G(x) > (y-x)^\beta$ and hence, by continuity of $\mathsf{E}[N^+(u)]$, for loads with asymmetric spectrum, $t \cdot d^+ < \mathsf{E}[D^{HM}(t)]$.

## 7.B SOME IMPORTANT SPECTRA

### 7.B.1 PIERSON-MOSKOWITZ SPECTRUM

Based on the upper bound expression for the energy of wind-generated deepwater gravity waves, derived by Phillips (1958), [155] and from empirical measurements in the North Atlantic, Pierson and Moskowitz (1964) [156] suggested the following spectrum for fully developed deep-water sea waves,

$$S^{PM}(\omega) = \frac{\alpha g^2}{\omega^5} \exp\left(-\beta\left(\frac{\omega_0}{\omega}\right)^4\right),$$

with constants $\alpha = 0.0081$, $\beta = 0.74$, and $\omega_0 = g/v$, where $v$ is the wind speed 19.5 m above still water sea level and $g$ is the gravity constant. By definition of $h_s$ and $t_z$ it follows

$$\frac{t_z^2}{h_s} = \frac{\pi^2\sqrt{\omega_0}}{\omega_2} = \frac{\pi^2\sqrt{\int_0^\infty S(\omega)\, d\omega}}{\int_0^\infty \omega^2 S(\omega)\, d\omega} \approx 13.$$

Hence, the Pierson-Moskowitz spectrum imposes a deterministic relation between the significant wave height and the mean period. A generalised form of the P-M spectrum, allowing for independent $h_s$ and $t_z$, was given by ISSC (1964),

$$S^{GPM}(\omega) = \frac{h_s^2}{4\pi}\left(\frac{2\pi}{t_z}\right)^4 \omega^{-5} \exp\left(-\frac{1}{\pi}\left(\frac{2\pi}{t_z}\right)^4 \omega^{-4}\right).$$

(The spectrum is also called Bretschneider spectrum.)

### 7.B.2 JONSWAP SPECTRUM

The drawback of the P-M spectrum is the limitation to deep-water waves and fully developed sea states. A result from the Joint North Sea Wave Project (JONSWAP) was the generalisation of the P-M spectrum also to include growing seas. The JONSWAP spectrum is a five-parameter spectrum, fitted to measurements in the North-Sea, for $\omega < \omega_c$,

$$S^J(\omega) = c_0 \frac{T_p h_s^2}{2\pi} \exp\left(-\frac{5}{4}\left(\frac{2\pi}{T_p\omega}\right)^4 + \log(\kappa)\exp\left(-\frac{(T_p\omega - 1)^2}{2\sigma^2}\right)\right), \quad (7.45)$$

with $\sigma = \sigma_a$ if $T_p\omega < 1$ and $\sigma = \sigma_b$ if $T_p\omega > 1$. Here, $h_s$ is the significant wave height defined as $h_s = 4\sqrt{\omega_0}$, $T_p$ is the peak period of the spectrum and $\omega_c$ is the cut-off frequency, i.e. a frequency such that $S(\omega) = 0$ for all $\omega > \omega_c$. Often, $\kappa = 3.3$, $\sigma_a = 0.07$, and $\sigma_b = 0.09$. The normalising constant $c_0$ is chosen in such a way that $\int_0^\infty S(\omega)\,\mathrm{d}\omega = h_s^2/16$. In general, the constant $c_0$ is a function of parameters $\kappa, \sigma_a, \sigma_b$. See Hasselmann et al. (1973) [81] for further details. Keeping $\omega_c$ constant for different values of $T_p$ means that the sea with smaller $T_p$ is more narrow-banded, since the peak frequency $2\pi/T_p$ is closer to $\omega_c$. (Theoretically for the JONSWAP spectrum $\omega_c = \infty$; however, this is somewhat inconvenient, since $\omega_4 = \infty$ and hence $T_4 = 0$, i.e. the intensity of rainflow cycles is infinite. In addition, the number of zero-crossings of $W(vt,t)$, where $W(x,t)$ denotes the wave height at position $x$ and time $t$ for a long-crested Gaussian sea having JONSWAP spectrum, that a ship encounters when sailing with velocity $v$, is infinite too. Consequently, for practical reasons, the cut-off frequency $\omega_c$ is introduced.)

### 7.B.3 BIMODAL SPECTRA

The JONSWAP spectrum was further developed by Torsethaugen (1993, 1996) [202, 203] to also include the contribution from swell generated at a far distance from the measuring point. This results in four different types of spectrum peaks: primary peak generated by local wind fields, primary swell peak, secondary wind peak and secondary swell peak. Based on measurements from the Norwegian sea, a bimodal model consisting of two JONSWAP spectra is suggested,

$$S^{Tor}(\omega) = S^J(\omega \mid c_{0,1}, T_{p,1}, h_{s,1}, \kappa_1, \sigma_1) + S^J(\omega \mid c_{0,2}, T_{p,2}, h_{s,2}, \kappa_2, \sigma_2).$$

The following bimodal spectrum is used in Larsen and Lutes (1991) [114],

$$S^W(\omega) = \frac{7793}{T_D^4\omega^5} \cdot \frac{\exp\left(-1948/(T_D\omega)^4\right)}{\left(1-\left(\frac{\omega}{\omega_N}\right)^2\right)^2 + 4\zeta^2\left(\frac{\omega}{\omega_N}\right)^2},$$

which is equal to the spectrum of a one-degree of freedom dynamic system with input load having P-M spectrum.

### 7.B.4 ROAD AND VEHICLE STRESS SPECTRA

A commonly used Gaussian road surface model is the one presented in LaBarre et al. (1970) [112],

$$S^{MIRA}(\xi) = \begin{cases} 10^{a_0}\left(\frac{\xi}{\xi_0}\right)^{-w_1} & , \quad \xi \in [0.01, 0.2], \\ 10^{a_0}\left(\frac{\xi}{\xi_0}\right)^{-w_2} & , \quad \xi \in [0.2, 10], \\ 0 & , \quad \text{otherwise}, \end{cases}$$

where the reference spatial frequency is $\xi_0 = 0.2\ \mathrm{m}^{-1}$ and with typical values of $a_0 = -5$, $w_1 = 3$, and $w_2 = 2$. An extension of this model accounting for non-stationary irregularities is given in Bogsjö and Rychlik (2009) [22]. The ISO-standard for road roughness can be found in [86].

For a vehicle model consisting of two masses, one "unsprung" mass, $m$, representing the mass of the wheel, etc. and one sprung mass, $M$, representing the body of the vehicle, joined together by a damper, $c$, and a spring, $k$, and with an additional damper, $c_t$, and spring, $k_t$, representing the damping of the tire, the transfer function can be written as

$$H(\omega) = \frac{M\omega^2(k_t + i\omega c_t)}{k_t - \frac{(k+i\omega c)\omega^2 M}{M\omega^2 - k - i\omega c} - m\omega^2 + i\omega c_t} \left(1 + \frac{M\omega^2}{k - M\omega^2 + i\omega c}\right),$$

with typical parameter values $M = 3400$, $m = 350$, $k = 270000$, $k_t = 950000$, $c = 6000$ and $c_t = 300$. Hence, the stress load spectrum for a vehicle at speed $v$ is given by

$$S^{vehicle}(\omega) = \frac{1}{v}|H(\omega)|^2 S^{MIRA}(\omega/(2\pi v)).$$

## 7.C COMPARISON OF DAMAGE INTENSITY APPROXIMATIONS FOR NINE SPECTRA

Explanation to Table 7.3, Comparison of the Different Damage Intensity Approximations for Spectra Given in Figure 7.2 for Different $\beta$.

Benasciutti-Tovo (BT)
Bishop-Sherratt (BS)
Dirlik (Dir)
d2($\beta$)-bound (Bound), d1($\beta$)
Zhao-Baker (ZB)
Frendahl-Rychlik (FR)
Jiao-Moan (JM)
Lutes-Larsen (LL)
Value from simulations (rfc)
Wirsching-Light (WL)

**Table 7.3**

**Comparison of the Different Damage Intensity Approximations for Spectra Given in Figure 7.2 for Different $\beta$**

| | $\beta$ | rfc | $d_2(\beta)$ | $d_1(\beta)$ | LL | WL | Dir | BT | ZB | JM | FR | BS |
|---|---|---|---|---|---|---|---|---|---|---|---|---|
| P-M | 1 | 0.621 | 0.627 | 0.586 | 0.627 | 0.726 | 0.712 | 0.627 | 0.627 | 0.619 | 0.635 | 0.627 |
| | 2 | 0.477 | 0.500 | 0.468 | 0.468 | 0.455 | 0.545 | 0.479 | 0.452 | 0.494 | 0.481 | 0.414 |
| | 3 | 0.441 | 0.470 | 0.440 | 0.431 | 0.393 | 0.506 | 0.436 | 0.409 | 0.464 | 0.444 | 0.356 |
| | 4 | 0.460 | 0.500 | 0.468 | 0.455 | 0.398 | 0.536 | 0.454 | 0.428 | 0.493 | 0.464 | 0.364 |
| | 5 | 0.531 | 0.588 | 0.550 | 0.531 | 0.447 | 0.628 | 0.525 | 0.499 | 0.578 | 0.536 | 0.422 |
| JONSWAP | 1 | 0.624 | 0.627 | 0.586 | 0.627 | 0.750 | 0.763 | 0.627 | 0.627 | 0.616 | 0.653 | 0.627 |
| | 2 | 0.481 | 0.500 | 0.468 | 0.468 | 0.451 | 0.575 | 0.482 | 0.443 | 0.491 | 0.493 | 0.395 |
| | 3 | 0.445 | 0.470 | 0.440 | 0.432 | 0.391 | 0.531 | 0.442 | 0.400 | 0.461 | 0.454 | 0.332 |
| | 4 | 0.462 | 0.500 | 0.468 | 0.456 | 0.397 | 0.560 | 0.462 | 0.418 | 0.490 | 0.475 | 0.338 |
| | 5 | 0.528 | 0.588 | 0.550 | 0.533 | 0.447 | 0.656 | 0.537 | 0.488 | 0.575 | 0.549 | 0.394 |
| Vehicle | 1 | 0.625 | 0.627 | 0.608 | 0.627 | 0.684 | 0.659 | 0.627 | 0.627 | 0.621 | 0.629 | 0.627 |
| | 2 | 0.492 | 0.500 | 0.485 | 0.485 | 0.464 | 0.515 | 0.486 | 0.470 | 0.495 | 0.492 | 0.454 |
| | 3 | 0.459 | 0.470 | 0.456 | 0.452 | 0.399 | 0.482 | 0.447 | 0.431 | 0.465 | 0.458 | 0.406 |
| | 4 | 0.484 | 0.500 | 0.485 | 0.479 | 0.401 | 0.512 | 0.466 | 0.453 | 0.495 | 0.482 | 0.419 |
| | 5 | 0.560 | 0.587 | 0.570 | 0.561 | 0.448 | 0.602 | 0.539 | 0.529 | 0.581 | 0.559 | 0.483 |
| Torseth. | 1 | 0.626 | 0.627 | 0.565 | 0.627 | 0.776 | 0.782 | 0.627 | 0.627 | 0.627 | 0.653 | 0.627 |
| | 2 | 0.471 | 0.500 | 0.451 | 0.451 | 0.448 | 0.584 | 0.472 | 0.435 | 0.554 | 0.481 | 0.375 |
| | 3 | 0.434 | 0.470 | 0.424 | 0.411 | 0.390 | 0.538 | 0.428 | 0.390 | 0.558 | 0.440 | 0.308 |
| | 4 | 0.451 | 0.500 | 0.451 | 0.430 | 0.397 | 0.567 | 0.445 | 0.408 | 0.627 | 0.459 | 0.313 |
| | 5 | 0.514 | 0.588 | 0.529 | 0.501 | 0.447 | 0.664 | 0.515 | 0.476 | 0.770 | 0.530 | 0.366 |
| Exp1 | 1 | 0.633 | 0.627 | 0.487 | 0.627 | 0.809 | 0.738 | 0.627 | 0.627 | 0.549 | 0.638 | 0.627 |
| | 2 | 0.425 | 0.500 | 0.389 | 0.389 | 0.445 | 0.498 | 0.422 | 0.425 | 0.530 | 0.412 | 0.354 |
| | 3 | 0.380 | 0.470 | 0.365 | 0.330 | 0.390 | 0.438 | 0.356 | 0.380 | 0.563 | 0.357 | 0.284 |
| | 4 | 0.400 | 0.500 | 0.389 | 0.332 | 0.397 | 0.454 | 0.354 | 0.397 | 0.655 | 0.370 | 0.287 |
| | 5 | 0.467 | 0.587 | 0.457 | 0.377 | 0.447 | 0.528 | 0.400 | 0.463 | 0.827 | 0.432 | 0.337 |
| Exp2 | 1 | 0.620 | 0.627 | 0.372 | 0.627 | 0.889 | 0.696 | 0.627 | 0.627 | 0.398 | 0.632 | 0.627 |
| | 2 | 0.306 | 0.500 | 0.296 | 0.296 | 0.441 | 0.356 | 0.325 | 0.403 | 0.354 | 0.328 | 0.314 |
| | 3 | 0.225 | 0.470 | 0.279 | 0.216 | 0.389 | 0.253 | 0.229 | 0.357 | 0.359 | 0.254 | 0.239 |
| | 4 | 0.218 | 0.500 | 0.296 | 0.201 | 0.397 | 0.231 | 0.205 | 0.372 | 0.404 | 0.256 | 0.239 |
| | 5 | 0.252 | 0.587 | 0.348 | 0.218 | 0.447 | 0.251 | 0.220 | 0.435 | 0.498 | 0.303 | 0.282 |
| 2JONSWAP | 1 | 0.627 | 0.627 | 0.577 | 0.627 | 0.735 | 0.720 | 0.627 | 0.627 | 0.621 | 0.637 | 0.627 |
| | 2 | 0.479 | 0.500 | 0.461 | 0.461 | 0.453 | 0.547 | 0.475 | 0.448 | 0.527 | 0.477 | 0.407 |
| | 3 | 0.442 | 0.470 | 0.433 | 0.422 | 0.392 | 0.507 | 0.430 | 0.405 | 0.518 | 0.438 | 0.346 |
| | 4 | 0.462 | 0.500 | 0.461 | 0.443 | 0.398 | 0.537 | 0.446 | 0.424 | 0.570 | 0.457 | 0.354 |
| | 5 | 0.532 | 0.587 | 0.541 | 0.516 | 0.447 | 0.629 | 0.515 | 0.495 | 0.689 | 0.526 | 0.411 |
| Bimodal | 1 | 0.614 | 0.624 | 0.469 | 0.624 | 0.819 | 0.734 | 0.624 | 0.624 | 0.477 | 0.624 | 0.627 |
| | 2 | 0.387 | 0.495 | 0.372 | 0.372 | 0.440 | 0.479 | 0.407 | 0.417 | 0.447 | 0.371 | 0.346 |
| | 3 | 0.341 | 0.463 | 0.348 | 0.315 | 0.384 | 0.412 | 0.336 | 0.370 | 0.465 | 0.308 | 0.274 |
| | 4 | 0.361 | 0.490 | 0.368 | 0.318 | 0.389 | 0.421 | 0.330 | 0.384 | 0.533 | 0.316 | 0.276 |
| | 5 | 0.429 | 0.572 | 0.430 | 0.363 | 0.435 | 0.485 | 0.369 | 0.446 | 0.664 | 0.370 | 0.325 |
| Wirsching | 1 | 0.606 | 0.623 | 0.446 | 0.623 | 1.09 | 0.908 | 0.624 | 0.624 | 0.607 | 0.809 | 0.627 |
| | 2 | 0.380 | 0.497 | 0.356 | 0.356 | 0.434 | 0.573 | 0.397 | 0.356 | 0.513 | 0.450 | 0.253 |
| | 3 | 0.346 | 0.467 | 0.334 | 0.305 | 0.387 | 0.503 | 0.340 | 0.306 | 0.502 | 0.368 | 0.175 |
| | 4 | 0.370 | 0.496 | 0.356 | 0.311 | 0.395 | 0.525 | 0.350 | 0.318 | 0.550 | 0.369 | 0.172 |
| | 5 | 0.434 | 0.583 | 0.418 | 0.358 | 0.444 | 0.612 | 0.406 | 0.370 | 0.663 | 0.419 | 0.206 |

# 8 Markov Loads

## 8.1 INTRODUCTION

In Chapter 7, we approximated the dependence structure of the sequence of turning points in Gaussian loads by means of Markov chain. As shown in Table 7.2, the computed rainflow damage intensity for the Markov chain agrees very well with the simulated one for the Gaussian load for the considered spectrum. Encouraged by the accuracy of this approximation procedure, we wish to further investigate the algorithms for computation of the expected rainflow damage for discrete valued time series that are functions of Markov chain, here called *Markov loads*. The class of Markov loads is very wide, and many real stationary loads can be approximated by members of this class. In addition, for Markov loads, there exist explicit formulas for the oscillation intensities $\mu^{osc}$; see Section 8.3.2.

A particular type of Markov loads called *Markov chain of turning points* (MCTP) are often used in practice. Those models postulate that the discretised sequence of turning points is a Markov chain. In Chapter 5, we also used MCTP to approximate the solution to the Fredholm equation (5.6). MCTP are also efficient means to generate random sequences of turning points for testing of components; see e.g. Agerskov [3].

Another interesting type of Markov load is the switching markov chain (SMC) introduced later on. This model can be used in situations when operation conditions change in an abrupt way. For example, for wave loads under changing sea-states or when road quality is changing or a vehicle is switching between heavy loaded and unloaded condition. SMC were first used for modelling fatigue loads in Johannesson [92] where algorithms to compute the rainflow damage were also presented. Use of the model in the automotive industry can be found in [94].

In this chapter, we shall briefly present the MC, MCTP, SMC, and SMCTP (*switching Markov chain of turning points*) models and also discuss the inverse problem, i.e. how to find the Markov load when only the oscillation intensity $\mu^{osc}$ is known. This is an important problem since often measured loads are saved in the form of rainflow matrices and hence one needs to find a simple load model $x$ having oscillations similar to the observed ones. Then, $x$ can be used to test components for fatigue strength for a load that have observed rainflow matrix.

## 8.2 CYCLE MODELS BASED ON MARKOV CHAIN

In this section, we present several alternatives for load models and load cycle models, all based on some form of Markov property. The alternatives are the following.

**Markov chain load:** In this model, the load process is a discrete Markov chain, or a Markov process in continuous time. It can seldom be used as a model for real loads and fatigue damage.

DOI: 10.1201/9781003019947-8

**Markov load:** In this book, a Markov load is defined as a function of a Markov chain. If the function is invertible, so each load value corresponds to a unique state in the Markov chain, a Markov load is also a Markov chain load. If several states in the Markov chain are mapped to the same state in the observed load process, a Markov load is not a Markov chain. States in the Markov chain are "lumped together" which makes it possible to represent dynamical effects in the load process while keeping an underlying Markov structure.

**Markov chain of turning points:** A Markov chain or Markov load such that each value is a local extremum is called a Markov chain of turning points. The requisite that each point is a local extremum destroys the simple one-step Markov property. One possible solution is to use a two-step Markov chain, $(X_{n-1}, X_n)$ with Markov transitions from a min-Max to a Max-min state and from Max-min to min-Max state. An equivalent and numerically more economic alternative is to use a Markov load where the underlying Markov chain keeps track of the status (maximum or minimum) of each observed $X_n$.

**Switching Markov chain load:** A switching load model is used when a system is governed by different laws during different time intervals. In a switching Markov chain load, the transition probability matrix changes randomly at times.

**Switching Markov chain of turning points:** This is a very general model, SM-CTP, that combines the simplicity of the MCTP with the flexibility of switching between different modes of operation.

### 8.2.1 BASIC PROPERTIES

First, we shall review some basic facts on Markov sequences with stationary transitions and countable state space, here simply called *Markov chain* MC.

A stochastic process $Z_0, Z_1, \ldots$ with values in $S = \{1, 2, \ldots, N\}$, $N$ can be infinite, is a Markov chain if there is a function $P$ defined on $S \times S$ ($P_{ij} \geq 0$ for all $i, j \in S$) such that for all $n > 0$,

$$\mathsf{P}(Z_{n+1} = j \mid Z_n = i, Z_{n-1} = k, \ldots, Z_0 = l) = P_{ij}. \tag{8.1}$$

This basic Markov condition says that if the value of the process at a time $n$, chosen independently of the process, is known *exactly*, $Z_n = i$, then the future of the process is completely independent of what has happened before time $n$. We assume that $S$ contains only the essential values, namely those for which $\mathsf{P}(Z_n = i) > 0$ for some $n$.

In the following, if not specified, we assume that all states in $S$ are positive recurrent, i.e. that the expectation of the least $n > 0$ with $Z_n = i$ given that $Z_0 = i$ is finite and equal $1/\pi_i$, where $\pi_i > 0$ for all $i \in S$. This means that the expected proportion of $i$-values in a long realisation of the process is $\pi_i$, and hence the row vector $\pi$ is a probability distribution on $S$, i.e. $\pi\mathbf{1} = \sum_{i \in S} \pi_i = 1$, where $\mathbf{1}$ is a column vector of ones. The probabilities $\pi$ are a solution to the following equation system $\pi = \pi P$, viz.

$$\pi_i = \sum_{j \in S} \pi_j P_{ji}. \tag{8.2}$$

The row vector $\pi$ is called a stationary distribution of $Z_n$, since if $\mathsf{P}(X_0 = i) = \pi_i$ then $\mathsf{P}(X_n = i) = \pi_i$ for all $n$, and

$$\lim_{n\to\infty} \mathsf{P}(Z_n = i) = \pi_i,$$

regardless of how the process has started.

In the following, we shall consider stationary Markov chain $Z_n$, i.e. $\mathsf{P}(Z_n = i) = \pi_i$. Denote the joint probabilities of $Z_0, Z_1$ by $F = [F_{ij}]$, where

$$F_{ij} = \mathsf{P}(Z_n = i, Z_{n+1} = j) = \pi_i P_{ij}. \tag{8.3}$$

If $F$ is symmetrical, then the chain $Z_n$ is time reversible. We shall discuss the time reversibility in the following sections.

A Markov process with continuous time obeys a condition analogous to (8.1).

### 8.2.2 MARKOV LOADS

A Markov chain has a very simple and local dependence structure, and it can seldom be used to model the variability of real loads generated in a dynamic mechanical system. Hence, we shall introduce a wider class of loads, named here Markov loads, which are more flexible and may fit a variety of real loads.

**Definition 8.1** (Markov load). *A Markov load $X_n$ with state space $\mathbf{x} = \{x_1 > x_2 > \cdots > x_J\}$ is a time series that is a function of a Markov chain, viz. $X_n = G(Z_n)$, where $Z_n$ is a stationary Markov chain taking values in $S = \{1, 2 \ldots, N\}$, $N \geq J$, and $G$ is a function from $S$ onto* $\mathbf{x}$.

The Markov loads form a very general class of models for discretised random loads $X_n^d$ since if $X_n^d$ is observed for $0 \leq n \leq t < \infty$, then one can find a Markov load $X_n = G(Z_n)$ such that a vector $X_0^d, X_1^d, \ldots, X_t^d$ has the same distribution as $X_0, X_1, \ldots, X_t$. (The so-called Markov chain Monte Carlo algorithm to simulate $\mathbf{X}^d = (X_0^d, X_1^d, \ldots, X_t^d)$ defines a Markov chain that has the same stationary distribution as $\mathbf{X}^d$.) However, to define a Markov load $X_n = G(Z_n)$ with a complicated distribution may require that $Z_n$ has a very large sample space $S$ which makes it numerically prohibitive. A Markov load is a simple form of a *partially observed Markov chain* or *Hidden Markov Models*, HMM. This class of time series models has found wide applications in data analysis, decision algorithms and general machine learning, see, e.g. [224].

**Remark 8.1.** *An important and useful property of Markov loads is that the rainflow filtered $X_n$ is again a Markov load.*

**Example 8.1.** *This example illustrates two aspects of Markov loads. Let $Z_n$ be a Markov chain taking four values* 1, 2, 3, 4 *and having transition matrix $P$ and stationary measure $\pi$. The process $Z_n$ jumps between a group of "upper" states and a*

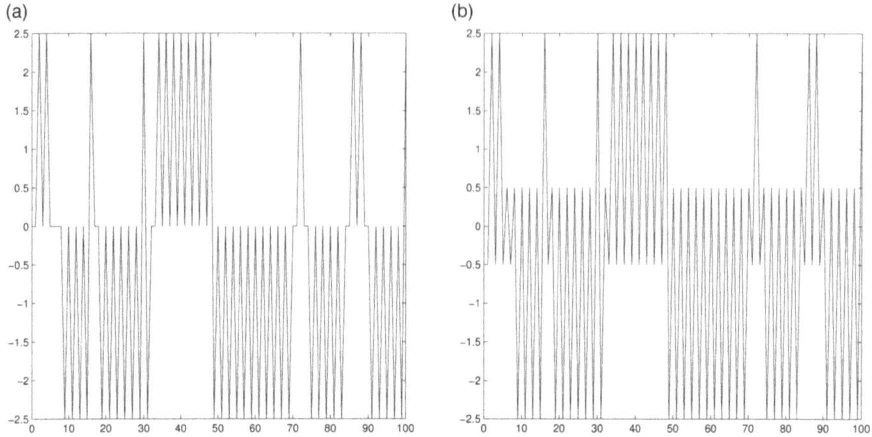


**FIGURE 8.1** Markov loads defined in Example 8.1; (a) $X_n = G(Z_n)$ with $G(1) = 2.5$, $G(2) = G(3) = 0$, $G(4) = -2.5$. (b) $Y_n = G(Z_n)$ with $G(1) = 2.5$, $G(2) = -0.5$ $G(3) = 0.5$, $G(4) = -2.5$.

*group of "lower" states.*

$$P = \begin{bmatrix} 0 & 0 & 0.9 & 0.1 \\ 0 & 0 & 0.1 & 0.9 \\ 0.9 & 0.1 & 0 & 0 \\ 0.1 & 0.9 & 0 & 0 \end{bmatrix}, \quad \pi = \begin{bmatrix} 0.25 & 0.25 & 0.25 & 0.25 \end{bmatrix}.$$

*The discrete time series $X_n = G(Z_n)$, where $G(1) = 2.5$, $G(2) = G(3) = 0$, $G(4) = -2.5$, is not an MC; two states, one from the upper and one from the lower group, are pooled into one observable value. Prediction of the future from an observation $X_n = 0$ is obviously much improved by the information of the value $X_{n-1}$. This is not a Markov chain but a Markov load. Note that this is not an MCTP either since $X_n$ can be constantly equal zero for a period of time.*

*However, $Y_n = G(Z_n)$, where $G(1) = 2.5$, $G(2) = -0.5$, $G(3) = 0.5$, $G(4) = -2.5$, gives a new stationary MC. A simulation is shown in Figure 8.1 right plot. By relabelling the states $\{1,2,3,4\}$ to $\{2.5,-0.5,0.5,-2.5\}$, we have even made it a chain of turning points, MCTP!*

*Both $X_n$ and $Y_n$ are Markov loads, but one is not an MC and the other is an MCTP. In Figure 8.1, a simulation of 100 values of these Markov loads is presented.*

### 8.2.3 MARKOV CHAIN OF TURNING POINTS

It is an empirical fact that the sequence of turning points in many load processes can be well approximated by a two-step alternating Markov chain. This means that, e.g., for an observed min-Max cycle $(m_n, M_n)$, the distribution of the following Max-min cycle $(M_n, m_{n+1})$ does not depend on previous maxima or minima, and similarly, for the next min-Max cycle. However, for reasonable accuracy, the dimension of the model has to be large. The MCTP model is a computationally efficient model which keeps enough of the Markov structure to be realistic under stationary circumstances.

A discrete time series $X_n$, with values in $\mathbf{x} = (x_1 > x_2 > \ldots > x_J)$, such that all $X_n$ are local extremes is a sequence of turning points. MCTP will be defined next.

**Definition 8.2** (Markov chain of turning points). *Let $P^{Mm}$, $P^{mM}$ be two $(J-1, J-1)$ dimensional upper and lower triangular matrices, called Mm- and mM-transition matrices, with transition probabilities from $x_1, \ldots, x_{J-1}$ to $x_2, \ldots, x_J$ and $x_2, \ldots, x_J$ to $x_1, \ldots, x_{J-1}$, respectively. Let $Z_n$ be a stationary MC having the transition matrix*

$$P = \begin{bmatrix} 0 & P^{Mm} \\ P^{mM} & 0 \end{bmatrix}.$$

*Suppose that all states are essential (if not the non-essential states should be removed). Let the G-function be defined as follows: $G(1) = x_1$, $G(k) = G(J+k-2) = x_k$, for $1 < k \leq J-1$, and $G(2(J-1)) = x_J$. Then, the time series $X_n = G(Z_n)$ will be called an MCTP with range $\mathbf{x}$.*

The dimension of $P$ is $2(J-1) \times 2(J-1)$ and the state space of $Z_n$ is $\{1, 2, \ldots, 2(J-1)\}$. The state space of $X_n$ is $\mathbf{x}$ of dimension $J$. Note that an MCTP is a function of a Markov chain, and it can be, but need not be, a Markov chain.

**Example 8.2.** *Let the Mm- and mM-transition matrices be*

$$P^{Mm} = \begin{bmatrix} 0.5 & 0.25 & 0.25 \\ 0 & 0.75 & 0.25 \\ 0 & 0 & 1 \end{bmatrix}, \quad P^{mM} = \begin{bmatrix} 1 & 0 & 0 \\ 0.25 & 0.75 & 0 \\ 0.25 & 0.25 & 0.5 \end{bmatrix},$$

*and construct the Markov transition matrix for $Z_n$ as*

$$P = \begin{bmatrix} 0 & P^{Mm} \\ P^{mM} & 0 \end{bmatrix}.$$

*It has stationary distribution $\pi = [0.1765, 0.2353, 0.0882, 0.0882, 0.2353, 0.1765]$ and hence the $F = [\pi_i \cdot P_{ij}]$ matrix is*

$$F = \begin{bmatrix} 0 & 0 & 0 & 0.0662 & 0.1176 & 0.0441 \\ 0 & 0 & 0 & 0 & 0.1765 & 0.0441 \\ 0 & 0 & 0 & 0 & 0 & 0.1765 \\ 0.1765 & 0 & 0 & 0 & 0 & 0 \\ 0.0441 & 0.1765 & 0 & 0 & 0 & 0 \\ 0.0441 & 0.1176 & 0.0662 & 0 & 0 & 0 \end{bmatrix}.$$

*As expected, $F$ is symmetric and hence $Z_n$ is time reversible.*

*Again, consider two loads, $X_n = G_x(Z_n)$ and $Y_n = G_y(Z_n)$ defined by*

| $k$ | 1 | 2 | 3 | 4 | 5 | 6 |
|---|---|---|---|---|---|---|
| $G_x(k)$ | 3 | 2 | 1 | −1 | −2 | −3 |
| $G_y(k)$ | 3 | 1 | −1 | 1 | −1 | −3 |

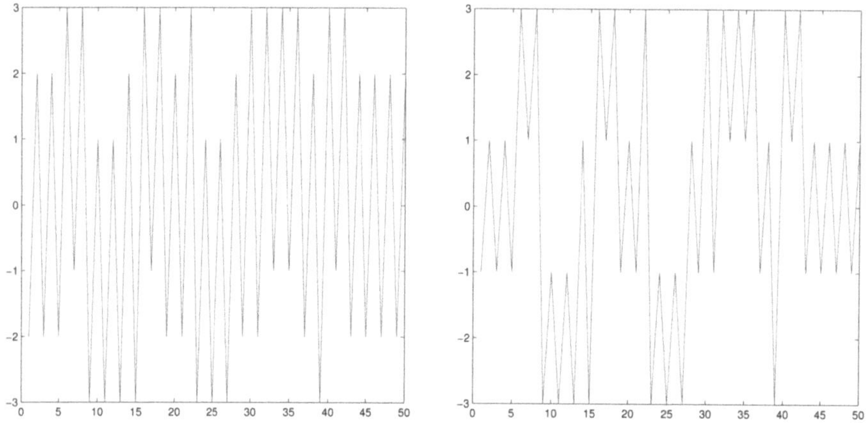


**FIGURE 8.2** Markov loads defined in Example 8.2; $X_n = G(Z_n)$ with $G(1) = 3$, $G(2) = 2$, $G(3) = 1$, $G(4) = -1$, $G(5) = -2$, $G(6) = -3$ (left) and $Y_n = G(Z_n)$ with $G(1) = 3$, $G(2) = G(4) = 1$, $G(3) = G(5) = -1$ $G(6) = -3$ (right) plot.

*Both $X_n$ and $Y_n$ are sequences of turning points. The time series $X_n$ is an MC, while $Y_n$ is not (it is a two-step MC). The series $X_n$ has a special structure that all local maxima are higher than the local minima; that's why it is MC and TP. Simulations of both time series are presented in Figure 8.2.*

Note that the MCTP has the correct long-term distribution of $M$, $m$, $(M,m)$ and $(m,M)$, but that it still can be a too simple model to have the oscillation intensity $\mu_{ij}^{osc}$ that agrees with the observed one. Bogsjö [20] contains a discussion of the validation problem that the model predicts well the rainflow damage accumulation if only one realisation of the load is available.

### 8.2.4 SWITCHING MARKOV CHAIN LOADS

There are situations with abruptly changing loading conditions, e.g. when a truck is driving on a highway and then on a gravel road, etc. Suppose that loads, under each of those conditions, could be adequately modelled as a Markov load. Since Markov loads are convenient for the generation of load sequences and for estimation of expected damage, one would like to include switching between different loading environments, also called regimes, into the more complex Markov load model (this is done by enlarging the state space $S$). Modelling and properties of loads with Markov regimes were discussed in depth in Johannesson 1999, [92], and here we give two examples of such models and consider only two regimes.

Suppose that one has $r = 2$ regimes. Under each of the regimes, the variability of the load can be well modelled by a Markov load $X_n^s$, $s = 1,2$, having the same range $\mathbf{x}$ (some states are maybe non-essential). Consider a special case when the two regimes have identical G-functions, $X_n^s = \widetilde{G}(Z_n^s)$, where $Z_n^s$, $s = 1,2$, are two Markov chain with $N \times N$-transition matrix $P^s$, respectively.

First, consider the transition matrix

$$P = \begin{bmatrix} P^1 & 0 \\ 0 & P^2 \end{bmatrix}.$$

of order $2N \times 2N$ and the function $G(i) = \widetilde{G}(\text{mod}(i,N)+1)$, where the modulo operation $\text{mod}(i,j)$ finds the remainder of division of the first number by the second. The Markov chain $Z_n$ with transition matrix $P$ is not positive recurrent since if it starts in regime $s$, then it will stay there forever.

Next, suppose that regimes switch at random times and let $R_n$ indicate the regime at time $n$. "Random swiches" then means that $R_n$ is a Markov chain with a transition matrix

$$Q = \begin{bmatrix} q_{11} & q_{12} \\ q_{21} & q_{22} \end{bmatrix}.$$

Finally, let $Z_n$ be a Markov chain having the transition matrix

$$P = \begin{bmatrix} q_{11}P^1 & q_{12}P^2 \\ q_{21}P^1 & q_{22}P^2 \end{bmatrix},$$

and define $G(i) = \widetilde{G}(\text{mod}(i,N)+1)$. Then, $X_n = G(Z_n)$ is the SMC load.

**Remark 8.2.** *The transitions in this model are occurring according to the following scheme. If, at time $n$, the process $R_n$ is in state $s$, it is first updated according to the transition matrix $Q$. If the new value is $R_{n+1} = w$, the new $Z_{n+1}$ is defined using the transition matrix $P^w$.*

**Remark 8.3.** *To simplify the computations of oscillation intensity, one is often reordering the states of $Z_n$ by means of a permutation $k(i))$ defined as follows:*

$$[k(1), k(2), k(3)\ldots, k(2N)] = [1, N+1, 2, N+2, \ldots, N, 2N].$$

*This implies that columns and rows in $P$ are permuted to a new transition matrix $\widehat{P}$ say. Let $\widehat{Z}_n$ be the Markov chain having transition matrix $\widehat{P}$. (Obviously in distribution $\widehat{Z}_n = k(Z_n)$). A new G-function is defined as $\widehat{G}(i) = G(k^{-1}(i)$.) Finally*

$$\widehat{X}_n = \widehat{G}(\widehat{Z}_n) = G(k^{-1}(k(Z_n))) = X_n,$$

*which means that we can use equivalently $X_n$ or $\widehat{X}_n$ as a Markov load.*

**Example 8.3.** *Consider a load taking three values $\mathbf{x} = \{1, 0, -1\}$ and that there are two regimes with transition matrices*

$$P^1 = \begin{bmatrix} 0 & 0.9 & 0.1 \\ 0.9 & 0 & 0.1 \\ 0.95 & 0.05 & 0 \end{bmatrix}, \quad P^2 = \begin{bmatrix} 0 & 0.05 & 0.95 \\ 0.1 & 0 & 0.9 \\ 0.1 & 0.9 & 0 \end{bmatrix}.$$

*Switching takes place with switching matrix*

$$Q = \begin{bmatrix} 0.95 & 0.05 \\ 0.1 & 0.9 \end{bmatrix}.$$

*The SMC is* $\widehat{X}_n = \widehat{G}(\widehat{Z}_n)$, *where* $\widehat{Z}_n$ *is a Markov chain with transition matrix*

$$\widehat{P} = \begin{bmatrix} 0\cdot 0.95 & 0\cdot 0.05 & 0.9\cdot 0.95 & 0.05\cdot 0.05 & 0.1\cdot 0.95 & 0.95\cdot 0.05 \\ 0\cdot 0.1 & 0\cdot 0.9 & 0.9\cdot 0.1 & 0.05\cdot 0.9 & 0.1\cdot 0.1 & 0.95\cdot 0.9 \\ 0.9\cdot 0.95 & 0.1\cdot 0.05 & 0\cdot 0.95 & 0\cdot 0.05 & 0.1\cdot 0.95 & 0.9\cdot 0.05 \\ 0.9\cdot 0.1 & 0.1\cdot 0.9 & 0\cdot 0.1 & 0\cdot 0.9 & 0.1\cdot 0.1 & 0.9\cdot 0.9 \\ 0.95\cdot 0.95 & 0.1\cdot 0.05 & 0.05\cdot 0.95 & 0.9\cdot 0.05 & 0\cdot 0.95 & 0\cdot 0.05 \\ 0.95\cdot 0.1 & 0.1\cdot 0.9 & 0.05\cdot 0.1 & 0.9\cdot 0.9 & 0\cdot 0.1 & 0\cdot 0.9 \end{bmatrix},$$

*and* $\widehat{G}(\cdot)$ *is defined as follows:* $\widehat{G}(1) = \widehat{G}(2) = 1$, $\widehat{G}(3) = \widehat{G}(4) = 0$ *and* $\widehat{G}(5) = \widehat{G}(6) = -1$. *(Note that* $\widehat{X}_n$ *is not a Markov chain.) The simulated values of* $\widehat{Z}_n$ *and* $\widehat{X}_n$ *are presented in Figure 8.3.*

*Finally, the stationary distribution of* $\widehat{Z}_n$ *is*

$$\widehat{\pi} = [0.3240, 0.0291, 0.2833, 0.1388, 0.0594, 0.1654],$$

*and we can see that all states are essential.*

*The probability mass function of* $\widehat{Z}_0, \widehat{Z}_1$, $F = [\widehat{\pi}_i \cdot \widehat{P}_{ij}]$, *is*

$$F = \begin{bmatrix} 0 & 0 & 0.2770 & 0.0008 & 0.0308 & 0.0154 \\ 0 & 0 & 0.0026 & 0.0013 & 0.0003 & 0.0249 \\ 0.2422 & 0.0014 & 0 & 0 & 0.0269 & 0.0127 \\ 0.0125 & 0.0125 & 0 & 0 & 0.0014 & 0.1124 \\ 0.0536 & 0.0003 & 0.0028 & 0.0027 & 0 & 0 \\ 0.0157 & 0.0149 & 0.0008 & 0.1340 & 0 & 0 \end{bmatrix}.$$

*Since F is asymmetric,* $\widehat{Z}_n$ *is not time reversible.*

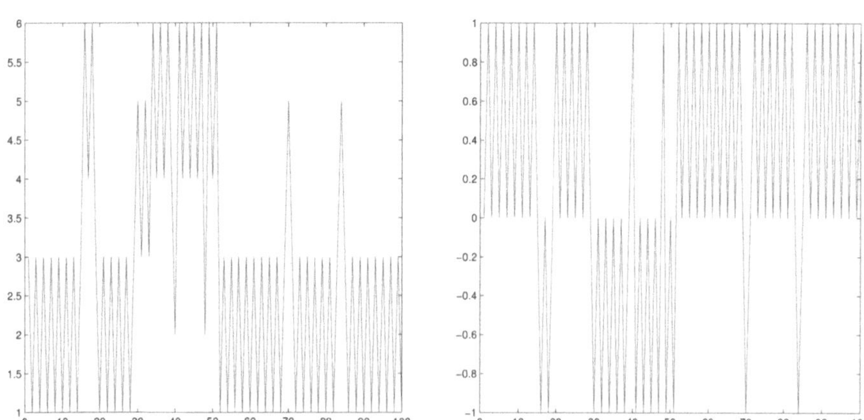


**FIGURE 8.3** The Markov chain $\widehat{Z}_n$ used in Example 8.3 (left) plot; the SMC $\widehat{X}_n = \widehat{G}(\widehat{Z}_n)$ with $\widehat{G}(1) = \widehat{G}(2) = 1$, $\widehat{G}(3) = \widehat{G}(4) = 0$, $\widehat{G}(5) = \widehat{G}(6) = -1$ (right) plot.

### 8.2.5 SWITCHING MARKOV CHAIN OF TURNING POINTS

The SMCTP belongs to the class of SMC, but since this special case has found applications in fatigue, we shall introduce some special notations here. Let the discretised load have $J$ levels.

Consider two regimes $r = 2$ and suppose that under a regime $s = 1,2$ the loads $X_n^s$ are MCTP. Then, there exists $(J-1) \times (J-1)$ dimensional upper and lower triangular matrices $P^{Mm1}, P^{Mm2}$ and $P^{mM1}, P^{mM2}$ such that the $P^s$ matrices are given by

$$P^1 = \begin{bmatrix} 0 & P^{Mm1} \\ P^{mM1} & 0 \end{bmatrix}, \quad P^2 = \begin{bmatrix} 0 & P^{Mm2} \\ P^{mM2} & 0 \end{bmatrix}.$$

Now, the rest is the same as for the general SMC given above.

## 8.3 OSCILLATION INTENSITY FOR MARKOV LOADS

In Section 4.8, we demonstrated the important equivalence (4.46) between the rainflow cumulative count intensity $\mu^{rfc}(u,v)$ and the intensity of oscillations across intervals $\mu^{osc}(u,v)$. By means of (4.31), the intensity of rainflow damage can therefore be estimated when the oscillation intensity (4.27) $\mu^{osc}(u,v)$ is known.

Here, we shall discuss computations of $\mu^{osc}(x_j,x_i;X_n)$ (note that for $x_{j+1} < u < x_j$ and $x_i < v < x_{i-1}$, $\mu^{osc}(u,v;X_n) = \mu^{osc}(x_j,x_i;X_n)$). For Markov chain the computation of $\mu^{osc}$ is particularly simple. However, often Markov loads $X_n = G(Z_n)$ are not Markov chain and one has to formulate a relation between the rainflow intensity of the Markov load $X_n$ by means of the oscillation intensity of $Z_n$. Consequently, we shall express $\mu^{osc}(x_j,x_i;X_n)$ using oscillations of $Z_n$.

In order to do so, a new concept of *oscillations intensity between sets* $\mu(U,V;Z_n)$, has to be introduced so that $\mu^{osc}(u,v;X_n) = \mu(U,V;Z_n)$, for $U = \{k \in S : G(k) < u\}$ and $V = \{k \in S : G(k) > v\}$. Of particular importance is the case when $u = x_j$ and $v = x_i$ when

$$\mu^{osc}(x_j,x_i;X_n) = \mu(U_j,V_i;Z_n), \tag{8.4}$$

where

$$U_j = \{k : G(k) < x_j\}, V_i = \{k : G(k) > x_i\}.$$

Characterisation of load variability by means of intensity of oscillations between sets was introduced in Beste et al. 1992 [16].

### 8.3.1 INTENSITY OF OSCILLATIONS BETWEEN SETS *U*, *V*

Let $Z_n$ be a time series taking only a countable (usually finite) number of values denoted by $S$. Without loosing generality, we assume that $S = \{1,\dots,N\}$.

For $Z_n$, $n = 0,1,\dots$, and $U$, $V$ two disjoint non-empty subsets of $S$, a crossing from $U$ to $V$ is said to start at time $n-k$ and end at $n$ $(0 \leq n-k < n)$ if the following conditions are satisfied:

1. $Z_{n-k} \in U$,
2. $Z_{n-k+l} \in W = S - U \cup V$ for $0 < l < k$,
3. $Z_n \in V$.

Let $\mathbf{1}_{UV}(n,n+k)$ be equal one if a crossing from $U$ to $V$ starts at $n$ and ends at $n+k$ and zero otherwise. Then, $\mathbf{1}_{UV}(n) = \sum_{k=1}^{\infty} \mathbf{1}_{UV}(n,n+k)$ is equal to one if there is a crossing from $U$ to $V$ starting at $n$.

Since there are no crossing overlaps the number of crossings from $U$ to $V$ that start in $\{0,1,\ldots,t\}$, denoted by $N_t(U,V)$, is equal to

$$N_t(U,V) = \sum_{n=0}^{t} \sum_{k=1}^{\infty} \mathbf{1}_{UV}(n,n+k) = \sum_{n=0}^{t} \mathbf{1}_{UV}(n).$$

For a stationary and ergodic time series $Z_n$, the intensity of $(U,V)$-oscillations is equal to the long-run average

$$\mu(U,V;Z_n) = \lim_{T\to\infty} \frac{1}{T} N_T(U,V) = \mathsf{E}[\mathbf{1}_{UV}(0)] = \sum_{k=1}^{\infty} \mathsf{E}[\mathbf{1}_{UV}(0,k)], \tag{8.5}$$

where

$$\mathsf{E}[\mathbf{1}_{UV}(0,1)] = \mathsf{P}(Z_0 \in U, Z_1 \in V),$$
$$\mathsf{E}[\mathbf{1}_{UV}(0,n)] = \mathsf{P}(Z_0 \in U, Z_j \in S - U \cup V \text{ for } 0 < j < n, Z_n \in V).$$

As before, for non-stationary $Z_n$, the long-run distribution of $Z_n$ can be used to evaluate the intensity. If $Z_n$ is a Markov chain, satisfying the general assumptions of this chapter, the long-run distribution means that the chain starts with the stationary distribution $\pi$.

### 8.3.2 ESTIMATION OF $\mu(U,V;Z_n)$ FOR A MARKOV CHAIN

As before, let $Z_0, Z_1, \ldots$, with values in $S$, be a Markov chain defined by the transition probabilities $P$ having the stationary distribution $\pi$, and let $F = [f_{ij} = \pi_i p_{ij}]$ be the Markov matrix of the stationary distribution of $(Z_n, Z_{n+1})$. We shall give formulas for $\mu(U,V;Z)$ and in order to simplify the presentation let us introduce some additional notation.

**Notation 8.1.** *a) We assume that for any subset $U \subset S = \{1,2,\ldots,N\}$, an order of elements is always specified, i.e. if $U$ has $k$ elements, then there is a function $u_i$ from $i \in \{1,2,\ldots,k\}$ onto $U$ so that the elements in $U$ can be ordered as $(u_1,u_2,\ldots,u_k)$. Since most of our results are independent of the order, we shall not distinguish in the notation between the set $U$ and the ordered sequence of its elements $(u_1,u_2,\ldots,u_k)$. However, in actual computations, the order is essential, and the complexity of calculations can depend on the particular choice of the order.*

*b) For any $N \times N$ matrix $R$, let $R(U,V)$ denote a sub-matrix of $R$, with elements defined by the orders of the sets $U$ and $V$. If the elements in $U$ and $V$ are ordered* *as follows* $(u_1, u_2, \ldots, u_k)$, $(v_1, v_2, \ldots, v_m)$, *then* $R(U,V)$ *is a* $k \times m$ *matrix with* $(i,j)$ *element given by* $R(u_i, v_j)$. *If any of the sets* $U, V$ *is empty, then* $R(U,V)$ *is a zero matrix with appropriate dimension.*

*c) Let* $\mathbf{1}$ *be an* $N \times 1$ *matrix of ones, i.e. a column vector of ones. Consequently,* $\mathbf{1}(U)$ *is a column vector of ones with the same dimension as* $U$. *The row vector* $\boldsymbol{\pi}$ *of stationary probabilities of the Markov chain* $Z$ *is a* $1 \times N$ *matrix, and hence* $\boldsymbol{\pi}(U)$ *is a row vector containing probabilities* $\pi(u_i)$ *for* $u_i \in U$. *(A simple consequence of this notation is the following formula:* $\mathsf{P}(Z_0 \in U) = \boldsymbol{\pi}(U)\mathbf{1}(U)$*).*

*d) Finally, for the vector of stationary probabilities* $\boldsymbol{\pi}$, *let* $\boldsymbol{\pi}(U,U)$ *be a matrix of the same dimension as* $U$ *with diagonal elements equal to the vector* $\boldsymbol{\pi}(U)$ *and with zeros outside the diagonal. Similarly, let* $\mathbf{1}(U,U)$ *be an identity matrix of the same dimension as* $U$.

Before we turn to transition between sets, $\mu(U,V) = \mu(U,V;Z)$, we introduce the intensity of transition between individual states, from state $i$ to state $j$, staying in a set $W$ in between. More precisely let, for any set $W \subset S$,

$$\begin{aligned}\mu_W(i,j) &= \mathsf{P}(Z_0 = i, Z_1 = j) \\ &\quad + \sum_{k=2}^{\infty} \mathsf{P}(Z_0 = i, Z_k = j, \text{ and } Z_n \in W \text{ for all } n = 1, \ldots, k-1). \qquad (8.6)\end{aligned}$$

If $U \cap V = \emptyset$ and $W = S - (U \cup V)$ then, summing over $i \in U, j \in V$, we obtain

$$\mu(U,V) = \sum_{i \in U, j \in V} \mu_W(i,j) = \mathbf{1}(U)^T \mu_W(U,V)\mathbf{1}(V).$$

To compute the elements in $\mu_W$, we need the following lemma, where $F_{ij}$ is the long-run (stationary) distribution of $Z_n, Z_{n+1}$, i.e.

$$F_{ij} = \pi_i \frac{\mathsf{P}(Z_0 = i, Z_1 = j)}{\mathsf{P}(Z_0 = i)} = \pi_i P_{ij}.$$

**Lemma 8.1.** *Let* $Z$ *be a Markov chain with transition matrix* $P$ *and stationary distribution* $\pi$. *Then, for any set* $W \subset S$ *and fixed states* $i, j \notin W$,

$$\begin{aligned}\mu_W(i,j) &= F(i,j) + F(i,W)[\mathbf{1}(W,W) - P(W,W)]^{-1} P(W,j) \\ &= F(i,j) + F(i,W)[\pi(W,W) - F(W,W)]^{-1} F(W,j). \qquad (8.7)\end{aligned}$$

*Proof.* We compute each term in the defining sum (8.6). Obviously, the first term is $F(i,j)$. Next, for any matrix $A$, let $A^0$ be the identity matrix of the same dimension as $A$. Using this notation,

$$\mathsf{P}(Z_0 = i, Z_2 = j, Z_1 \in W) = F(i,W)P(W,W)^0 P(W,j).$$

Similarly, for $k > 2$,

$$\begin{aligned}\mathsf{P}(Z_0 =& i, Z_k = j, \text{ and } Z_n \in W \text{ for all } n = 1, \ldots, k-1) \\ &= F(i,W)P(W,W)^{k-2} P(W,j),\end{aligned}$$

and hence

$$\mu_W(i,j) = F(i,j) + \sum_{k=2}^{\infty} F(i,W)P(W,W)^{k-2}P(W,j)$$
$$= F(i,j) + F(i,W)(I(W,W) - P(W,W))^{-1}P(W,j).$$

Now, $(I(W,W) - P(W,W))^{-1}P(W,j)$ is the solution to the linear equation system

$$X = P(W,j) + P(W,W)X, \tag{8.8}$$

which is equivalent to

$$\pi(W,W)X = F(W,j) + F(W,W)X, \tag{8.9}$$

which finishes the proof. □

Note that solutions of Equation (8.8) do not depend on which distribution for $Z_0$ is used; hence, $Z_n$ need not be stationary.

#### 8.3.2.1 Main Rainflow Markov Load Theorem

We are now ready to formulate the main rainflow Markov load theorem. Let $X_n = G(Z_n)$ be a stationary Markov load and $G$ a decreasing function, and set

$$U = \{k \in S; G(k) < u\}, \quad V = \{k \in S; G(k) > v\}, \quad W = S - U \cap V.$$

Then, for $u = x_j, v = x_i$.

$$\mu_{ji} = \mu(x_j, x_i) = \mathbf{1}(U)^T F(U,V)\mathbf{1}(V) \tag{8.10}$$

is the intensity of direct crossings from below $x_j$ to above $x_i$ by $X_n$, while

$$\mathscr{F}\mu_{ji} = \mathbf{1}(U)^T F(U,W)[\pi(W,W) - F(W,W)]^{-1}F(W,V)\mathbf{1}(V) \tag{8.11}$$

is the intensity of indirect crossings with landings in $W$.

**Theorem 8.1** (Main rainflow Markov load theorem). *Assume the general assumptions of this section.*
*a) If $U \cap V = \emptyset$ and $W = S - (U \cup V)$ then,*

$$\mu(U,V;Z) = \mu^{osc}(U,V;Z) = \sum_{i \in U, j \in V} \mu_W(i,j) = \mathbf{1}(U)^T \mu_W(U,V)\mathbf{1}(V)$$
$$= \mathbf{1}(U)^T F(U,V)\mathbf{1}(V)$$
$$+ \mathbf{1}(U)^T F(U,W)[\pi(W,W) - F(W,W)]^{-1}F(W,V)\mathbf{1}(V). \tag{8.12}$$

*b) For the Markov load $X_n = G(Z_n)$,*

$$\mu^{osc}(u,v;X_n) = \mu^{osc}(U_u, V_v; Z_n). \tag{8.13}$$

For future use, we write part (a), with reference to (8.10) and (8.11), as

$$\mu^{osc}(U,V;Z_n) = \mu + \mathscr{F}(\mu), \tag{8.14}$$

where $\mu$ is the suitably reduced part of the Markov matrix $F$.

**Corollary 8.1.** *If $X_n$ is a Markov chain of turning points with Markov matrix $F^{mM}$, the rainflow matrix is given by*

$$F^{rfc} = F^{mM} + \mathscr{F}(F^{mM}) = (I+\mathscr{F})(F^{mM}). \tag{8.15}$$

Equation (8.14) is the starting point to the inverse problem, i.e. to find $F$ when $\mu^{osc}$ of the Markov chain is known.

Finally, we comment on the assumption of time reversibility of the Markov chain. Consider a Markov chain $Z_n$ having joint probability mass matrix $F$. If $F_{ij} = F_{ji}$, then $Z_n$ is stationary, time reversible and $\pi_i = F_i = \sum_j F_{ij}$. However, if $\pi_i \neq F_i$ for some $i$, then $Z_n$ is not stationary, and we need to study its long-run crossing properties to compute its oscillations intensity. But this is equivalent to considering the stationary chain $Z_n$ having transition probabilities $P_{ij} = F_{ij}/F_i$ and $\mathsf{P}(Z_0 = i) = \pi_i$. The chain $Z_n$, $n = 0,1,\ldots N$, does not need to be time reversible.

Let us introduce the time reversed Markov chain $\widehat{Z}_n = Z_{N-n}$ ($N$ is large and we will let $N$ tend to infinity). Then, $\widehat{Z}_n$ is irreducible, has transition probabilities matrix $\widehat{P}$ satisfying

$$\pi_j \widehat{P}_{ji} = \pi_i P_{ij}, \qquad \widehat{F}_{ji} = F_{ij},$$

and has stationary distribution $\pi$. Consequently, $\widehat{F} = F^T$. Since $\mu(U,V)$ is the same for $Z_n$ and for the reversed chain $\widehat{Z}_n$, we notice that there can be two different asymmetric matrices $F$ and $\widehat{F} = F^T$ which will give the same $\mu(U,V)$. Consequently, Equation (8.14) ($\mu^{osc} = \mu + \mathscr{F}(\mu)$) cannot be uniquely solved for $\mu$ if only $\mu^{osc}$ is known. (Consult also Johannesson [92] for a more detailed discussion of this subject.)

However, since by (8.12),

$$\begin{aligned}\mu(U,V) &= \mathbf{1}(U)^T \widehat{F}(U,V)\mathbf{1}(V)\\ &\quad + \mathbf{1}(U)^T \widehat{F}(U,W)[\pi(W,W) - \widehat{F}(W,W)]^{-1}\widehat{F}(W,V)1(V),\end{aligned}$$

and hence, with $F^s = 0.5(F + \widehat{F})$,

$$\begin{aligned}\mu(U,V) &= \mathbf{1}(U)^T F^s(U,V)\mathbf{1}(V)\\ &\quad + \mathbf{1}(U)^T F^s(U,W)[\pi(W,W) - F^s(W,W)]^{-1} F^s(W,V)\mathbf{1}(V).\end{aligned} \tag{8.16}$$

Our claim is that $\mu^{osc} = \mu + \mathscr{F}(\mu)$ can be inverted under the additional assumption that $Z_n$ is time reversible.

**Example 8.4** (Continuation of Example 8.1). *Consider the Markov chain $Z_n$ defined in Example 8.1. Obviously, $F = 0.25 \cdot P$ and by (8.7), the oscillation intensity $\mu(\{1\},\{4\};Z_n) = \mu_{\{2,3\}}(4,1)$ is given by*

$$0.25 \cdot 0.1 + 0.25 \cdot \begin{bmatrix}0.9 & 0\end{bmatrix} \cdot \left( \begin{bmatrix}1 & 0\\ 0 & 1\end{bmatrix} - \begin{bmatrix}0 & 0.1\\ 0.1 & 0\end{bmatrix} \right)^{-1} \cdot \begin{bmatrix}0\\ 0.9\end{bmatrix} = 0.0455.$$

There exists an unexpected relation between the distribution of oscillations in a Markov chain and the currents in an electric circuit, see Rychlik (1999) [178]. This method was presented in Chapter 4 and used to evaluate oscillation intensities for random walk, and also in Appendix 5.A to Chapter 5 for the Gaussian Ornstein-Uhlenbeck process.

In this section, we presented an algorithm to evaluate the oscillations intensity $\mu^{osc}(u_i,v_i)$. Obviously, the oscillation intensity is equal to the expected rainflow count intensities $\mu^{rfc}(u_i,v_i)$, and the master-formula can be used to compute the expected rainflow damage. In this section, we consider loads that take a finite numbers of levels, i.e. oscillation and rainflow count intensities are $(N,N)$ matrices. Thus, we can use formula (4.32) to get rainflow cycles intensity $f_{ji}^{rfc}$, and then the intensity of rainflow damage is evaluated using the discrete master-formula, (4.33), i.e.

$$d^{rfc} = \sum_{ji} f_{ji}^{rfc} f(u_j,u_i).$$

A natural question is why not directly compute the $[f_{ji}^{rfc}]$ matrix instead of $\mu^{osc}(u_j,u_i)$. For discrete Markov chain it is possible to do so, and actually it was originally proposed by Rychlik 1988 [171], and Krenk and Gluver 1989 [110]. Bishop and Sherratt 1990 [17] proposed to use the Kowalewsky kernel (6.41) from Section 6.5.3.1 to define a Markov chain for the $(min,Max)$ pairs. However, for general Markov loads, it is easier to estimate the $\mu^{rfc}$ matrix and then use the master-formula to get damage intensity. This approach, for MCTP, was introduced by Frendahl and Rychlik 1993 [62].

**Example 8.5** (Rfc-distribution from MCTP). *We illustrate how one can obtain the rfc distribution $f_{M,m_{rfc}}$ from the Max-min distribution $f_{M,m}$ by means of an MCTP transformation. We chose a Gaussian load with Pierson-Moskowitz spectrum with $H_s = 7$ m and peak period $t_p$ 11 sec. We obtain the exact Max-min pdf $f_{M,m}$ by means of* WAFO *routines and the Kowalewski approximation $f_{M,m}^{Kow}$ by* (6.41)*. They form the transmission matrix (Markov matrix) for an MCTP model. From the Markov matrix, we computed the rainflow matrix (RFC cycle pdf).*

*The Markov and the RFC matrix are shown in Figure 8.4 with exact and approximate matrices in left and right columns.*

### 8.3.3 THE INVERSE PROBLEM

Since using Markov loads has many advantages – it is easy to simulate and compute the expected rainflow damage and the oscillation matrix $\mu_{ij}^{osc}$ – one would like to find the Markov chain $Z_n$ and the function $G$ so that $\mu^{osc}(x_j,x_i) = \mu_{ij}^{osc} = \mu(U_j,V_i;Z_n)$ as defined in (8.4).

In the case when one knows that $X_n = G(Z_n)$ is MC and the function $G$ is decreasing, then the distribution of $X_0, X_1$ and the intensity $\mu$ (defining the matrix $F$) can often be found by means of the fixed point algorithm

$$\mu = \mu^{osc} - \mathscr{F}(\mu),$$

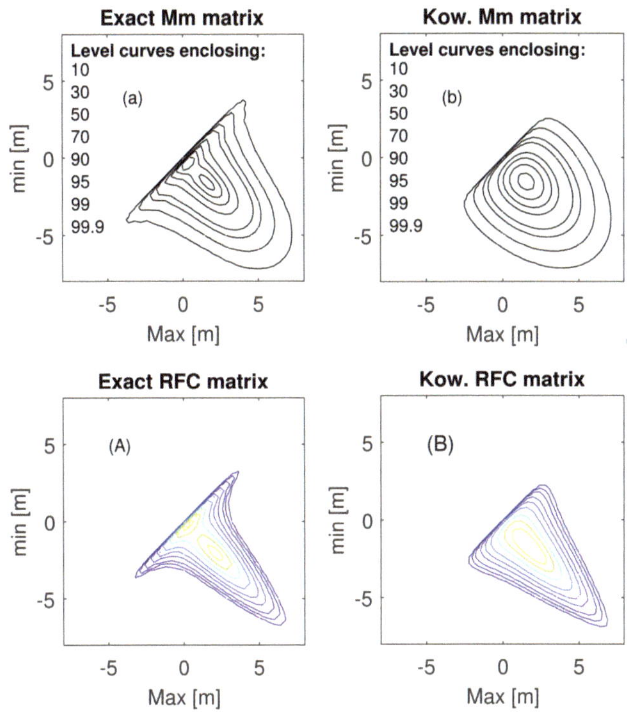


**FIGURE 8.4** Top row shows (a) exact and (b) Kowalewski approximation of pdf for Max-min cycles. Bottom row shows pdf for rfc-cycles by MCTP based on Max-min pdf. (A) and (B) based on (a) and (b).

see (8.14). Often, one has to filter the matrix $\mu$ by setting intensity of crossings of small intervals to zero, i.e. letting some few off diagonal elements to be equal zero.

An alternative to use a Markov load model for the complete load process is to model the sequence of turning poins as a Markov chain, an MCTP. Then, the relation between the Markov matrix $F^{mM}$ and the rainflow matrix $F^{rfc}$ is given by (8.15),

$$F^{rfc} = F^{mM} + \mathscr{F}(F^{mM}) = (I + \mathscr{F})(F^{mM}),$$

with the inverse

$$F^{mM} = (I + \mathscr{F})^{-1}(F^{rfc}). \tag{8.17}$$

**Remark 8.4.** *An iterative algorithm to solve* (8.17) *was presented by Rychlik 1996 [175]. An exact inversion algorithm from the full sequence of measured oscillations to the full sequence of turning points was presented by Dreßler et al. 1997 [58].*

**Example 8.6** (Inversion of the rainflow matrix)**.** *This example is taken from the original article by Rychlik 1996 [177] on the inversion of the rainflow matrix. The load* *is a Duffing oscillator*

$$\ddot{x}(t) + 2\zeta\dot{x}(t) - x(t) + \alpha x(t)^3 = \dot{w}(t), \quad (8.18)$$

*with* $\zeta = 0.25$, $\alpha = 0.5$, *and* $\dot{w}$ *Gaussian white noise. The sample paths of* $x(t)$ *oscillate between two distinct equilibria, and therefore the cycle ranges are either small-/moderately small or large, when the process jumps between the equilibria.*

*Figure 8.5, upper row, shows part of a simulation (a), and the observed rainflow cycles (b) together with contour lines of the smoothed rainflow matrix* $F^{rfc}$.

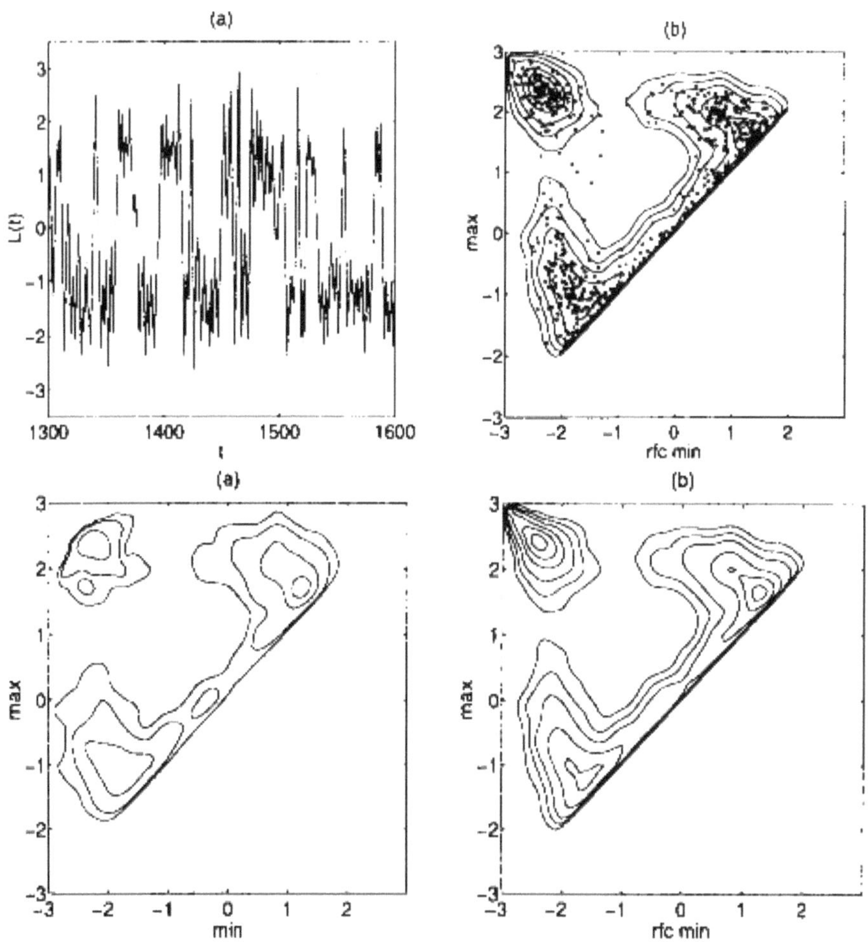


**FIGURE 8.5** Upper row: (a) Simulated sample path of a Duffing oscillator, (b) its rainflow cycles and rainflow matrix $F^{rfc}$. Lower row: (a) Reconstructed Markov matrix $F = (I+\mathscr{F})^{-1}(F^{rfc})$. (b) Corresponding rainflow matrix $G = (I+\mathscr{F})(F)$. (Reproduced with permission from [177]. Copyright © 1996 Elsevier.)

*The rainflow matrix* $F^{rfc}$ *was inverted by means of the function* $(I+\mathscr{F})^{-1}$, *to give a Markov matrix* $F=(I+\mathscr{F})^{-1}(F^{rfc})$, *which is shown as (a) in the lower row in the figure. The function* $(I+\mathscr{F})$ *returns the rainflow matrix, and the result* $G=(I+\mathscr{F})(F)$, *shown as (b) in the lower row.*

*The result is not exactly the same as the original, owing to the smoothing – the smoothed rainflow matrix cannot be inverted to a Markov matrix.*

# 9 Laplace Moving Average (LMA)

## 9.1 INTRODUCTION

The stationary Gaussian load process has many advantages over general time series models and Markov models, in the ease by which it is parametrised by its power spectral density and its stability in linear dynamic systems. Available numerical tools for computation of fatigue damage are easy to use. The main disadvantage is its "normality" that restricts its use to symmetric load distributions with modest extreme tails. In Chapter 10, we will present a number of ad hoc methods to mitigate that problem by the transformation of the load signal.

In this chapter, we will present the Laplace moving average (LMA) load, a non-normal model that takes better control of the more extreme parts of the distribution, while it keeps the spectral properties of the Gaussian model. The deviation from normality is governed by an excess kurtosis parameter $\kappa_e = \kappa - 3$, where 3 is the kurtosis in the normal distribution. In a hybrid model, it will be possible to combine the Gaussian load with occasional short but dangerous episodes of Laplace loads without destroying the spectral structure. Several examples will show the flexibility of the Laplace model that considerably widens the class of usable load models.

LMA as a versatile model for fatigue loads was presented by Åberg, Podgórski and Rychlik (2009) in [2], from which the basic LMA-model in this book is taken, together with a major simulation experiment. Applications and extensions of the technique are also taken from [111] and [21]. Estimation for such models has been presented in [159]. Introduction to road surface roughness models has been initiated in [21]. Theory for the LMA-model can be found in [104] and [1] with multivariate extensions in [109]. These models have been applied to multivariate fatigue analysis in [100]. Natural extensions to asymmetric loads were discussed in [6] and are also treated in this chapter.

To avoid possible language barriers, we first give some details on notation.

**Notation 9.1** (Fourier and inverse Fourier transform). *The power spectral density $S(\omega)$ and covariance function $r(\tau)$ of a stationary random load $X(t)$ form a Fourier transform pair. In Section 2.2.2.3, we used the name-standard common in probability theory that the Fourier transform generates a time-function from a frequency representation,*

$$\text{Fourier transform:} \qquad r(\tau) = \int_{-\infty}^{\infty} e^{i\tau\omega} S(\omega)\,\mathrm{d}\omega,$$

$$\text{Inverse Fourier transform:} \qquad S(\omega) = \frac{1}{2\pi}\int_{-\infty}^{\infty} e^{-i\omega\tau} r(\tau)\,\mathrm{d}\tau.$$


DOI: 10.1201/9781003019947-9

*In applications, signal theory, random vibration [215], etc., other conventions are used. In Definition 9.1, we will introduce the Fourier transform pair $g(t), \widehat{g}(\omega)$ to generate Gaussian and Laplace moving averages:*

$$\begin{aligned} \textit{Fourier transform:} \quad & \widehat{g}(\omega) = \int_{-\infty}^{\infty} e^{i\omega t}\, g(t)\,\mathrm{d}t, \\ \textit{Inverse Fourier transform:} \quad & g(t) = \frac{1}{2\pi}\int_{-\infty}^{\infty} e^{-i\omega t}\widehat{g}(\omega)\,\mathrm{d}\omega, \\ \textit{Parceval/Plancherel's theorem:} \quad & \int_{-\infty}^{\infty} |g(t)|^2\,\mathrm{d}t = \frac{1}{2\pi}\int_{-\infty}^{\infty} |\widehat{g}(\omega)|^2\,\mathrm{d}\omega. \end{aligned} \tag{9.1}$$

## 9.2 GAUSSIAN MOVING AVERAGE, GMA-LOAD

In fatigue applications, one often use stationary Gaussian processes to model unpredictably varying loads. The loads are defined by the Fourier pair power spectrum or covariance function. A particularly useful property of Gaussian loads is that the output of a linear filter with Gaussian input is a Gaussian process too.

A stationary Gaussian load is defined by its expected value and its power spectrum $S(\omega)$, $-\infty < \omega < \infty$. As before, we write $\omega_k = \int_{-\infty}^{\infty} \omega^k S(\omega)\,\mathrm{d}\omega$ for the spectral moments. The moments $\omega_0, \omega_2$ are the variance of the process and of its derivative, respectively. The mean-level upcrossing frequency for a Gaussian process is

$$\mu_0^+ = \frac{1}{2\pi}\sqrt{\frac{\omega_2}{\omega_0}}.$$

For simplicity of presentation, we assume that the process has mean zero and variance one, which means that the spectrum is normalised to have $\omega_0 = \omega_2 = 1$.

Given power spectrum $S(\omega)$, the Gaussian load can be well approximated by a sum of independent harmonics having Rayleigh distributed amplitudes and uniformly distributed phases,

$$X^{appr}(t) = \sum_{k=1}^{N} A_k \cos(\omega_k t - \phi_k), \tag{9.2}$$

with frequencies $\omega_k = k\Delta\omega, k = 1,\ldots,N$; see Section 2.2.4.1 for more details.

An alternative way to generate Gaussian processes is to filter Gaussian white noise to get a Gaussian moving average (GMA). In an approximate sense, a general moving average process is the convolution of a *kernel function* $g(t)$ with an infinitesimal "white noise" process having variance equal to the discretisation step, say $dt$. The kernel $g(t)$ is normalised so that its square integrates to one, which is equivalent to saying that the variance of $X(t)$ is one. In particular, the standardised (mean zero and variance one) Gaussian moving average (GMA) $X(t)$ can be approximated for $t \in [-T,T]$ by

$$X_N(t) \approx \sum_{i=-N}^{N} g(t - t_i)\sqrt{\Delta t}\, Z_i, \tag{9.3}$$

where $t_i = i\Delta t$ are discretisation points and $Z_i$ are independent standard normal variables. The choice of appropriate length of the increment $\Delta t$ is related to the smoothness of the kernel, while $N$ is chosen so that the approximation is accurate at the endpoints of $[-T, T]$. For example, if the kernel is restricted (or truncated) to the interval $[-S, S]$, one should take $N \geq (T + 4S)/\Delta t$ and remove the initial and final part of the generated series.

**Definition 9.1** (Gaussian Moving Average). *The Gaussian moving average $X(t)$, $-\infty < t < \infty$, is the limit of $X_N(t)$ in* (9.3) *as $T, N$ both tend to infinity with $\Delta t = T/2N \to 0$. It is a stationary Gaussian process with mean zero. Since the $Z_i$ are independent normalised variables, the spectrum of $X(t)$ can be computed using the spectrum rule* (2.55) *for a linear filter,*

$$S_X(\omega) = \frac{1}{2\pi} |\widehat{g}(\omega)|^2, \tag{9.4}$$

*where $\widehat{g}(\omega)$ stands for the Fourier transform of the kernel $g(t)$.*

*Necessary conditions for the existence of the limits $X(t)$ and the derivative $\dot{X}(t)$ are that the spectral moments are finite; i.e. by* (9.4) *and* (9.1),

$$\omega_0 = \int_{-\infty}^{\infty} S_X(\omega)\, \mathrm{d}\omega = \int_{-\infty}^{\infty} g(t)^2\, \mathrm{d}t < \infty, \tag{9.5}$$

$$\omega_2 = \int_{-\infty}^{\infty} \omega^2 S_X(\omega)\, \mathrm{d}\omega = \int_{-\infty}^{\infty} \dot{g}(t)^2\, \mathrm{d}t < \infty, \tag{9.6}$$

*respectively. Then,* $\mathsf{Var}[X(t)] = \omega_0$ *and* $\mathsf{Var}[\dot{X}(t)] = \omega_2$.

*In the following, we consider standardised loads, i.e. when the variance of the process and its derivative $\omega_0 = \omega_2 = 1$.*

**Remark 9.1.** (Symmetric and asymmetric kernels). *Note that* (9.4) *does not allow for unique identification of the kernel $g$. Only if we limit ourselves to kernels symmetric around zero, i.e. such that $g(-t) = g(t)$, they can be defined through their Fourier transform by*

$$\widehat{g}(\omega) = \sqrt{S_X(\omega) 2\pi}. \tag{9.7}$$

*Obviously, there are many asymmetric kernels that give the same spectrum, and hence the same Gaussian load. Of special interest among these are the causal kernels, i.e. those with $g(t) = 0$ for $t < 0$.*

### 9.2.1 KERNELS FOR THE LINEAR GAUSSIAN OSCILLATOR

The linear damped oscillator $X(t)$ driven by white noise has a power spectrum

$$S_X(\omega) = m^{-2} \frac{1}{(\omega_0^2 - \omega^2)^2 + 4\alpha^2\omega^2}, \tag{9.8}$$

and it is a solution to the standardised differential equation,

$$\ddot{x}(t) + 2\zeta\omega_0\dot{x}(t) + \omega_0^2 x(t) = m^{-1}\dot{W}(t), \tag{9.9}$$

where $\dot{W}(t)$ is white noise. Here, we will consider Gaussian and Laplace white noise as “driving force”.

Equation (9.7) can be used to find the symmetric kernel $g_1(t)$, say. The causal, asymmetric, kernel is given by the filter's impulse response function,

$$g_2(t) = m^{-1}\widetilde{\omega}_0^{-1}e^{-\alpha t}\sin(\widetilde{\omega}_0 t), \quad t \geq 0,$$

and zero otherwise, with $\alpha = \zeta\omega_0$, $\widetilde{\omega}_0 = \omega_0\sqrt{1-\zeta^2}$. The parameter $\omega_0$ is the undamped angular frequency, while $\zeta$ is the damping ratio. The Fourier transform of $g_2(t)$ is

$$\widehat{g_2}(\omega) = m^{-1}\frac{1}{\omega_0^2 - \omega^2 + 2\zeta i\omega_0\omega}.$$

The moving averages defined by means of the symmetric kernel $g_1$ and the causal $g_2$ have the same spectrum and hence GMA processes defined using the two kernels are equivalent, or rather statistically identical.

**Example 9.1.** *Figure 9.1 illustrates the symmetric and causal kernels needed to generate the Gaussian oscillator process in a system with damping $\zeta = 0.1$, undamped frequency $\omega_0 = 1.012$ and mass $m = 2.75$. The results are shown in Figure 9.2.*

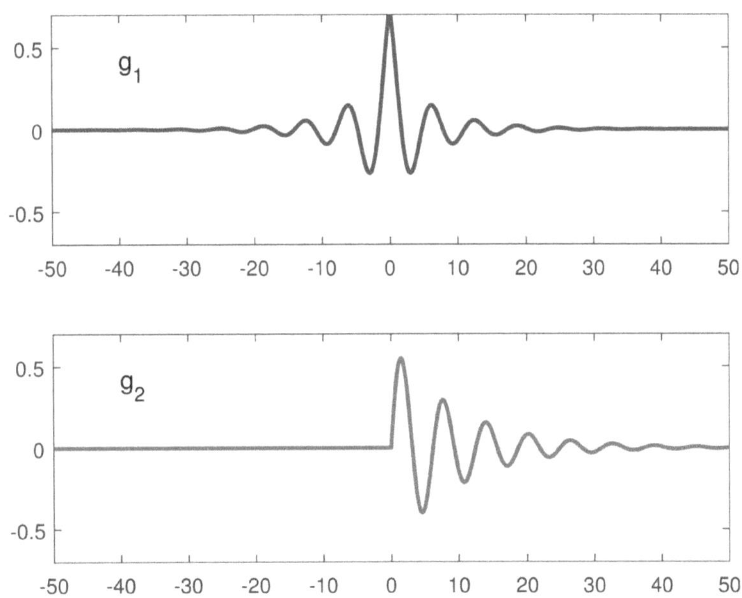


**FIGURE 9.1** Symmetric ($g_1$) and causal ($g_2$) kernels for the Gaussian moving average in Example 9.1.

## 9.3 LAPLACE MOVING AVERAGE, LMA-LOAD

### 9.3.1 THE NEED FOR ALTERNATIVES TO GAUSSIAN LOADS

The statistical properties of a stationary Gaussian load model are entirely determined by the frequency distribution of its total energy, as expressed by the power spectrum

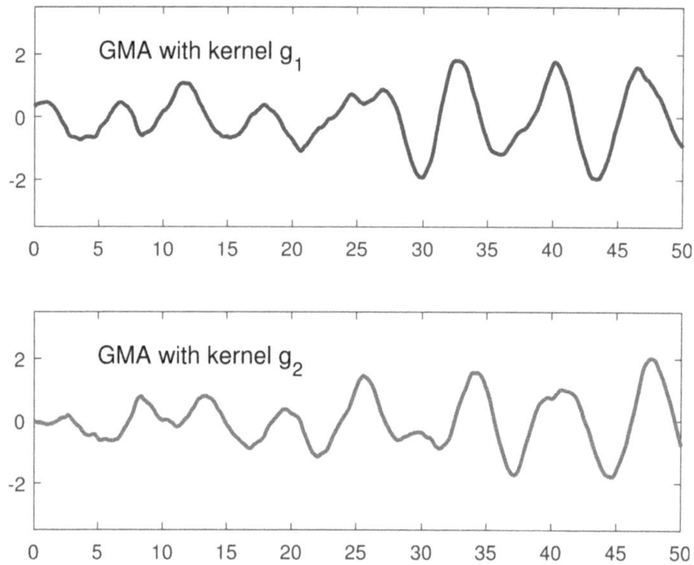


**FIGURE 9.2** Simulated GMA for Example 9.1 with symmetric ($g_1$) and causal ($g_2$) kernels.

density $S(\omega)$. The Fourier representation as a sum of independent smooth random harmonics together with the Gaussian marginal and multivariate distributions makes them ideal for loads in linear systems excited by environmental loads, like sea or wind loads, under moderately irregular conditions. In the previous section, we presented the GMA-load model as a tool to generate Gaussian loads with prescribed energy spectrum, (9.4) and (9.7).

Under certain environment conditions, the Gaussian load model will be too optimistic, since it will not catch extraordinary, abnormal disturbances in an otherwise "normal" irregular load. Then, the LMA, the LMA-load, offers an alternative. By a simple modification of the GMA-model (9.3), it can generate a load process with skewed marginal distribution and with heavier tails than in the normal distribution.

The skewness and kurtosis measures $\kappa_3$ and $\kappa_4$ were defined in (2.6), as measures of asymmetric and general heavy tails, respectively. For clarity, we will use the notations $\gamma = \kappa_3$ and $\kappa = \kappa_4$, respectively. For the LMA-process, the important measure will be the excess kurtosis $\kappa_e = \kappa_4 - 3$ that measures the tail heaviness above normality. When the kurtosis of a load exceeds level 3, it may have larger rainflow cycles than the Gaussian model predicts. The LMA is a generalisation of the Gaussian model in the sense that it contains the Gaussian case as a limit when its kurtosis approaches the value 3.

We will first deal with the symmetric LMA model in some detail. By a rather small modification of the construction, one can generate skewed Laplace loads. Details about such models are found in Section 9.5.

Applications of LMA processes, in particular in wave models, can be found in [65, 158, 159, 163]. Technical details on general LMA models can be found in [104].

### 9.3.2 THE EFFECT OF SKEWNESS AND KURTOSIS ON FATIGUE

Non-Gaussian skewness and kurtosis and unusual spectral content can have systematic effect on fatigue damage and fatigue damage prediction, see [218]. Åberg et al. [2] studied the effect of skewness and kurtosis on the accumulated damage rate and came to the conclusion that for fixed skewness, the damage increases with kurtosis, while for fixed kurtosis, the damage decreases when the absolute value of the skewness decreases. This result is valid when damage function is symmetric in the rainflow minima and maxima.

Wang and Sun (2005) [209] studied the effect of mean stress and noted that one can, by including the mean stress in the damage function, get a damage that is an asymmetric function of skewness with the highest damage for some positive value of the skewness parameter.

As the kurtosis increases, the damage will also increase. However, this effect can to some extent be reduced by also increasing the skewness. Therefore, a Laplace model with small kurtosis and relatively large skewness will give rise to similar damage as a Gaussian model. In studied examples, this could be explained by the fact that the rainflow range distribution is less dependent on skewness than the joint distribution of maximum and rainflow minimum. Our conclusion is that it is not a good idea to use a Gaussian models when the load has large kurtosis.

Further, a stationary Gaussian model may severely underestimate the damage rate if the load contains "transients", e.g. road surfaces with potholes which may introduce transient responses (vibrations) when a vehicle passes a pothole. Such loads can be modelled in various ways, e.g., one can add potholes to the Gaussian model located randomly according to a Poisson point process. This type of model will be named the *hybrid model* and is considered in Section 9.3.5.

We will describe several types of Laplace models for load with more than normal kurtosis. The first model, the standard LMA model, is a generalisation of the GMA with a kernel $g(t)$ chosen to give a certain predetermined spectrum $S(\omega)$. In the second model, the kernel represents independent random transients that show up in the load at random occasions. A hybrid model is a mixture of a transient LMA model and a GMA or a standard LMA. The last model is the asymmetric LMA with non-zero skewness and positive excess kurtosis.

Before we go to details, we present some visual and numerical differences between the three basic moving average models, the Gaussian, the symmetric Laplacian and the asymmetric Laplacian, in particular regarding the rainflow cycles.

**Example 9.2** (Visual and numerical differences). *We continue with the oscillator model from Example 9.1 but limit the analysis to the symmetric kernel $g_1$ in Figure 9.1. The load is the output of a resonant system excited by Gaussian, Laplace and skewed Laplace noise. We get three loads, all with spectrum* (9.8)*, variance and variance of the derivative equal to one, i.e. the spectrum is standardised so that spectral moments $\omega_0 = \omega_2 = 1$. By definition, the mean of the LMA is zero, see (9.41).*

*Thus, we have three models: GMA with skewness $\gamma = 0$ and excess kurtosis $\kappa_e = 0$, symmetric LMA with $\gamma = 0$, $\kappa_e = 7.8$, and asymmetric LMA with $\gamma = 1.3$, $\kappa_e = 8$. We*

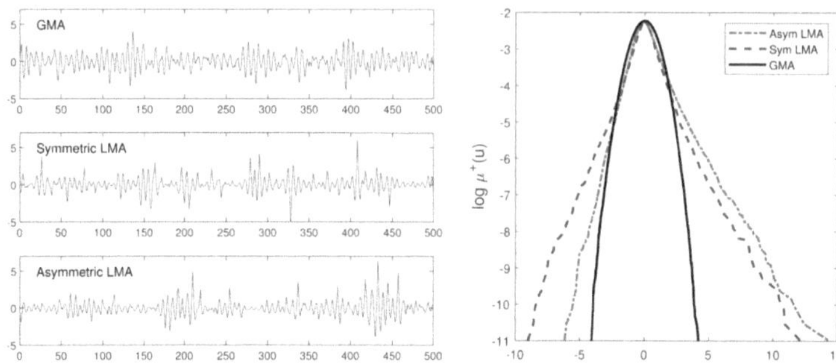


**FIGURE 9.3** Samples and logarithmic crossing spectra $\log \mu^+(u)$ for GMA, symmetric LMA and asymmetric LMA in Example 9.2.

*simulated 25 000 seconds of loads with sampling frequency 100 Hz. Parts of the loads are shown in Figure 9.3. A characteristic feature of the shown LMA loads is that they have large extremes not present in the GMA, and that the shape of the transient responses is visually dominating. Rainflow plots for the three loads are shown in Figure 9.4.*

*In the symmetric case, the amplitudes of transients can be positive and negative with the same distribution of their absolute values. However, the shape of the transient $g_1$, see Figure 9.1, dies out fast and hence the extreme maxima will often (but not always) be paired with local minima that are smaller in absolute value. Obviously, also extreme minima may be paired with less extreme maxima.*

*For symmetric loads, the result is that the rfc cycle count gets a wider spread around the diagonal with increasing range. This is not the case for Gaussian loads where extreme maxima are paired with extreme minima, which in this narrow-band case are of similar size.*

*The third rainflow cycle plot represents variability of skewed LMA. Here, positive skewness implies that the intensity of high positive transients amplitudes is much higher than the size (in absolute value) of the negative transients, and as a consequence, the plot becomes bent.*

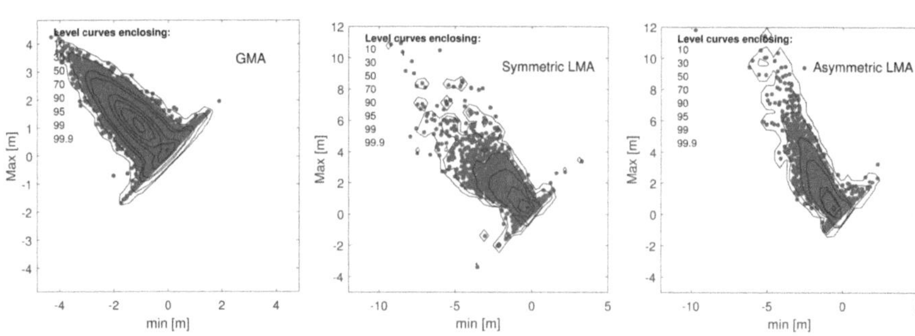


**FIGURE 9.4** RFC for the three oscillating loads in Example 9.2.

**Table 9.1**
**Damage Intensity and the Upper Bound $d^+$ (4.41)**

| | GMA | Symmetric LMA | Asymmetric LMA |
|---|---|---|---|
| rfc est, $\beta = 1$ | 0.402 | 0.344 | 0.367 |
| $d^+, \beta = 1$ | 0.402 | 0.344 | 0.367 |
| rfc est, $\beta = 3$ | 4.59 | 8.58 | 8.67 |
| $d^+, \beta = 3$ | 4.8 | 10.6 | 9.16 |
| rfc est, $\beta = 4$ | 19.3 | 79.9 | 74.6 |
| $d^+, \beta = 4$ | 20.4 | 117 | 81.5 |
| rfc est, $\beta = 5$ | 89.4 | 968 | 832 |
| $d^+, \beta = 5$ | 95.4 | 1124 | 907 |

*The rainflow damage intensities are estimated from the long simulations and are shown in Table 9.1 in rows 1, 3, 5 and 7 for the damage exponent $\beta$ = 1, 3, 4, 5 and three loads. In rows 2, 4, 6, and 8, we compare the estimated rainflow damages with the upper bound $d^+$ (4.41) for the rainflow damage intensity $d^{rfc}$. For LMA loads with symmetric crossing intensity, we could use the Holm-de Maré damage intensity $d^{HM}$, which is equal to $d^+$ for LMA with skewness zero. However, for skewed LMA, $d^{HM}$ becomes a very conservative bound, for example $d^{HM}(5) > 2.5\,d^{rfc}(5)$.*

### 9.3.3 SYMMETRIC LAPLACE MOVING AVERAGE

#### 9.3.3.1 Generation and Definition

In the simplest terms, a symmetric LMA model is obtained by replacing the discretisation step $\Delta t$ in the definition of Gaussian moving average defined in (9.3) by positive random variables which squared all have mean $\Delta t$. This is done using independent gamma random variables with shape $k > 0$ and scale $\nu > 0$. The parameters $k$ and $\nu$ are chosen in such a way that LMA will have the same mean and variance as GMA but have excess kurtosis $\kappa - 3 > 0$ proportional to $\nu$.

**Definition 9.2** (Laplace distributed r.v.). *A standard gamma variable $K$ with shape parameter $k > 0$ has density function*

$$f_K(x) = \frac{x^{k-1}e^{-x}}{\Gamma(k)}, \quad x > 0, \tag{9.10}$$

*where $\Gamma(k)$ is the gamma function. It has mean and variance both equal to $k$, $\mathsf{E}[K] = \mathsf{Var}[K] = k$.*

*Further, if $Z$ is a standard Gaussian variable, independent of $K$, then $\Lambda = \sqrt{K}Z$ has the generalised Laplace distribution with shape parameter $k$ and characteristic* *function*

$$\phi_\Lambda(u) = \left(1 + \frac{u^2}{2}\right)^{-1/k}. \tag{9.11}$$

A symmetric LMA process with kernel $g(t)$ and kurtosis $\kappa > 3$ will be approximated in similar way as GMA (9.3) with $\sqrt{\Delta t}Z_i$ replaced by Laplace distributed variables $\sigma\Lambda_i$ with the same mean square, $\sigma^2\mathsf{E}[\Lambda_i^2] = \Delta t$ and with gamma-shape parameter $k$ depending on an "excess-kurtosis" parameter $\nu$, defined by (9.15).

The asymmetric LMA process will be introduced in Section 9.5. It needs an extra parameter, in the literature denoted $\mu$, to govern the degree of skewness.

**Definition 9.3** (Symmetric LMA). *The LMA process $X(t)$, $-\infty < t < \infty$, with kernel $g(t)$ defined on $[-S,S]$ and excess-kurtosis parameter $\nu$ and scale parameter $\sigma$, is the limit, as $N,T$ tend to infinity and $\Delta t \to 0$, of the approximation*

$$X_N(t) = \sum_{i=-N}^{N} g(t-t_i)\Lambda_i\sigma, \quad -T \le t \le T, \tag{9.12}$$

*with $t_i = i\Delta t$, $N \ge (T+4S)/\Delta t$ and $\Lambda_i = \sqrt{K_i}Z_i$. Here, the gamma-shape parameter $k = \Delta t/\nu$ tends to zero with $\Delta t$.*

#### 9.3.3.2 Basic Properties

It can be shown [2] that the symmetric LMA $X(t)$ is a stationary process with mean zero and variance

$$\mathsf{Var}[X(t)] = \frac{\sigma^2}{\nu}\int g(t)^2\,\mathrm{d}t. \tag{9.13}$$

Its excess kurtosis $\kappa_e = \kappa - 3$ is related to the excess-kurtosis parameter $\nu$,

$$\kappa_e = 3\nu\frac{\int g(t)^4\,\mathrm{d}t}{\left(\int g(t)^2\,\mathrm{d}t\right)^2} > 3, \tag{9.14}$$

$$\nu = \kappa_e\frac{\left(\int g(t)^2\,\mathrm{d}t\right)^2}{3\int g(t)^4\,\mathrm{d}t}. \tag{9.15}$$

Thus, if we want to generate a standardised LMA with prescribed kurtosis $\kappa$, we have to choose $\nu$ by (9.15) and the scale parameter $\sigma = \sqrt{\nu}$, by (9.13).

The spectrum and spectral moments of the standardised LMA follow the same rules as those for GMA. Thus, $S(\omega)$ is given by (9.4) regardless of the value of $\nu$. The variance of $\dot{X}(t)$ is equal to $\int \dot{g}(t)^2\,\mathrm{d}t$, according to (9.6).

If not stated otherwise, we assume in the future that the kernels are normalised and have

$$\int g(t)^2\,\mathrm{d}t = \int \dot{g}(t)^2\,\mathrm{d}t = 1. \tag{9.16}$$

**Summary of notation and properties of symmetric LMA:** The parameters in the approximation $X_N(t)$ and the limiting LMA $X(t)$ are linked together as follows:

- The normalised kernel $g(t)$ determines the spectrum and the dynamics of the load. The GMA and LMA processes (9.3), (9.12) are zero-mean stationary processes having the same power spectrum, as given by (9.4).
- The discretisation $\Delta t$ determines the time resolution of the approximation.
- The excess-kurtosis parameter $\nu$ determines how "heavy" the tails of the load distribution are.
- Scale parameter $\sigma = \sqrt{\nu}$ gives $\mathsf{Var}[X(t)] = 1$.
- The time step $\Delta t$ and $\nu$ determine the gamma shape parameter $k = \Delta t/\nu$.
- In the approximation, the random $\Lambda_i = \sqrt{K_i}Z_i$ are built by iid gamma variables $K_i$ with shape $k$ and iid standard normal variables $Z_i$.
- The $N \geq (T+4S)/\Delta t$ is large enough to make the approximation accurate after removal of initial and final data.

**Remark 9.2** (Non-timereversible LMA). *Note that given a kernel $g$, the Gaussian and LMAs have the same spectrum. Spectrum and kurtosis equal 3 defines the GMA. Similarly, for timereversible LMA with symmetric kernel $g(-t) = g(t)$, defined in (9.7), the LMA is defined by kurtosis and spectrum. If we accept non-timereversible models, there are many asymmetric kernels that give the same spectrum but which have different value of the integral $\int g(t)^4\,dt$, and therefor also different kurtosis.*

**Example 9.3.** *Consider the oscillator in Example 9.1 loaded by Gaussian white noise. In Figure 9.1, the symmetric $g_1$ and causal kernels $g_2$ are shown. The integrals $\int g_i(t)^4\,\mathrm{d}t$ are not equal as seen in Table 9.2. The integral for the symmetric kernel is equal 0.25, while the causal kernel has 0.15. The two Gaussian loads in Figure 9.2 have the same statistical properties.*

**Table 9.2**
**$\int g(t)^4\,\mathrm{d}t$ Is Different for Symmetric and Causal Kernel**

| | | |
|---|---|---|
| | 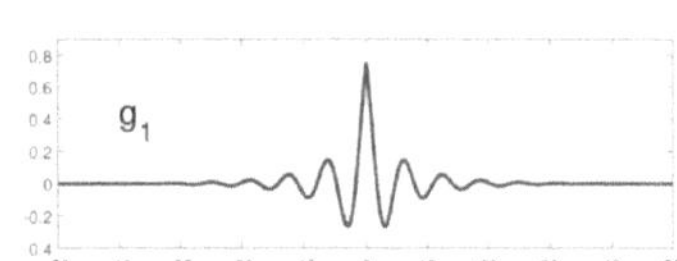  | 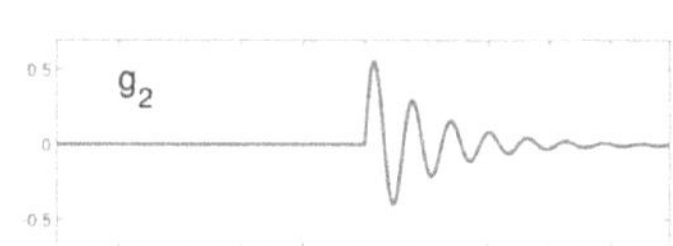  |
| $\int g(t)^4\,\mathrm{d}t$ | 0.25 | 0.15 |

**Example 9.4** (LMA with oscillator spectrum). *We want to generate an LMA-process with the same spectrum as in Example 9.1 and with kurtosis $\kappa = 10$. Formula (9.15) gives $\nu_1 = 9.34$ for the symmetric kernel and $\nu_2 = 15.6$ for the causal kernel.*

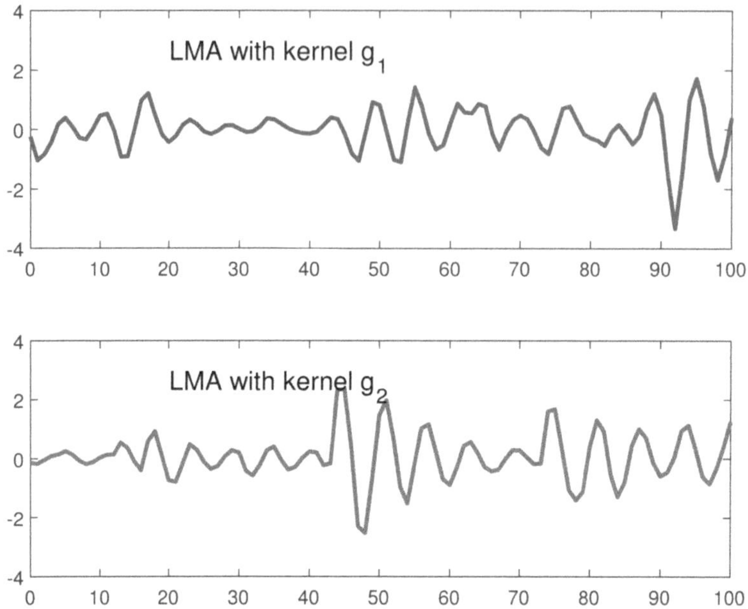


**FIGURE 9.5** Simulated LMA for Example 9.4 with $g_1$ and $g_2$ kernels.

*Figure 9.5 shows parts of LMA-loads for the two kernels $g_1$ and $g_2$, simulated with the correct $\nu$-values, 9.34 and 15.6, respectively. Figure 9.6 shows histograms of observed values in the full 5000-samples-long loads, together with the observed kurtosis for the samples, which are quite close to the target value 10.*

*To illustrate the effect of choosing the wrong $\nu$-values for the two kernels, we interchanged the $\nu$-values and simulated 100 replicates of the LMA with kernels $g_1, g_2$ and with correct and wrong values. The table shows the average kurtosis for the 100 replicates for each of the combinations. The combinations in the diagonal are almost correct (standard errors in parenthesis).*

| *kernel* | $\nu = 9.34$ | $\nu = 15.6$ |
|---|---|---|
| $g_1$ | *9.91(0.22)* | *14.5(0.39)* |
| $g_2$ | *7.2(0.14)* | *9.85(0.24)* |

*As mentioned in Remark 9.2, a symmetric kernel gives timereversible loads, a causal kernel does not. The $g_2$-load in Figure 9.5 seems to be clearly non-timereversible, confirming the claim. In fact, for high kurtosis values, the kernel shape will be visible in the LMA, leading to non-timereversibility for causal kernels.*

*The shape of the causal kernel $g_2$ as well as simulations give the impression that LMA with kernel $g_2$ is more narrow-band than the LMA with kernel $g_1$. This is more visible in the rainflow cycle plots derived from simulations of the two LMA processes. We return to this in the section on crossing intensities.*

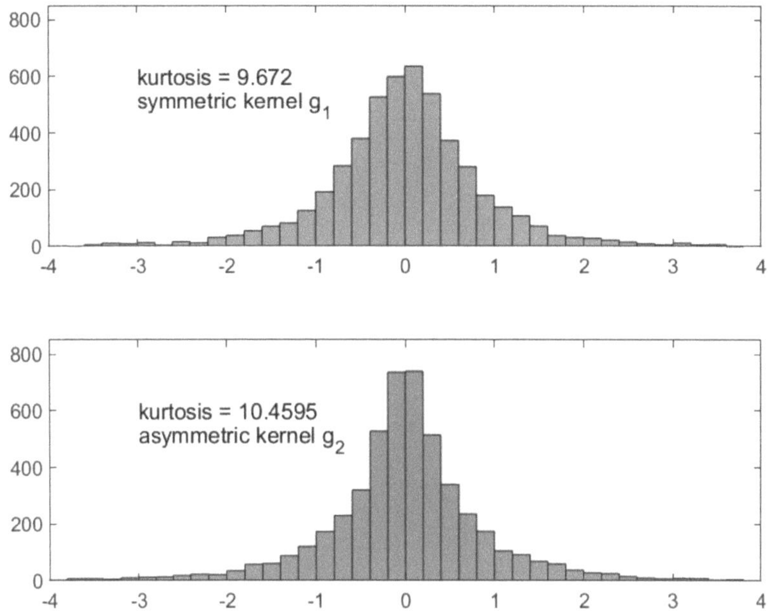


**FIGURE 9.6** Histograms for 5000 sample points of the LMA in Example 9.4, target kurtosis $\kappa = 10$.

**Remark 9.3.** *For a given spectrum $S(\omega)$, there are many kernels that can be used to construct an LMA with that spectrum, and the choice of the kernel is a separate issue. We have restricted ourselves to choose between a symmetric and a causal kernel, denoted $g_1(t)$, $g_2(t)$. As noted in Example 9.4, these require different $\nu$-values in the simulation model; allowing non-causal asymmetric kernels widens the possibilities further. Having kernel and kurtosis the LMA is uniquely defined, since the parameter $\nu$ can be computed using (9.15).*

**Remark 9.4** (LMA in a linear filter). *A convenience with the class of Laplace driven moving average models is its invariance under the passage through a linear filter. Let $h(x)$ be the impulse response of a linear filter. A filtered LMA with kernel $g$ is again an LMA but now with kernel $h * g$, i.e. the convolution between $h$ and $g$. This means e.g. that if one can compute crossing intensities for the original LMA, and thereby obtain an upper bound* (4.43) *for the damage, one can also do it in the same manner for linearly filtered Laplace MA.*

*Note that when the process is filtered, the spectrum is multiplied by $|H(\omega)|^2$ as for Gaussian loads but the parameter $\nu$ defining $\Lambda_i$ remains unchanged. However, kurtosis of the filtered process changes since it is a function of the kernel which in the filtered process is $h * f$.*

#### 9.3.3.3 Distribution and Characteristic Function

Unfortunately, no explicit expression is known for the marginal distribution of LMA, but its pdf can be expressed as the inverse Fourier transform of its characteristic function, c.f.,

$$\phi(u) = \mathsf{E}[e^{iuX}] = \int_{-\infty}^{\infty} e^{iux} f_X(x)\,\mathrm{d}x,$$
$$f_X(x) = \frac{1}{2\pi}\int_{-\infty}^{\infty} e^{-iux}\phi(u)\,\mathrm{d}u. \tag{9.17}$$

For a symmetric LMA without trend, with standardised kernel $\int g(t)^2\,\mathrm{d}t = 1$, and excess kurtosis $\kappa_e = \kappa - 3$, the distribution depends on the fourth power of $g(t)$ through the parameter

$$\nu = \frac{\kappa - 3}{3\int g(t)^4\,\mathrm{d}t}.$$

The distribution of the LMA in (9.12) has the characteristic function

$$\phi_X(u) = \exp\left\{-\frac{1}{\nu}\int_{-\infty}^{\infty} \log\left(1 + \frac{\sigma^2 u^2 g(t)^2}{2}\right)\mathrm{d}t\right\}, \tag{9.18}$$

where $\sigma^2$ can be replaced by $\nu$ for a standardised load.

Note that the $\nu$-parameter is dependent on the type of kernel even if the difference between symmetric and asymmetric kernel can be small; see Example 9.5 and Table 9.3.

For this and more details on the LMA distribution, see [2]. We will use the characteristic function in Section 9.4.3.2 to evaluate Rice's formula for level crossings. In the next example, we utilise simulations to find the distributions.

**Table 9.3**
**High Quantiles for GMA and Four LMA for the Oscillator in Example 9.4**

| | 95% | 99% | 0.999% | 95% | 99% | 0.999% |
|---|---|---|---|---|---|---|
| | Symmetric kernel $g_1$ | | | Asymmetric kernel $g_2$ | | |
| GMA | 1.648 | 2.332 | 3.092 | 1.647 | 2.334 | 3.088 |
| $\kappa = 4$ | 1.635 | 2.508 | 3.710 | 1.628 | 2.501 | 3.709 |
| $\kappa = 8$ | 1.553 | 2.799 | 4.919 | 1.564 | 2.817 | 4.895 |
| $\kappa = 12$ | 1.504 | 2.941 | 5.583 | 1.505 | 2.949 | 5.523 |
| $\kappa = 16$ | 1.462 | 2.980 | 5.835 | 1.471 | 3.120 | 6.281 |

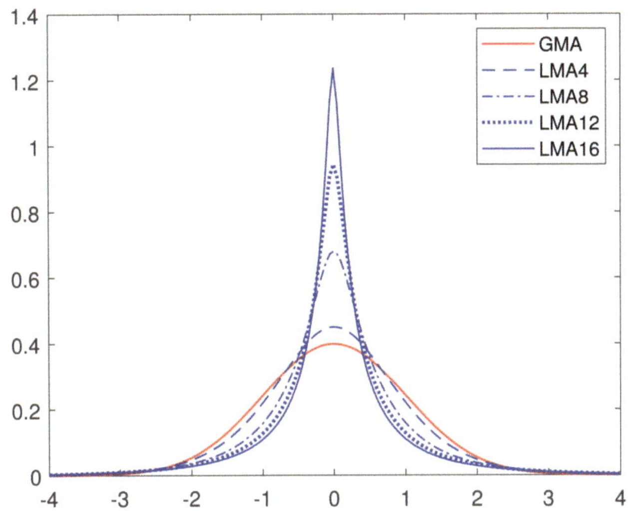


**FIGURE 9.7** Pdf of GMA and LMA with the narrowing of Pdf with increasing $\kappa$.

**Example 9.5** (Effect of high kurtosis). *What are the consequences of changing the value of the parameter* $\nu$*, in particular, how are the amplitudes affected?*

*Figure 9.7 shows the central part of the distribution for one GMA and four LMAs with kurtosis* $\kappa = 4, 8, 12, 16$. *The narrowing of the pdf with increasing* $\kappa$ *is clearly visible. The kernel is the symmetric "Cauchy-kernel",* $g(t) = 2/\sqrt{\pi}/(1+(2t)^2)$, *with the exponential spectrum* $S(\omega) = 0.5\exp(-|\omega|)$*; see also* (9.34).

*Table 9.3 shows the increase in extreme quantiles for the harmonic oscillator for the symmetric and asymmetric kernels* $g_1, g_2$. *Note that all five load distributions have variance (and standard deviation) one.*

*The effect can be easiest understood from an alternative definition of Laplace noise which will represent the LMA is an infinite sum of transients all having shape of a kernel g. This model is presented in Section 9.3.4.*

**Remark 9.5.** *When modelling fatigue loads, it can happen that it is not kurtosis that is the essential property to be correctly modelled. For example, in the model of road surface irregularity, it may be important to describe the frequency and sizes of the large "chocks" or "transients" which are due to potholes or other obstacles. The LMA can also be used for this purpose and hence we will concentrate now on occurrence of transients in LMA. It will be shown that the parameter* $\nu$ *has also another role, not only to determine kurtosis, but also to describe sizes of transients. This will be shown in the next section where we give another representation for LMA showing that this process essentially represents the effects of jumps (shocks) as the resulting load is the sum of transients caused by the jumps. The transients have the shape of the kernel that is scaled by the jump sizes.*

### 9.3.4 LMA AS A SUM OF SORTED TRANSIENTS

#### 9.3.4.1 An Alternative Definition of LMA

The approximations GMA and LMA are sums of transients with independent, identically distributed amplitudes, $g(t-t_i)Z_i\sqrt{dt}$ and $g(t-t_i)\Lambda_i\sqrt{\nu}$, respectively, at the discretisation points $t_i$; where $\Lambda_i = \sqrt{K_i}Z_i$. $\Lambda_i = \sqrt{K_i^*}Z_i$. However, as opposed to the Gaussian case, the random scaling $K_i$ can be large which makes $g(t-t_i)$ visible as transients in the moving average process.

Following [2, 24, 111], we will present another representation of LMA as a sum of randomly located transients with amplitudes stochastically ordered, from the largest to successively vanishing of minute order. We call that model the *sorted transient LMA*. Bondesson (1982) [24] used shot-noise processes as a tool to simulate random variables from infinitely divisible distributions. The interest in [2, 111] lies in the randomness of the occurrence and size of large transients.

To modify the approximation $X_N(t)$ (9.12) of an LMA process $X(t)$, for $0 \le t \le T$, we will use a sorted transient approximation, as a sum of a potentially infinite number of transients. Keeping in mind how (9.13) allows the use of $\nu$ as an excess-kurtosis parameter and, simultaneously, $\sqrt{\nu}$ as a scale parameter, the transient LMA process is defined as the sum,

$$X(t) = \sum_{i=1}^{\infty} g(t-U_i)\sqrt{\nu K_i^*}Z_i \tag{9.19}$$

$$= \lim_{N\to\infty} \sum_{i=1}^{N} g(t-U_i)\sqrt{\nu K_i^*}Z_i. \tag{9.20}$$

Here, $U_i$ are independent, uniformly on $[0,T]$, distributed locations of transients with shape $g(t)$ and dependent sizes $Z_i\sqrt{\nu K_i^*}$. The random variables $\{K_i^*\}_{i=1}^{\infty}$ are defined as a stochastically decreasing sequence as follows:

$$K_i^* = W_i e^{-\nu\eta_i/T}, \tag{9.21}$$

where $W_i$ are iid standard exponential variables independent of $Z_i$ and $U_i$. Finally, the decrease is governed by the variables $\eta_i$, which are the location of the $i$-th point in a Poisson process, i.e. $\eta_i = \sum_{j=1}^{i} G_j$, where $G_j$ are independent standard exponential random variables, independent of all previously introduced variables.

Approximation (9.20) of LMA resembles the one in (9.12), but it differs in two aspects. First, the $t_i$'s that are regularly spaced over $[0,T]$ have been replaced by the $U_i$'s which are independently and uniformly distributed over the same interval. Second, the independent sizes of transients $\sqrt{K_i}Z_i$ at regularly spaced and ordered points $t_i$ have been replaced by dependent amplitudes $\sqrt{K_i^*}Z_i$ at random points $U_i$. Their sizes are statistically decreasing with increasing index $i$. But note that the $U_i$'s are not ordered, so the size of the transients is not ordered over time.

As for the GMA and LMA (9.12), the width $[-S,S]$ of the kernel will affect the initial and final parts of the simulated approximating time series, which therefore should be removed in simulation experiments.

**Remark 9.6.** *Note that the only parameters in the approximation (9.20) are the interval length $T$, the kernel/kurtosis-dependent parameter $\nu$ and the number $N$ of used transients. In computer simulations, the transients are located at points $U_i$ that can be selected from a large fixed grid of equidistant points at locations $t_k = k\,\Delta t$, i.e. at a random permutation of $N$ points selected at random with equal probabilities from the grid.*

#### 9.3.4.2 Influential and Non-Influential Transients – Choice of $N$

The infinite sum (9.19) converges, and the terms decrease at a geometric rate in average. In this section, we will verify this statement and also investigate the approximation error if the sum is truncated after a finite number $N$ of transients.

We first consider the average amplitude of the individual term

$$\sqrt{\nu K_i^*}Z_i = \sqrt{\nu W_i}\prod_{j=1}^{i} e^{-\nu G_j/2T} Z_i.$$

Since all involved variables are independent, the evaluation is straightforward,

$$\begin{aligned}\mathsf{E}[\sqrt{\nu K_i^*}|Z_i|] &= \sqrt{\nu}\,\mathsf{E}[\sqrt{W_i}] \times \prod_{j=1}^{i}\mathsf{E}[e^{-\nu G_j/2T}] \times \mathsf{E}[|Z_i|]\\ &= \frac{\sqrt{\nu/2}}{(1+\nu/2T)^i} \approx \sqrt{\nu/2}\,e^{-\nu i/2T}.\end{aligned} \tag{9.22}$$

The average amplitude of transients is thus exponentially decaying at a rate determined by the parameter $\nu$. An LMA process with large $\nu$ is therefore locally dominated by a few transients with very large amplitudes, while low values of $\nu$ mean that the sizes of transients are less variable and that the LMA is closer to a Gaussian model.

**Example 9.6** (LMA with oscillator spectrum, cont.). *This example illustrates the rapid decrease in transient amplitudes, thereby separating influential and non-influential transients. We take the linear oscillator as reference model and use the same symmetric and causal kernels $g_1, g_2$ as in Example 9.4. The goal is an LMA with kurtosis $\kappa = 10$, which requires the parameter $\nu$ equal to 9.36 and 15.83, for symmetric and causal kernels, respectively. Note that these $\nu$-values are slightly larger than those used in Example 9.4, which is due to a stricter truncation of the spectrum, here at $|\omega| > 5$.*

*In this example the load duration $T$ is $10^4$ seconds, i.e. almost three hours of load. In a Gaussian load, there would on average be about 2000 rainflow cycles. We may think that three hours is the assumed length of a mission in harsh environment (that is why we have chosen such high kurtosis value). Then in order to schedule a plan for inspections, one needs both the average damage growth rate and the risk for extreme damage increments. For this, we simulated 1000 missions and looked on the difference between the symmetric and causal kernels, both for average damage growth and for the distribution of damage increments per mission.*

*The sampling frequency was 10 Hz, which means that the sample step* $\Delta t = 0.1$ *seconds, and there is (on average) 50 data points between consecutive local maxima. Consequently, by (9.12), one needs* $N = T/\Delta t = 10^5$ *independent Laplace distributed random numbers* $\Lambda_i$ *in (9.12) to simulate a sample of the LMA load for* $0 \le t \le T$. *This is also the number* $N = 10^5$ *of transients we chose to use to generate the LMA load by means of (9.20). We will show that this is far too much, and here it is sufficient to have* $N^{tr} = 10^4$ *transients placed at random at the* $10^5$ *possible locations.*

*We first illustrate the decreasing amplitudes* $\sqrt{\nu K_i^*} Z_i$ *of the available transients. Figure 9.8 shows one replicate of simulated amplitudes for the two kernels. In the top plot,* $\nu = 9.42$ *giving kurtosis 10 for the symmetric kernel* $g_1$*, while in the bottom plot we have amplitudes when* $\nu = 15.83$ *for the causal kernel* $g_2$*. We can see that the "essential" amplitudes (not negligible) are fewer in the bottom plot. In the plots, the solid lines show the expected absolute values of the transient amplitudes, (9.22).*

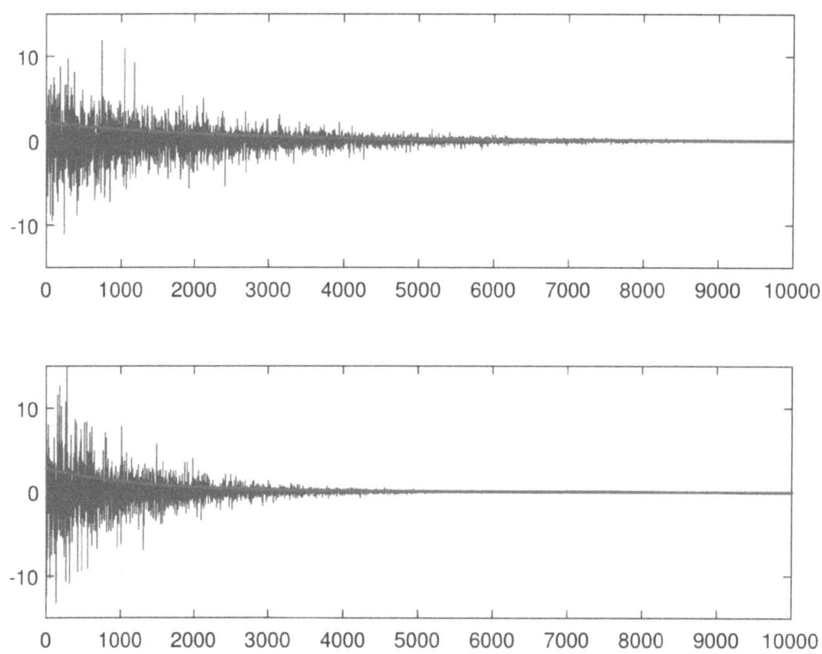


**FIGURE 9.8** Amplitudes for available transients and expected absolute amplitude. Upper plot: Kernel $g_1$, lower plot: $g_2$.

The transient LMA process $X(t)$ is the infinite sum (9.19). As seen in the example, transients in the series rapidly become negligible, which makes it possible to limit the number of transients to $N^{tr} < \infty$ without loosing the accuracy of the approximation. This is an important property of the transient LMA. We introduce an additional parameter for the number of transients per time unit $\mu = N^{tr}/T$. In the example, it has a value $\mu = 10$ Hz.

Let us use the variance of

$$X_N(t) = \sum_{i=1}^{N} g(t-U_i)\sqrt{\nu K_i^* Z_i}$$

to find the intensity of transients $\mu = N/T$ needed to get the fraction $1-p$ of the total variance of $X(t)$ (9.19). We start with the variance of individual transients,

$$\mathsf{Var}\left[g(t-U_i)\sqrt{\nu K_i^* Z_i}\right] = \frac{\nu}{T}\left(\frac{T}{T+\nu}\right)^i \int_0^T g(t-s)^2\,\mathrm{d}s,$$

with $\mathsf{Var}[X(t)] = \int_0^T g(t-s)^2\,\mathrm{d}s = 1$, which is just the variance of LMA and GMA. Now if one removes the influential transients, the remainder has variance

$$\mathsf{Var}[X(t) - X_N(t)] \approx \left(1 - \frac{\nu}{T+\nu}\right)^N \approx e^{-\nu\mu}, \tag{9.23}$$

for large $T$.

Thus, $p = \exp(-\mu)$ is the fraction of the total process variance that is not accounted for by the first $N = \mu T = -\log(p)/\nu$ transients. As a consequence, the smaller the $\nu$, the more terms you need in order to account for $(1-p)100\%$ of the total load variance. In Example 9.6, we generated $N = 10^5$ transients over $T = 10^4$. To account for 99% of the variation, only $N = -10^4 \log 0.01/\nu$ = 4920 and 2909, respectively, had been needed.

### 9.3.5 THE HYBRID MODEL WITH GMA BASE

In some applications, it can be hard to find a single type of model that can reproduce occational large variations in environment conditions. Then, one can use a mix of GMA and LMA models. The result, called a hybrid model, has been used to model road surface roughness; Bogsjö et al. (2012) [21].

In that application, GMA is used to model an "undamaged" road surface to which one is adding "potholes" and other types of irregularities. Since such an approach is physically appealing, we propose to alter the LMA model so that it will resemble this general scheme. The model will be called the hybrid model and is a mixture of a GMA, $Z_0(t)$, and a transient LMA, $X_n(t)$,

$$X(t) = Z_0(t) + \sum_{i=1}^{n} g(t-U_i)\sqrt{\nu K_i^* Z_i}, \quad 0 \le t \le T, \tag{9.24}$$

where $n$ is much smaller than the number of simulated values of LMA, $N = 1 + T/dt$. The frequency of potholes, called "essential" transients, will be denoted by $\mu = n/T$.

The process $Z_0(t)$ is a GMA (9.3),

$$Z_0(t) = \sqrt{p}\sum_{i=1}^{N} g(t-t_i)\sqrt{dt} Z_i. \tag{9.25}$$

Under our assumption that $\int g(t)^2 dt = 1$ the variance of $Z_0(t)$ is $p$. Consequently, the variance of the transient LMA-part, $1-p$. This is estimated as follows.

Let us use the variance of

$$X_n(t) = \sum_{i=1}^{n} g(t-U_i)\sqrt{\nu K_i^*}Z_i$$

to find the intensity of transients $\mu = n/T$ needed to get the fraction $1-p$ of the total variance of $X(t)$ (9.24). We start with the variance of individual transients,

$$\mathsf{Var}\left[g(t-U_i)\sqrt{\nu K_i^*}Z_i\right] = \frac{\nu}{T}\left(\frac{T}{T+\nu}\right)^i \int_0^T g(t-s)^2\,\mathrm{d}s,$$

and $\mathsf{Var}[X(t)] = \int_0^T g(t-s)^2\,\mathrm{d}s = 1$, which is just the variance of LMA and GMA. Now, if one removes the influential transients, then, as in (9.23),

$$\mathsf{Var}[X(t) - X_n(t)] = \left(1 - \frac{\nu}{T+\nu}\right)^n \approx e^{-\nu\mu}. \tag{9.26}$$

Since $p = \exp(-\nu\mu)$ and hence the intensity of transients $\mu = -\log(p)/\nu$, the smaller the $\nu$, the more terms you need to account for $(1-p)100\%$ of the total load variance.

**Example 9.7** (LMA with oscillator spectrum, cont.). *Let us again consider the response of a damped oscillator having the symmetric kernels $g_1$ and the asymmetric causal kernel $g_2$. The response will be modelled as the hybrid process (9.24). In that model, part of signals variability is described by a Gaussian process (9.25) having spectrum $S(\omega)$ and variance p. If $p = 0$, it is an LMA process, and if $p = 1$ ($\mu = 0$), it becomes GMA. We are interested in how the damage rate depends on the variance p. Obviously, the Gaussian part of this model is independent of kernel type, but LMA is not.*

*As in the previous example, we assume that the LMA component has kurtosis 10 and the defining parameter $\nu$ equal 9.4 and 15.8, for symmetric and causal kernel, respectively. The 200 missions each three hour long (described in the previous example) were used to estimate the damage rates for $\beta = 3,4,5$.*

*In Figure 9.9, the results for the $g_1$ kernel are shown with solid line, while dashed lines show results for causal kernel $g_2$. On the x axis, we have the proportion p. Having p and $\nu$, we can compute the rate of transients $\mu = -log(p)/\nu$ if $p > 0$, and if $\mu = 0$, then $p = 0$. From the figure, we can see that estimates of kurtosis $\kappa$ and damages $d^{rfc}$ are decreasing functions of p; however, for $p \leq 0.3$, the hybrid and LMA give similar damage rates. Recall that the parameter $\nu$ for LMA is a function of excess kurtosis and kernel g,*

$$\kappa_e(LMA) = 3\nu\int g(t)^4\,\mathrm{d}t.$$

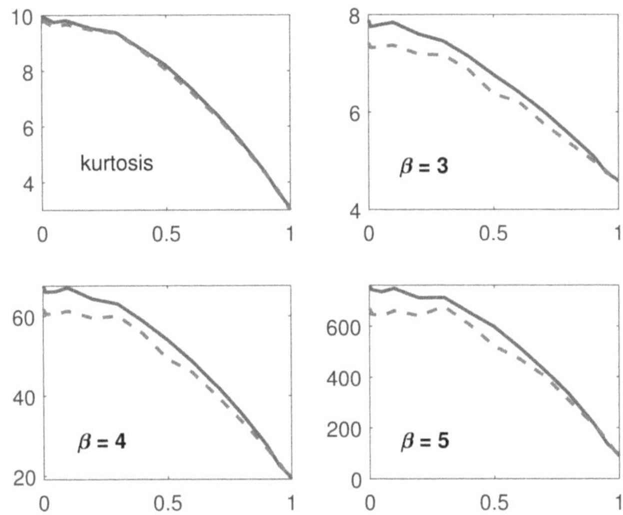


**FIGURE 9.9** Upper left plot: Kurtosis for hybrid model as function of GMA proportion $p$. Next three plots: Damage intensities for hybrid linear oscillator. On the horizontal axis $p =$ the proportion of the Gaussian part. Solid line: $g_1$-kernel. Dashed line: $g_2$-kernel.

## 9.4 COMPUTATION OF DAMAGE INTENSITY FOR LMA

### 9.4.1 SCALE INVARIANT MODELS

Power spectral density (psd) is an important characteristic of stationary responses. For stationary Gaussian responses, the fatigue damage rate is a function of psd alone. Even for LMA, the power spectrum remains an important characteristic, together with the parameters $\sigma$ and $\nu$, as e.g., in (9.12) and (9.13).

In many applied areas, it is common to use special parametric families of spectra to model load processes. For example, one assumes a frequency/time-scale invariant family of models with spectral densities satisfying the relation

$$S_a(\omega) = a S(a\omega), \quad a > 0, \tag{9.27}$$

with a normalised standard spectrum $S(\omega)$ generated by a standard kernel $g(t)$. The kernel is scaled as

$$g_a(t) = g(t/a)/\sqrt{a}. \tag{9.28}$$

**Example 9.8** (Some frequency-scale invariant spectra). *Scale invariant spectra like (9.27) are often used to describe environmental loads, e.g. the Pierson-Moskowitz spectrum for* sea surface elevation,

$$S(\omega) = h_s^2 \frac{5}{\omega_p} \left(\frac{\omega_p}{\omega}\right)^5 \exp(-\frac{5}{4}(\omega_p/\omega)^4),$$

*where the scale* $a \sim 1/\omega_p$ *is basically the average wind speed 19.5 m above sea surface. The ISO spectrum for encountered* road surface roughness *[86] is also of scale invariant type, with scaling parameter equal to vehicle velocity.*

*The Davenport spectrum for* wind gustiness *[46] is obtained as white noise of unit standard deviation filtered by a Davenport filter [69]. This model is used for wind turbulence causing dangerous slow random vibration on bridges and high buildings as well as fast small-scale vibration on radio astronomical antennas, [69]. The Davenport spectrum depends on average wind speed* $v$ *and terrain roughness and is given by*

$$S_v(\omega) = c_0 \frac{\beta\omega}{(1+\beta^2\omega^2)^{4/3}}, \qquad \omega > 0,$$

*with a space scale* $c_0$ *and a time/frequency scale* $\beta = 600/\pi v$.

*An exponential spectrum for stress and bending moments on a cultivator frame was used in [111], and results from that study will be presented later on.*

*All the mentioned standard spectra contain a multiplicative constant c that relates to a physically interpretable quantity, e.g., the significant wave-height factor* $H_s^2$ *in the Pierson-Moskowitz spectrum.*

In the following, we sometimes will write $g(t) = g_1(t)$ to emphasise that it is a fixed chosen kernel. Obviously, for a normalised spectral density $S(\omega)$, one can find many kernels, but only one symmetric. Each of them could be chosen as $g_1$, the one defining the family. We will next show the advantage of using the scale invariant family of psd (kernels) for computing the damage intensity in a non-Gaussian model.

### 9.4.2 LMA IN A FREQUENCY/TIME INVARIANT FAMILY

It follows from (2.47) that Gaussian moving averages, generated by spectra in a frequency invariant family $S_a(\omega) = a\,S(a\omega)$, only represent time-scaled versions of each other,

$$\mathsf{Cov}[X_a(s), X_a(t)] = \mathsf{Cov}[X(s/a), X(t/a)].$$

As a consequence, fatigue damage intensity, mean damage per time unit, scales with the factor $a$.

While a stationary Gaussian process is completely defined by its second-order covariance structure, the LMA is more complex, with its focus on skewness and kurtosis, third and fourth order moments. How do the parameters in the LMA definition Definition 9.3 change and how is the LMA affected by a time scale modulation of the kernel $g(t)$? And perhaps most important, how does the damage intensity change?

We start with the standardised kernel $g(t)$ satisfying

$$\int_{-\infty}^{\infty} g(t)^2 \, \mathrm{d}t = 1, \quad \int_{-\infty}^{\infty} g(t)^4 \, \mathrm{d}t = g_4,$$

to get the relation between excess kurtosis and the excess kurtosis parameter $\nu$,

$$\kappa_e = 3\nu g_4. \tag{9.29}$$

The symmetric LMA load $X(t)$ is defined as a limit of

$$X_N(t) = \sum_{i=-N}^{N} g(t-t_i)\Lambda_i\sigma, \quad -T \leq t \leq T, \tag{9.30}$$

with $t_i = i\Delta t$, and $\Lambda_i = \sqrt{K_i}Z_i$. The gamma-variables $K_i$ have parameter $k = \Delta t/\nu$, and the $Z_i$ are standard normal. The variance is $\mathsf{Var}[X(t)] = \int g(t)^2\,\mathrm{d}t = \sigma^2/\nu$.

Now, we change kernel to $g_a(t) = g(t/a)/\sqrt{a}$ and construct an LMA $X_a(t)$ with the same variance and kurtosis as the standardised $X(t)$.

**Theorem 9.1** (LMA as a scale invariant family). *The family of LMA with spectrum $S(\omega)$ and kernel $g(t)$ is frequence/time scale invariant in the parameter $\nu$ and the damage intensity scales as*

$$d_\beta^{rfc}(X_a) = d_\beta^{rfc}(\nu/a)/a, \tag{9.31}$$

*where $d_\beta^{rfc}(s)$ is the rainflow damage intensity of LMA having kernel $g_1(t)$ and the LMA scale $s = \nu/a$.*

*Proof.* With $g_a(t) = g(t/a)/\sqrt{a}$, one finds $\int g_a(t)^2\,\mathrm{d}t = \int g(t)^2\,\mathrm{d}t = 1$ and $\int g_a(t)^4\,\mathrm{d}t = 1/a\int g(t)^4\,\mathrm{d}t = g_4/a$. Thus, by (9.29), the new LMA scale is $\nu_a = a\nu$, and the space scale in (9.30) is $\sigma_a = \sqrt{a}\sigma$. □

Like for the GMA, a frequency/time transformation of an LMA does not change the cycle structure, only the damage per time unit.

We conclude this section with some consequences of the scale invariance properties. We use the exponential spectrum for road roughness classifications, where $a$ is the velocity of a vehicle travelling on the road, while $S(\omega)$ depends on the linear filter that is used to model responses and the spectral properties of a road profile, see [21].

**Example 9.9** (LMA with exponential spectrum). *In this example, we demonstrate the very convenient LMA $X_{a,\sigma}(t)$ having the exponential spectrum*

$$S_a(\omega) = 0.5\sigma^2 a\exp(-a|\omega|), \quad a > 0. \tag{9.32}$$

*The spectral moments are*

$$\omega_0 - \sigma^2 = \mathsf{Var}[X_{\sigma,a}(t)], \quad \omega_2 = 2\sigma^2/a^2 = \mathsf{Var}[\dot{X}_{\sigma,a}(t)].$$

*We assume $\sigma = 1$ and analyse the role of the frequency/time scale $a$.*

*There is a single-parameter family of kernels for which moving average process will have the spectrum (9.32). The kernels depend on the real parameter $-1 < b < 1$ and are defined by their Fourier transforms,*

$$\mathscr{F}g_a(\omega;b) = \sqrt{0.5a}\exp\{-0.5a|w|\cdot(1+ib\,\mathrm{sgn}(w)\ln(|w|)2/\pi)\}, \tag{9.33}$$

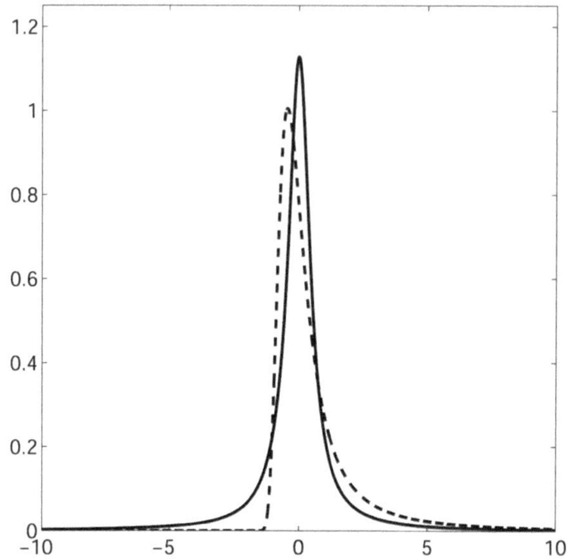


**FIGURE 9.10** Two kernels with Fourier transform (9.33); symmetric kernel, $a = 1, b = 0$ (solid line); asymmetric kernel, $a = 1, b = -1$ (dashed line).

*where* $i = \sqrt{-1}$ *and* $\operatorname{sgn}(x)$ *is the sign of* $x$. *For* $b = 0$, *the kernel is symmetric and given by*

$$g_a(t) = \frac{2}{\sqrt{\pi}} \frac{a^{-1/2}}{1 + (2t/a)^2}, \tag{9.34}$$

*and for* $b \neq 0$, *it is asymmetric. In this example, we use two kernels: Symmetric* $g$ ($a = 1$, $b = 0$ *in (9.33)) and the most asymmetric* $g$ *with* $a = 1$, $b = -1$. *The kernels are shown in Figure 9.10 as solid and dashed lines, respectively.*

*In contrast to Gaussian modelling, the parameters in the LMA model and the damage index depend not only on psd but also on the kernel function* $g(t)$. *The two kernels will be used to demonstrate this dependence.*

*The LMA with kernel (9.34) was used in several examples in [21] and [111], e.g., to model the variability of stresses on a cultivator frame. This model will be used in the following examples. The kernel is explicit and for symmetric LMA defined by the one-parameter symmetric kernel (9.34) together with a shape parameter* $\nu$ *for excess kurtosis* $\kappa_e$, *given by*

$$\nu = 0.4189\,\kappa_e a. \tag{9.35}$$

*This relation follows from (9.15) and*

$$\int g_a(t)^n \,\mathrm{d}t = a^{1-n/2} \frac{\Gamma(n-1/2)}{\Gamma(n)} \left(\frac{2}{\sqrt{\pi}}\right)^{n-1}.$$

*The relation (9.35) is very convenient because if we can find the relation between rainflow damage intensity and scale parameter* $\nu$ *for the symmetric kernel (9.34)* *with $a = 1$, then we can compute the damage intensity for any $a$ and excess kurtosis of the load.*

*Using non-linear regression and Monte Carlo simulations of the LMA with kernel (9.34), one obtained in [111] an approximate relation between $\nu$ and the damage rate ($\beta = 3$)*

$$d_3^{rfc}(\nu) = 4.84 + 0.06\nu + 8.32\sqrt{\nu} - 5.15\,\nu^{1/3}.$$

*The regression curve was estimated using the symmetric kernel with parameters $a = 1, b = 0$, by simulating different damage intensities for various values of the parameter $\nu$.*

*Similarly, using the asymmetric kernel with $a = 1, b = -1$, the following regression curve was estimated*

$$d_3^{rfc}(\nu) = 4.84 + 0.37\nu + 3.80\sqrt{\nu} - 2.23\,\nu^{1/3}.$$

*This type of relations is useful when one wishes to estimate damage in a variable slowly changing environment. The environment can be described by means of the distribution of parameters $\sigma, a, \nu$. This resembles the structure of a hierarchical model. In such a model, one assumes that the parameters come from a certain distribution obtained experimentally that can be used for statistical verification of the assumptions on the parameter distributions.*

*Note that for GMA, the parameter $\nu = 0$ and the damage $d_3^{rfc}(0) = 4.84$ which is higher than we had for GMA with standardised oscillator kernel. The reason for this is that spectrum $S_1(\omega)$ (9.32) has spectral moments $\omega_0 = 1$ and $\omega_2 = 2$ while for the oscillator spectrum $\omega_0 = \omega_2 = 1$.*

## 9.4.3 DAMAGE BOUNDS FOR LMA, RICE'S FORMULA

### 9.4.3.1 Damage Bound and Crossing Spectrum

To quantify the damage caused by an LMA load, one can use the approximations and bounds described in earlier chapters that are based on master-formula (4.31). The exact formula contains the intensity of oscillations over intervals, $\mu^{osc}(u,v)$, and this has to be obtained by numerical simulation. The upcrossing spectrum $\mu^+(u)$ comes as a by-product of the simulation, but it can also be computed exactly by a generalisation of Rice's formula, to be derived in Section 9.4.3.2.

For LMA with narrow-band spectrum, one can use the narrow-band approximation by Bendat [12] from Section 4.7.1. When the load process has only one local maximum between zero up- and downcrossings, the distribution of the maximum height is $\mathsf{P}(H > u) = \mu^+(u)/\mu^+(0)$. Assuming symmetric load and damage function $f(u,v) = (v-u)^\beta$, the narrow-band approximation of the expected damage intensity is, according to (2.5),

$$d^{rfc} \approx \mu^+(0)\mathsf{E}[(2H)^\beta] = \mu^+(0)\,2^\beta \int_0^\infty u^{\beta-1}\mu^+(u)\,\mathrm{d}u. \tag{9.36}$$

Note that (4.35) was derived for Gaussian loads, while (9.36) is general.

For symmetric LMA, both the narrow-band and the Holm-de Maré approximation can be used. They are equal to the general upper bound $d^+$ based on the minimal crossing spectrum $k^+(u,v) = \min_{u \le z \le v} \mu^+(z)$, for $\beta > 2$,

$$d^{rfc} \le d^+ = \int_{-\infty}^{+\infty} \int_{-\infty}^{v} \beta(\beta-1)(v-u)^{\beta-2} k^+(u,v) \, \mathrm{d}u \, \mathrm{d}v. \tag{9.37}$$

For skewed LMA, the narrow-band approximation can be both unconservative, if the load has negative skewness, and very conservative, if it has positive skewness. The H-M approximation is always conservative, and the $d^+$-bound is the least conservative.

#### 9.4.3.2 Rice's Formula for LMA Loads

In order to compute the narrow-band bound for the rainflow damage intensity, the upcrossing intensity has to be computed e.g., by means of Rice's formula (2.61). For the LMA model, we will use the basic form

$$\mu^+(u) = \int_0^{\infty} z f_{X(0),\dot{X}(0)}(u,z) \, \mathrm{d}z, \tag{9.38}$$

with the joint density of $X(0)$ and $\dot{X}(0)$. But unfortunately, for the LMA, no explicit formula for this density is known.

However, one can find one for the joint characteristic function, [1]. In this section, we consider the standardised and symmetric LMA, i.e. variances of the load and its derivative is one, skewness is zero and kurtosis $\kappa > 3$ which means that parameter $\nu > 0$,

Again, $g(t)$ is the kernel, such that $\int g(t)^2 \, \mathrm{d}t = \int \dot{g}(t)^2 \, \mathrm{d}t = 1$. In order to find the characteristic function $\phi(u_1,u_2)$ of $(X(0),\dot{X}(0))$, we will compute the characteristic function of $u_1 X(0) + u_2 \dot{X}(0)$,

$$\phi_{X(0),\dot{X}(0)}(u_1,u_2) = \mathsf{E}\left[e^{i(u_1 X(0) + u_2 \dot{X}(0))}\right] = \phi_{u_1 X(0) + u_2 \dot{X}(0)}(1).$$

First, we consider the approximation

$$u_1 X_N(0) + u_2 \dot{X}_N(0) = \sum_{i=-N}^{N} \sqrt{\nu K_i} Z_i (u_1 g(-t_i) + u_2 \dot{g}(-t_i)),$$

and observe that it has exactly the same structure as the approximation (9.12) but with kernel $u_1 g + u_2 \dot{g}$.

Similar arguments as for the characteristic function of $X(0)$ will give the following result,

$$\phi_{X(0),\dot{X}(0)}(u_1,u_2) = \exp\left(-\frac{1}{\nu}\int_{-\infty}^{\infty} \log\left(1 + \frac{\nu}{2}(u_1 g(x) + u_2 \dot{g}(x))^2\right) \mathrm{d}x\right). \tag{9.39}$$

Note that as $\nu$ tends to zero, the characteristic function converges to

$$\phi_{X(0),\dot{X}(0)}(u_1,u_2) = \exp\left(-\frac{1}{2}\int_{-\infty}^{\infty}(u_1 g(x) + u_2\dot{g}(x))^2\,\mathrm{d}x\right),$$

which, since $\int g(t)\dot{g}(t)\,\mathrm{d}t = 0$, simplifies to

$$\phi_{X(0),\dot{X}(0)}(u_1,u_2) = \exp\left(-\frac{u_1^2+u_2^2}{2}\right).$$

Thus, in the limit, as $\nu$ tends to zero $X(0)$, $\dot{X}(0)$ become independent standard Gaussian variables and the upcrossing intensity approaches $\mu^+(u) = \exp(-u^2/2)/2\pi$.

In the case when $\nu$ is significantly positive, the characteristic function is given by the full expression (9.39), and we obtain the upcrossing intensity by combining Rice formula (9.38) with the two-dimensional inverse Fourier transform,

$$\mu^+(u) = \frac{1}{(2\pi)^2}\int_0^{\infty}\int_{-\infty}^{\infty}\int_{-\infty}^{\infty} z e^{-i(v_1 u + v_2 z)}\phi_{X(0),\dot{X}(0)}(v_1,v_2)\,\mathrm{d}v_1\,\mathrm{d}v_2\,\mathrm{d}z. \quad (9.40)$$

The upcrossing intensity is not given by an explicit formula and thus must be evaluated numerically. For example, one could apply the two-dimensional fast Fourier transform algorithm combined with the trapezoidal method. The narrow-band bound for the rainflow damage intensity can then be obtained by means of the general bound (4.41) or the simpler (9.37), which have to be computed numerically.

**Example 9.10.** *This example illustrates in Figure 9.11 the accuracy of the numerical computation of the crossing spectrum. The kernel is a Cauchy kernel,*

$$g(t) = \sqrt{2/\pi}/(1+(\sqrt{2}t)^2).$$

*Ten replicates of the LMA were simulated over* $T = 10^5$ *seconds and a common crossing spectrum estimated. The two-dimensional characteristic function for* $X(t), \dot{X}(t)$ *was inverted over* $(u_1,u_2) \in [10:0.1:10]^2$ *and the theoretical crossing spectrum obtained by numerical integration as* (9.40).

*In the figure, the triangular solid curve is the simulated crossing spectrum with the dots the result of Rice's formula. The solid parabola is the crossing spectrum for GMA.*

**Example 9.11** (LMA with oscillator spectrum, cont.). *This example is focused on the crossing spectrum by Rice's formula and the narrow-band damage bounds. In Figure 9.12, we compare the computed crossing spectra as dashed lines with solid lines for simulated spectra. Symmetric and asymmetric kernels give virtually identical results.*

*To evaluate the damage intensities, we simulated 2000 missions of LMA, each of duration* $T = 10.8 \cdot 10^3$ *seconds, i.e. three hours of load; thus, the setup is similar to that in Example 9.7.*

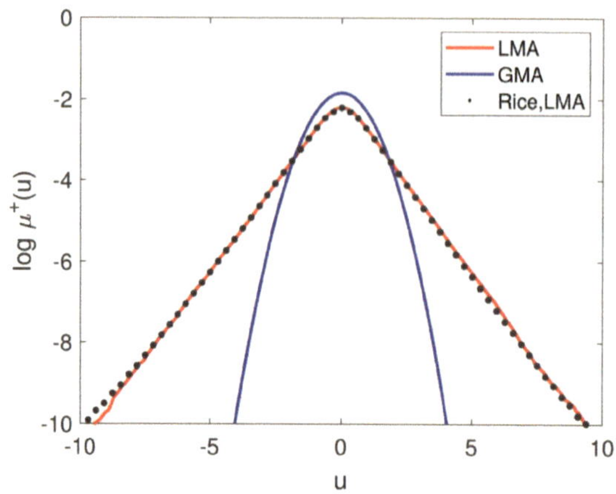


**FIGURE 9.11** Log crossing spectra. Solid triangular: LMA with Cauchy kernel; solid parabolic: GMA; dots: LMA computed by Rice's formula.

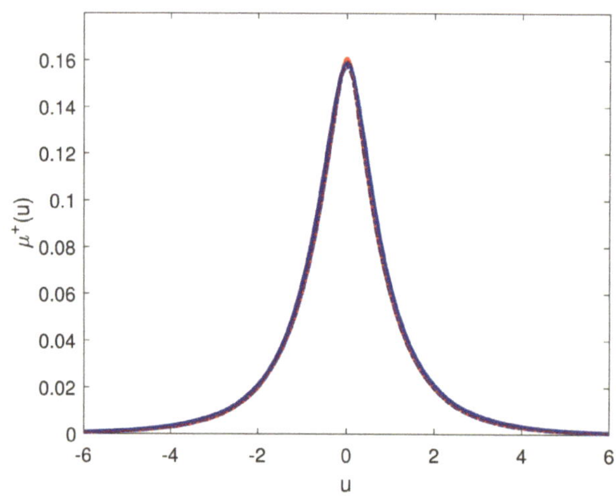


**FIGURE 9.12** Upcrossing spectra for oscillation LMA. Symmetric kernel $g_1$; Asymmetric kernel $g_2$. Solid; simulated, dashed: By Rice's formula.

*Table 9.4 shows the average damage of 2000 replicates of three hours missions, each containing about 2100 rainflow cycles. The damage function is $f(u,v) = (v-u)^\beta$, $\beta = 3,4,5$. For comparison, we give in parenthesis the narrow-band bound.*

*We can see that the expected damage rate for the symmetric kernel is higher than for the causal one. However, as can be seen in Table 9.5, containing quantiles for rainflow damage intensity, when $\beta = 5$, the risk to experience an extremely damaging mission is higher when the causal kernel is used for modelling of responses.*

*Important conclusions: a) With high kurtosis, it is hard to estimate parameters and damage intensities. b) The LMA with symmetric kernel is more damaging than LMA with causal kernel.*

**Table 9.4**
**Simulated Average Damage from 2000 Replicates of 3 Hours Missions, Each Containing About 2100 Rainflow Cycles**

| | | | |
|---|---|---|---|
| $g_1$ | 7.79 (9.04) | 66.4 (77.21) | 742.5 (917.1) |
| $g_2$ | 7.40 (8.26) | 62.1 (66.43) | 678.2 (742.5) |

Damage function $f(u,v) = (v-u)^\beta$. $\beta = 3,4,5$. Narrow-band bound in parenthesis.

**Table 9.5**
**Quantiles in Damage Distribution for Oscillator LMA with Symmetric and Asymmetric Kernel**

| | Quantile | 0.25 | 0.5 | 0.75 | 0.975 | 0.99 |
|---|---|---|---|---|---|---|
| $g_1$ | $\beta = 3$ | 7.17 | 7.73 | 8.35 | 9.80 | 10.24 |
| | $\beta = 4$ | 56.7 | 64.5 | 73.5 | 101.4 | 110.4 |
| | $\beta = 5$ | 562.8 | 684.1 | 849.1 | 1503 | 1735 |
| $g_2$ | $\beta = 3$ | 6.65 | 7.31 | 8.04 | 9.75 | 10.25 |
| | $\beta = 4$ | 51.1 | 59.30 | 70.3 | 102.6 | 114.5 |
| | $\beta = 5$ | 479.3 | 604.5 | 792.6 | 1531 | 1839 |

**Remark 9.7.** *Filtering an LMA load with kernel $f$, one gets back an LMA but now with kernel $h * f$. This means e.g. that if one can compute crossing intensity for the Laplace MA, and thereby the upper bound* (4.41) *for the damage, one can also do it in the same manner for linearly filtered LMA load.*

*Note that as the process is filtered, the spectrum is multiplied by $|H(\omega)|^2$ as for Gaussian loads, but kurtosis of the filtered process changes since by virtue of* (9.14) *this quantity is a function of the kernel which in the filtered process is $h * f$.*

**Remark 9.8.** *There exist some explicit formulas for the approximation of the crossing intensity for symmetric LMA. The method employs the so-called saddle point method, introduced in statistics by Daniels (1954), [45]. Since this approach requires some additional concepts, we refer to the papers by Butler, Machado and Rychlik (2009) [34] and Galtier (2011) [65].*

## 9.5 ASYMMETRIC LMA

A LMA load with skewed marginal distribution can be generated by a convolution equation like (9.12) where the gamma variable $K_i$ has gotten an extended role. The following description and the extended simulation experiment are taken from [2] and [180] with theoretical details from [1]. We use the parametrisation introduced in [180], which differs from [1, 2].

### 9.5.1 GENERATION AND BASIC FACTS

**Definition 9.4** (Asymmetric LMA). *The standardised asymmetric LMA process $X(t)$ with kernel $g(t)$, $\int g(t)^2\,\mathrm{d}t = 1$, defined on $[-S,S]$, relative skewness parameter $q$, $-1 \le q \le 1$, excess-kurtosis parameter $\nu$ and scale parameter $\sigma = \sqrt{\nu}$, is the limit, as $N,T$ tend to infinity and $\Delta t \to 0$, of the approximation*

$$X_N(t) = \sum_{i=-N}^{N} g(t-t_i)(q\sqrt{\nu}(K_i - \Delta t/\nu) + \sqrt{1-q^2}\sigma\sqrt{K_i}Z_i), \quad -T \le t \le T, \tag{9.41}$$

*with $t_i = i\Delta t$, $N \ge (T+4S)/\Delta t$. The $K_i$ are independent gamma variables with shape parameter $k = \Delta t/\nu$ and $Z_i$ are standard normal variables.*

The standardised asymmetric LMA has mean zero and variance one and the same spectrum as the symmetric LMA, $S(\omega) = \frac{1}{2\pi}\left|\mathscr{F}g(\omega)\right|^2$. It has

$$\text{skewness: } \gamma = \sqrt{\nu}q(3-q^2)\int g(t)^3\,\mathrm{d}t, \tag{9.42}$$

$$\text{excess kurtosis: } \kappa_e = 3\nu\left(2-(1-q^2)^2\right)\int g(t)^4\,\mathrm{d}t. \tag{9.43}$$

To invert (9.42) and (9.43) to get suitable $q$- and $\nu$-values for simulation requires numerical tools. Figure 9.13 shows contour lines of $\gamma$ and $\kappa_e$ as functions of $q$ and $\nu$.

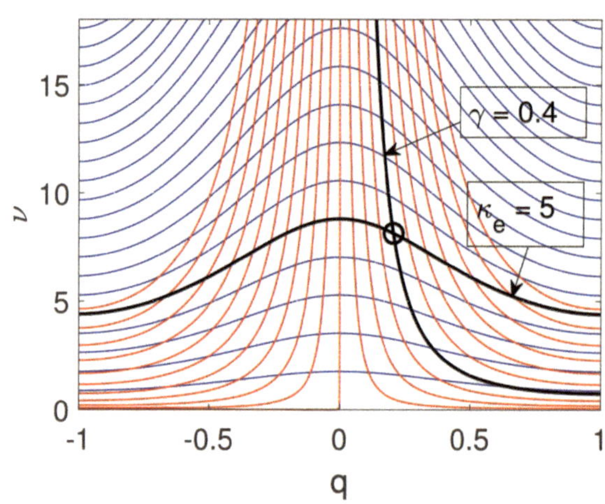


**FIGURE 9.13** Contour lines of overlaid plots of $\gamma$ and $\kappa_e$ as functions of $q$ and $\nu$. "Vertical" : Contour lines for $|\gamma| = 0:0.1:0.9$; $\gamma < 0$ for $q < 0$. "Horizontal" : Contour lines for $\kappa_e = 1:20$. Thick black contours for $(\gamma,\kappa_e) = (0.4,5)$ cross at $(q,\nu) = (0.2044, 8.1333)$.

### 9.5.2 SIMULATION EXPERIMENT

In order to illustrate the new model and also to evaluate the performance of the upper bound for the damage, given in (4.43), the following simulation experiment is done.

#### 9.5.2.1 Stiff Structure

In the first version of the experiment, the structure on which the load is acting is supposed to be stiff with stress proportional to the applied load. For simplicity, the coefficient of proportionality is set to one so the load coincides with the stress. The load $Y(t)$ is modelled as a LMA with parameters $\nu$, $\sigma$ and $q$ and a symmetric kernel $g(t)$ corresponding to a Pierson-Moskowitz type spectrum with significant wave height 7 m and peak period 11 s, see Figure 9.14. When spectrum is truncated at $\omega_c = 3$, the load variance is 1.9. The same spectrum is used throughout the experiment, and in all examples, the signals are centred to mean zero.

The damage intensity on the structure, that is the accumulated damage per time unit, can be estimated in two different ways: Either by direct simulation of the response and computing the observed rainflow damage according to the master-equation (4.31) or by direct computation of the upper bound (4.43) for the damage without making any simulations. In Table 9.6, the simulated damage intensity and the upper bound (4.43) for the Laplace driven MA are shown for different values of skewness and kurtosis. The simulated values are computed as a mean over 1000 independent simulations, each consisting of 10000 data points sampled at 1.85 Hz. For comparison, the simulated damage and the upper bound in the Gaussian model are also computed. Simulations are given in plain text and the upper bounds in bold text. The last two columns give the corresponding values for a Gaussian load.

The Gaussian model clearly causes much less damage than the LMA for large values of the load skewness $\gamma$ and excess kurtosis $\kappa_e$ in combination with high fatigue exponent $\beta$, mainly due to the fact that the Gaussian distribution has too light tails and thus does not include enough big cycles. Moreover, the upper bound seems to follow the simulated values closely, both for the Laplace and the Gaussian model; see also Figure 9.15. This has to do with the fact that the spectrum is fairly narrow-banded, at least from a fatigue application point of view.

As skewness and excess kurtosis are close to zero, the simulations from the two models are close, as can be expected since the Laplace model converges to a Gaussian process as $\kappa_e \to 0$. However, the Laplace upper bound is not in perfect agreement with the Gaussian upper bound because of numerical problems when computing the crossing intensity. These problems arise since one has to choose a finite grid to

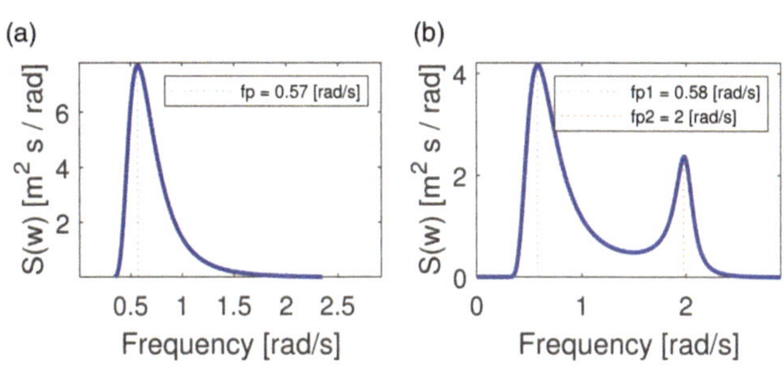


**FIGURE 9.14** (a) Spectrum of the input process. (b) Spectrum of the load filtered by the dynamic structure.

**Table 9.6**

**Estimated Fatigue Damage Intensity for a Stiff Structure Subjected to a Laplace Driven Moving Average Load for Different Values**

| | Laplace model | | | | | | Gaussian model | |
|---|---|---|---|---|---|---|---|---|
| $\beta$ | $\gamma = 0,\ \kappa_e = 0.01$ | | $\gamma = 0.3,\ \kappa_e = 1$ | | $\gamma = 0.6,\ \kappa_e = 2$ | | $\gamma = 0,\ \kappa_e = 0$ | |
| 2 | 2.89 | **3.16** | 2.93 | **3.16** | 2.90 | **3.08** | 2.90 | **3.06** |
| 3 | 18.6 | **19.6** | 21.0 | **21.8** | 22.1 | **22.4** | 18.7 | **20.1** |
| 4 | 135 | **146** | 180 | **190** | 207 | **211** | 135 | **149** |
| 5 | 1080 | **1200** | 1790 | **1940** | 2300 | **2350** | 1080 | **1230** |
| 6 | 9430 | **10,700** | 20100 | **22,500** | 29,600 | **30,400** | 9380 | **11,000** |

Different values of skewness $\gamma$, excess kurtosis $\kappa_e$ and fatigue exponent $\beta$; upper bound $d^+$ (4.43) in bold.

evaluate the characteristic function before computing the crossing intensity. If the grid is not fine enough, the crossing intensity is underestimated and thereby also the upper bound.

#### 9.5.2.2 Dynamic Structure

In the second version of the experiment, the load is supposed to act on a dynamic structure described by a linear damped oscillator, i.e. having dynamics that is described by the differential equation

$$Z''(t) + 2zw_0 Z'(t) + w_0^2 Z(t) = \frac{1}{m} Y(t),$$

where $Z(t)$ is the response and $Y(t)$ is the load. The parameters have the following interpretation: $w_0$ is the resonance frequency of the undamped system, $z$ is a damping coefficient and $m$ equals the mass that is accelerated. The frequency response $H(\omega)$ is given by

$$H(\omega) = \frac{1/m}{-\omega^2 + 2z\omega_0 i\omega + \omega_0^2}.$$

If the load $Y(t)$ is a Laplace-driven MA, then due to the properties of this class of models, the response $Z(t)$ is also a Laplace-driven MA. However, the kernel of $Z(t)$ is $h * g(t)$, where $g(t)$ is the kernel of $Y(t)$ and $h = \mathscr{F}^{-1}H$ is the causal impulse response of the linear oscillator. Moreover, the parameters $\sigma$, $\nu$ and $q$ remain unchanged, whereas the kurtosis is changed according to (9.14).

The same spectrum is used for the load as in the first version of the experiment, namely a truncated Pierson-Moskowitz spectrum with significant wave height 7 m and peak period 11 sec. and variance 1.9. The parameters of the oscillator are, for

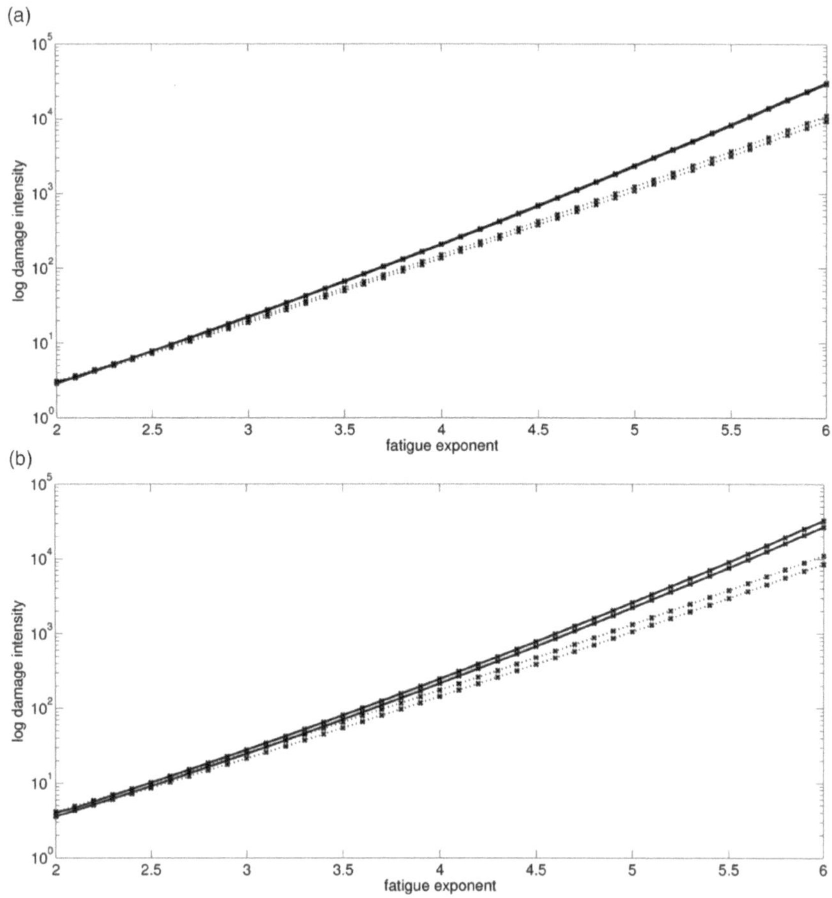


**FIGURE 9.15** Log-damage intensity vs. fatigue exponent $\beta$ for (a) a stiff structure and (b) a linear oscillator (right) driven by Laplace MA: Simulated (solid), upper bound in Laplace MA (dashed) and Gaussian (dotted) models.

illustration reasons, set to $m = 0.37$ kg, $\omega_0 = 2$ rad/s and $z = 0.05$. The resulting spectrum of the output stress process $Z(t)$ is shown in Figure 9.14 (variance of the stress is 4.0). The first peak can be thought of as the stress induced by, e.g. ocean waves and the second peak as the stress induced by resonances in the structure.

In Table 9.7, the damage intensity is shown for the filtered process. Again, the values correspond to direct simulation and the upper bound (4.43) for the Laplace-driven MA, and the corresponding quantities for a Gaussian process have the same spectral density function. The values of $\gamma$ and $\kappa_e$ refer in this case to the skewness and excess kurtosis of the process driving the oscillator, while $\widetilde{\gamma}$ and $\widetilde{\kappa}_e$ correspond to skewness and excess kurtosis for the stress response. As before, simulations are given in plain text and the upper bounds in bold text. The last two columns in the table give the corresponding values for a Gaussian load.

**Table 9.7**
**Estimated Fatigue Damage Intensity for a Linear Oscillator Subjected to a Laplace Driven MA Load for Different Values**

| | Laplace model | | | | | | Gaussian model | |
|---|---|---|---|---|---|---|---|---|
| $\beta$ | $\gamma=0,\ \kappa_e=0.01$ | | $\gamma=0.3,\ \kappa_e=1$ | | $\gamma=0.6,\ \kappa_e=2$ | | $\gamma=0,\ \kappa_e=0$ | |
| 2 | 3.62 | **4.12** | 3.61 | **4.10** | 3.59 | **4.06** | 3.63 | **4.13** |
| 3 | 21.3 | **24.3** | 23.3 | **26.4** | 24.7 | **27.9** | 21.4 | **25.2** |
| 4 | 143 | **167** | 182 | **197** | 215 | **246** | 144 | **174** |
| 5 | 1060 | **1270** | 1660 | **1970** | 2220 | **2610** | 1060 | **1330** |
| 6 | 8540 | **10,500** | 17100 | **21,100** | 26700 | **32,500** | 8520 | **11,100** |

*Note*: Different values of skewness $\gamma$, excess kurtosis $\kappa_e$, corresponding values $\widetilde{\gamma}$ and $\widetilde{\kappa}_e$ for the stress response, and fatigue exponent $\beta$; upper bound $d^+$ (4.43) in bold.

The biggest difference in this case, compared to the stiff structure in Table 9.6, is that the upper bound differs more from the simulated values. This is due to the broadband character of the spectrum. Keeping in mind that the upper bound (4.43) has the interpretation of being a narrow-band approximation, at least in the Gaussian case, this is not so surprising. In Figure 9.15, the simulated damage is compared to the Laplace MA upper bounds for excess kurtosis $\kappa_e = 2$ and skewness $\gamma = 0.6$.

## APPENDICES

### 9.A WHY LMA CONVERGES TO GMA AS $\nu$ GOES TO ZERO

By (9.12), the variable $X_N(0)$ is a sum of independent components $g(t_i)\sqrt{\nu}\Lambda_i$, which have characteristic functions $\phi_i(u) = \phi(\sqrt{\nu}g(t_i)u)$. In the definition of the LMA-process, $\Lambda = \sqrt{\nu}\sqrt{KZ}$ and $k = \Delta t/\nu$, hence

$$\phi_i(u) = \phi(\sqrt{\nu}g(t_i)\,u) = \left(1+\frac{\nu g(t_i)^2 u^2}{2}\right)^{-\nu/\Delta t} = \exp\left(-\frac{\ln(1+\nu u^2 g(t_i)^2/2)}{\nu}\Delta t\right).$$

For very small values of $\nu$, one can employ the Taylor expansion of the function $\ln(1+x) = x - x^2/2 + x^3/3 + \ldots$, to get

$$\phi_i(u) \approx \exp\left(-\frac{u^2}{2}g(t_i)^2\Delta t\right).$$

Finally, the characteristic function of $X_N(0)$, which is a product of $\phi_i(u)$, can be written as follows:

$$\phi(u) = \exp\left(-\frac{u^2}{2}\sum_{i=-N}^{N} g(t_i)^2\Delta t\right) \approx \exp(-u^2/2), \tag{9.44}$$

since $\sum_{i=-N}^{N} g(t_i)^2 \Delta t = 1$. Since $\exp(-u^2/2)$ is a characteristic function of a standard Gaussian variable, the claim that LMA $X_N(t)$ tends to GMA as $\nu$ tends to zero is motivated.

## 9.B RECIPE FOR SIMULATION OF LMA

A simple algorithm for simulating the Laplace-driven moving average with kernel $f$ is given by:

1. Pick $m$, and $dx$ so that $f$ is well approximated by its values on $0 < dx < \cdots < m \cdot dx$.
2. Pick $n \gg 2m - 2$ so the $k = n - 2m + 2$ values of $Y$ will be generated at $0 < dx < 2 \cdot dx < \cdots < k \cdot dx$.
3. Simulate $n$ iid $\Gamma(dx/\nu, 1)$ random variables and store them in a vector $G = [G_j]$.
4. Simulate $n$ iid zero-mean standard normal random variables and store them in a vector $Z$.
5. Compute $Y = \gamma \int f(x)\, dx + \mu f * G + \sigma f * (\sqrt{G} \cdot Z)$, where $\sqrt{G} \cdot Z = [\sqrt{G_j} \cdot Z_j]$, $*$ denotes convolution (we extend to infinite sequences by setting zero to the entries that have not been set otherwise), and the integral $\int f(x)\, dx$ is computed by some numerical method.
6. Remove the first $m - 1$ and last $m - 1$ entries of the vector $Y$. The resulting vector $Y$ is a simulation of the Laplace-driven moving average with parameters $\mu$, $\nu$, $\sigma$ and drift $\gamma$.

# 10 Non-Linear Load Transformations

## 10.1 INTRODUCTION

Markov loads introduced in Chapter 8 essentially constitute a sequential time series $X_n$ and are functions of Markov chains $X_n = G(Z_n)$. The class of Markov loads is broad and can be used as models for many loading time histories that are observed in practical situations. This class of models is flexible enough to fit to many types of variability, measured by means of observed oscillations between sets in measured sequences. For fatigue life prediction, the assumption of the loading following Markov approximation is suitable, since the fatigue accumulation is assumed to be *rate independent*, independent of time transformations in signals, and is only dependent on the sequence of turning points. The situation is however different when we are modelling external loads that act on structures. Real-life structures are not rigid, and consequently, the dynamical effects have to be included in the computation of stresses at fixed locations in the structure. In these situations, the frequency content of the external signal is important. This information is not preserved in Markov models for the loadings and hence is unsuitable for these applications.

Chapters 6 and 7 focussed on Gaussian random processes $x(\tau)$, having power spectral density $S(\omega)$. Gaussian random processes are commonly used models for many naturally occurring processes. However, in many practical situations, the probability density of $x(\tau)$ is markedly non-Gaussian. Assuming the processes to be Gaussian in such situations has a potential significant influence on the conclusions emerging from the analysis.

We start the discussions in this chapter by considering processes that can be expressed as a simple marginal transformation $y(\tau) = h(x(\tau))$ and explain how the rainflow fatigue damage due to the time history $y(\tau)$ is related to the damage calculated from the "parent" load $x(\tau)$. The relations are not coupled to any particular load model but are valid for all types of uni-dimensional loads. In particular, the technique applies to Markov loads, as transformed Markov chain. It is also applicable to transformed Gaussian loads, where the crossing properties are computable, based on the power spectrum of the parent load.

The idea with the transformation is that one should be able to combine well understood technique and properties of the parent process with the fatigue causing characteristics of an observed load. Here, several methods have been suggested, and we will describe the most common ones. The method suggested by Winterstein (1988) [212, 213] and Ochi and Ahn (1994) is based on skewness and kurtosis. A non-parametric transformation was presented by Rychlik et al. (1997) [182]. It generates a smooth transformed Gaussian load that has the same crossing spectrum as observed

 DOI: 10.1201/9781003019947-10

data. The transformed Gaussian loads are compared with the Laplace moving average loads with excess kurtosis.

Next, we give examples of so-called quadratic loads which are functions of vector valued Gaussian loads. Here, an approximate method to compute the upper bound $d^+$ for the rainflow damage intensity is discussed. These discussions are based on the studies carried out by Hagberg (2005) [75], Baxevani et al. (2005) [5] and Gupta and Rychlik (2007) [74]. There are stationary loads that may locally resemble a realisation of a Gaussian process, and hence one would like to model them as a transformed Gaussian process $y(\tau) = h(x(\tau))$. However, such models may fail to accurately predict the fatigue damage, especially when the stress is a dynamical response to the load. The problem is that there are loads which may have variable frequency contents over time. Such effects are difficult to model using a memoryless function $h$. Hence, sometimes one needs to use transformations that varies with time. An example of such analysis will also be presented here.

We also consider non-Gaussian processes which conditionally are Gaussian but with variable variance. Three typical cases will be presented with the following specific applications in mind: (a) wave loads with seasonally changing storm intensities, (b) response of a linear system subjected to Gaussian load with added transients at random instances, and finally, (c) loads with regime switching mean. The results corresponding to these applications are based on the works by Baxevani and Rychlik (2007) [7], Bengtsson and Rychlik (2009) [15], Bogsjö and Rychlik (2007) [22], and Karlsson (2007) [101], respectively.

## 10.2 RAINFLOW DAMAGE IN TRANSFORMED LOADS

### 10.2.1 DAMAGE RELATION BETWEEN PARENT AND TRANSFORMED LOAD

We first consider the rainflow fatigue damage due to a load $y(\tau)$, $0 \le \tau \le t$, obtained as a non-linear transformation of load $x(\tau)$, expressed as

$$y(\tau) = h(x(\tau)).$$

Here, typically $h(\cdot)$ is a non-linear transformation function. We assume that $h(\cdot)$ is a differentiable function with finite number of local extremes. We consider a particular time history $x(\tau)$ and assume that its rainflow pairs $(m_i^{\text{rfc}}, M_i)$ or the functions for crossing of intervals, given by $N^{osc}(u,v)$, are found for $x$. Formally, we would need to use both upcrossings and downcrossings of an interval $[u,v]$ (those can differ most by one), and hence for loads $x$ (having a large number of cycles), we let $N^+(u,v) = N^{osc}(u,v)$, while $N^-(u,v)$ denotes the number of downcrossings of the interval $[u,v]$.

As before, we denote the damage function by $f(u,v)$, $u \le v$, with $f_{12}(u,v) = \frac{\partial^2 f(u,v)}{\partial u \partial v}$ and, for simplicity of presentation, we assume that $f_2(v,v) = 0$ for all $v$ and $f(v,v) = 0$. The most important case is $f(u,v) = (v-u)^\beta$.

The rainflow damage can be computed when the rainflow cycles $(m_i^{\text{rfc}}, M_i)$ or interval up- downcrossings are known. Here, it is natural to assume that the cycles and crossing are counted in $x(\tau)$, while we would like to know the damage of $y(\tau) = h(x(\tau))$ for a chosen transformation $h$. It is simple if transformation is monotonic since then for differentiable non-decreasing transformation maxima and rainflow minima are just transformed by means of $h$, rainflow cycles in $y$ are $(h(m_i^{rfc}), h(M_i))$, and

$$D^{rfc}(y) = \sum_{i=1}^{n} f\left(h(m_i^{rfc}), h(M_i)\right).$$

Similarly, if the transformation has negative derivative, then

$$D^{rfc}(y) = \sum_{i=1}^{n} f(h(M_i), h(m_i^{rfc})).$$

If the interval crossings are known, we could use the master-formula to derive the damages for the transformed load. One convenient way is to just say that the damage function $f(u,v) = f(h(u),h(v))$ if $h$ is increasing and similarly $f(u,v) = f(h(v),h(u))$ if the transformation has negative derivative.

Employing the master-formula, we get the formula

$$D^{\mathrm{rfc}}(h(x)) = -\int_{-\infty}^{\infty}\int_{-\infty}^{v} h'(u)^+ h'(v)^+ f_{12}(h(u),h(v)) N^+(u,v)\,\mathrm{d}u\,\mathrm{d}v. \qquad (10.1)$$

Similarly, if $h(x)$ is decreasing, i.e. $h'(x) < 0$, one needs to use the downcrossings of the intervals instead of the upcrossings. This leads to the expression

$$D^{\mathrm{rfc}}(h(x)) = -\int_{-\infty}^{\infty}\int_{-\infty}^{v} h'(u)^- h'(v)^- f_{12}(h(v),h(u)) N^-(u,v)\,\mathrm{d}u\,\mathrm{d}v, \qquad (10.2)$$

where $x^- = \max(0,-x)$.

The case of non-monotonic transformations can be treated in a similar way by splitting the transformation into monotonic segments. The result is presented in Theorem 10.1, which basically leeds to two transformations, one increasing and one decreasing, and we get (10.5).

In this general case, it is not obvious how to write the rainflow damage of $y$ from the rainflow count of $x$. Obviously, it is possible since from the rainflow count, we can get the oscillations intensity and then apply (10.5). However, one can again use the master-formula and define the damage function $G(u,v)$ as given in (10.8). In the following subsection, we give two examples of non-monotonic transformations.

**Theorem 10.1.** *Let us assume that $g(\cdot)$ is a differentiable transformation having finite number of local extremes. We define*

$$g^+(u,v) = -f_{12}(h(u),h(v))h'(u)^+h'(v)^+, \qquad (10.3)$$

*if* $\max_{u\le z\le v} h(z) \le h(v)$ *and* $\min_{u\le z\le v} h(z) \ge h(u)$, *and* $g^+(u,v) = 0$ *otherwise, and*

$$g^-(u,v) = -f_{12}(h(v),h(u))h'(u)^-h'(v)^-, \qquad (10.4)$$

*if* $\max_{u\le z\le v} h(z) \le h(u)$ *and* $\min_{u\le z\le v} h(z) \ge h(v)$, *and* $g^-(u,v) = 0$ *otherwise. If* $f_2(v,v) = 0$, *then Eq. (10.2) can be written as*

$$D^{\mathrm{rfc}}(h(x)) = \int_{-\infty}^{\infty}\int_{-\infty}^{v} \{g^+(u,v)N^+(u,v) + g^-(u,v)N^-(u,v)\}\,\mathrm{d}u\,\mathrm{d}v. \tag{10.5}$$

Since $N^+(u,v)$ and $N^-(u,v)$ can differ at most by one, then for $x(\tau)$ having a large number of cycles $(u_i^{\mathrm{rfc}}, v_i)$, $i = 1,\ldots,n$ and for large $n$, in practical situations, one can assume that there are equally many upcrossings as there are downcrossings of interval $[u,v]$. This leads to a simplification of Eq. (10.5) of the form

$$D^{\mathrm{rfc}}(y) = \int_{-\infty}^{\infty}\int_{-\infty}^{v} g(u,v)N^{osc}(u,v)\,\mathrm{d}u\,\mathrm{d}v. \tag{10.6}$$

Here, $g(u,v) = g^+(u,v) + g^-(u,v)$.

Finally, using again the master-formula (backwards), it can be shown that an alternative formula for $D^{\mathrm{rfc}}(y)$ is given by

$$D^{\mathrm{rfc}}(h(x)) = \sum_{i=1}^{n} G(m_i^{\mathrm{rfc}}, M_i), \tag{10.7}$$

where

$$G(u,v) = \int_{u}^{v}\int_{u}^{z} g(t,z)\,\mathrm{d}t\,\mathrm{d}z, \tag{10.8}$$

since $g(v,v) = 0$ for all $v$ and $n$ is the total number of rainflow cycles for the parent load $x(\tau)$.

Consequently, for stationary $x(\tau)$, the rainflow damage intensity of $y(\tau)$ is given by

$$d^{\mathrm{rfc}}(y) = \int_{-\infty}^{\infty}\int_{-\infty}^{v} g(u,v)\mu^{osc}(u,v)\,\mathrm{d}u\,\mathrm{d}v. \tag{10.9}$$

Obviously, if $\mu^{osc}(u,v)$ is not known, it can be bounded by the minimal crossing spectrum $k^+(u,v) = \min_{u\le z\le v} \mu(z)$ and by means of (7.26)

$$d^+(y) = \int_{-\infty}^{\infty}\int_{-\infty}^{v} g(u,v)k^+(u,v)\,\mathrm{d}u\,\mathrm{d}v. \tag{10.10}$$

**Example 10.1** (The $g(u,v)$, $G(u,v)$ functions)**.** *Let the transformation be* $h(x) = |x|$ *and consider a damage function* $f(u,v) = (v-u)^\beta$. *Following Eqs. (10.3–10.4), one gets*

$$g(u,v) = \beta(\beta-1)\begin{cases} (v-u)^{\beta-2} & \textit{if } u \ge 0 \textit{ or } v \le 0,\\ 0 & \textit{if } u < 0 < v.\end{cases} \tag{10.11}$$

*The function* $G(u,v)$ *is obtained by double integration (see Eq. (10.8)) which leads to*

$$G(u,v) = \begin{cases} (v-u)^\beta & \textit{if } u \ge 0 \textit{ or } v \le 0,\\ (-u)^\beta + (v)^\beta & \textit{if } u < 0 < v.\end{cases}$$

## 10.3 TRANSFORMED GAUSSIAN LOADS

In Chapters 6 and 7, we have studied fatigue damage estimation for Gaussian loads. If the Gaussian load is stationary, its probabilistic properties are fully described by the mean $\mathsf{E}[x(0)] = m$ and the spectrum $S(\omega)$. The important property is that if loads are Gaussian and structure linear, then even local stresses that are essential for crack growth, i.e. fatigue, are Gaussian too. The Gaussian case has been extensively studied, and hence many approximations for the computations of damage intensity are available.

However, real loads are usually non-Gaussian and using Gaussian models can lead to underestimation of the expected damage. In such a case, one could model $x(\tau)$ by means of a Markov or Gaussian load together with a suitable transformation, viz. $y(\tau) = h(x(\tau))$. Now, if the oscillation intensity of $x$ is known, then the expected damage of $y$ can be computed using Theorem 10.1. Computations of the oscillation intensity for Markov loads were discussed in Chapter 8, where we also mentioned that for stationary Gaussian loads, the MCTP approximation often gives accurate damage estimates and hence could be employed even here.

In randomly vibrating large engineering structures, the load effect which causes fatigue damage, at a particular location in the structure, usually has contributions from a large number of components. Often, the resulting load effect is obtained as a non-linear combination of these individual components and is thus non-Gaussian, even if the external loads are Gaussian. An example is when weakly non-linear systems have to be used to analyse responses to Gaussian loads. Here, the responses are often approximated by means of quadratic transfer functions, which also leads to stresses which are non-Gaussian but as a function of a vector of Gaussian processes, viz.

$$y(\tau) = h(x_1(\tau), \ldots, x_n(\tau)), \tag{10.12}$$

where $h(\cdot)$ is a non-linear deterministic function (defined by means of mechanical laws), $\{x_i(t)\}_{i=1}^n$ are stationary Gaussian processes, and hence $y(t)$ is the non-Gaussian process that causes fatigue damage. Obviously, crossing of intervals by $y(\tau)$ can be translated to crossings between sets in $n$ dimensions. However, often $n$ is very large, and hence this property is hard to use in practice.

Consequently, one may instead evaluate the bound $d^+$, see (7.26) for the rainflow damage intensity of $y$. For this, the crossing intensity $\mu^+(u)$ of $y$ needs to be computed. For high (low) levels $u$, which are the most important in this problem, often the asymptotic formula, due to Breitung (1988) [26], see also Hagberg (2005) [75], can be employed for a wide class of transformations $h$. For low levels, the formula is less accurate, but it can still be used in many cases. (If the accuracy of the approximation is too poor, one could simulate $y(\tau)$ for a relative short period of time and estimate the oscillation intensity for small and medium ranges of $u, v$.) For larger $u, v$, the bound $\mu^{osc}(u,,v) \leq \min_{u \leq z \leq v} \mu^+(z)$ can be used, where $\mu^+(z)$ is computed using Breitung's method. (Note that it still may happen that $y(\tau) = h(x(\tau))$ is a sufficiently accurate model for (10.12).)

### 10.3.1 CHOOSING THE TRANSFORMATION $h$ AND SPECTRUM $S$

#### 10.3.1.1 Physically Based Transformations

There are situations when both transformation and spectrum $S$ can be derived by means of physical considerations. For example, let us consider the expected rainflow fatigue damage due to wind loading fluctuation. The force due to wind loadings is given by $F(t) = c(V(t)+m_v)^2$ (see Nigam and Narayanan (1994) [150] for a review), where $V(t)$ is a zero-mean Gaussian random process describing variability of wind velocity around the mean $m_v$. For illustrating the proposed method, we ignore the effect of structural vibrations and assume that the fatigue damage is due to load $F(t)$. In this case, the load is

$$y(\tau) = c(x(\tau)+m)^2$$

That is, $h(u) = c(u+m)^2$. Again using Eqs. (10.3–10.4), we obtain that

$$g(u,v) = 4c^{\beta+2}\beta(\beta-1)\times$$
$$\begin{cases} (v+m)(u+m)((v+m)^2-(u+m)^2)^{\beta-2} & \text{if } -m \le u \text{ or } v \le -m, \\ 0 & \text{if } u < -m < v, \end{cases} \tag{10.13}$$

and the corresponding form for $G(u,v)$ is given by

$$G(u,v) = c^\beta \begin{cases} ((v+m)^2-(u+m)^2)^\beta & \text{if } -m \le u \text{ or } v \le -m, \\ (-m-u)^{2\beta}+(v+m)^{2\beta} & \text{if } u < -m < v. \end{cases} \tag{10.14}$$

Similar reasoning when wave loads are considered will give a more complex function. Let $y(\tau)$ (see Skjong and Madsen (1987) [190]) be the stress process at a particular location of an offshore structure where one wants to evaluate the fatigue damage caused by wave forces (here the water particle movement). A finite element analysis is used to compute the stresses and, in general, for an element the stress can be approximated by

$$y(t) = cx(t)|x(t)| + \dot{x}(t),$$

where $x(t)$ is a zero-mean Gaussian process obtained through linear filtering, with transfer function $H(\lambda)$, of sea surface elevation $W(t)$. In Appendix 7.B, several parametric families of spectra $S_W(\omega)$ used in offshore engineering are presented. This is a simple example for $y(\tau) = h(x_1(\tau), x_2(\tau))$.

#### 10.3.1.2 Empirically Based Transformations

When no physical argument is at hand, one can rely on measured data $y$ from load experiments. If the distribution of the measured data is slightly non-Gaussian, then one may look for a transformation $h(\cdot)$ and a spectrum $S(\omega)$ such that the load is described by $y(\tau) = h(x(\tau))$, where $x$ is a Gaussian process with the spectrum $S$. Often, the transformation $h$ is chosen first and then the inverse transformed signal $h^{-1}(y(\tau))$ is used to estimate the spectrum $S$. There are several transformations proposed in the literature, [151, 212].

Generally, one is looking for a transformation $h$ such that $h(x(0))$ has probability density as close as possible to the observed one. In Rychlik et al. (1997) [182], a different approach was taken. It was proposed to use the expected upcrossing intensity $\mu^+(u)$ to choose the transformation. If $\mu^+(u)$ is not known, then the observed intensity $N^+(u) = N^{osc}(u,u)$ can be used to estimate the upcrossing intensity $\widehat{\mu}^+(u)$ of level $u$ by the $y$-process. If $\widehat{\mu}^+(u)$ is unimodal, then there is a transformation $h$ such that $h(x(\tau))$ has the same crossings as $y(\tau)$, see (10.19). The advantage of this approach is that for damage function $f(u,v) = v - u$, the model $h(x(\tau))$ and $y(\tau)$ have the same damage intensity. Also, the upper bound $d^+(h(x(\tau)))$ is the same as $d^+(y)$. In Machado (2002) [135], several methods to estimate the transformation were compared and further developed.

**Example 10.2** (Moment-based and crossings-based transformations)**.** *We use the Winterstein Hermite transformation [213] and the Rychlik crossing-based transformation [182] to illustrate how one can relate the Gaussian linear oscillator to the Laplace moving average with the same spectrum but with different kurtosis.*

*The spectrum* (9.8) *is*

$$S_X(\omega) = m^{-2} \frac{1}{(\omega_0^2 - \omega^2)^2 + 4\alpha^2\omega^2},$$

*and we compare the Gaussian solution with symmetric LMA with kurtosis* $\kappa = 5$.

*The Winterstein transformation* $y = h_w(x)$ *is a cubic expression in the Gaussian process* $x$ *with coefficients that are functions of mean, standard deviation, skewness and kurtosis of the target process. The Rychlik transformation* $y = h_r(x)$ *is a smoothed non-parametric transform that approximates the crossing spectrum of* $y$ *as close as possible to that of the target process.*

*The procedure is as follows:*

- *Simulate long realisations of a GMA and a symmetric LMA with spectrum* $S_X(\omega)$ *and kurtosis* $\kappa = 5$ *and* 10.
- *Compute the first four moments of the LMA and find* $h_w$.
- *Compute the crossing spectrum of LMA and find* $h_r$.
- *Compute* $LMA_w = h_w(GMA)$ *and* $LMA_r = h_r(GMA)$ *and find their crossing spectra.*
- *Plot the three crossing spectra and approximate the rainflow damage intensity by* $d^+$, *for symmetric crossing spectra equal to* $d^{nb}$.

*As seen in Figure 10.1, both transforms work well and give approximately the same crossing spectrum as the LMA and the same rainflow damage intensity. The* $h_r$*-transformation is designed to fit the crossing spectrum of data, i.e. the LMA, and obviously it succeeds doing that. The* $h_w$*-transformation is based on the empirical moments of data, and for the LMA with high kurtosis* $\kappa = 10$ *that becomes statistically more difficult. Together with the cubic transformation that can explain the less good fit in the right plot for that transformation.*

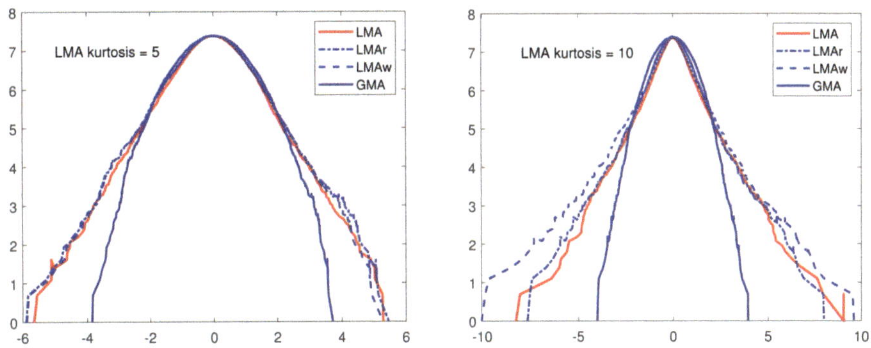


**FIGURE 10.1** Crossing spectrum for LMA and GMA (solid) and GMA transformed to LMAw and LMAr.

#### 10.3.1.3 Examples for Evaluating $d^+$ for Transformed Gaussian Load

We now illustrate the estimation of $d^{\mathrm{rfc}}$ by means of the upper bound $d^+$, (10.10), for the transformed Gaussian loads.

Assume that $x(t)$ is a zero-mean, stationary, Gaussian process with apparent frequency $f_z$ and significant amplitude $h_s = 4\sqrt{\omega_0}$, ($\omega_0$ is the variance of $x$). In Lemma 7.1, the upper bound for rainflow damage intensity $d^+$ was computed for $x$ and $f(u,v) = (v-u)^\beta$. From the lemma, it follows that

$$d^+(x) = f_z h_s^\beta 2^{-\beta} \mathsf{E}[R^\beta],$$

where $R$ is standard Rayleigh distributed range. Under assumptions of this section, if $y(\tau) = h(x(\tau))$, then with $G$ defined by (10.7),

$$d^+(y) = f_z \mathsf{E}[G(-h_s R/4, h_s R/4)].$$

**Example 10.3.** *We first consider the transformed load process $h(x(t)) = c|x(t)+m|$ and assume first that $m = 0$. It can be shown that $G(-h_sR/4, h_sR/4) = 2(ch_sR/4)^\beta$, and the upper bound can be computed explicitly, from Eq.* (7.14), *viz.*

$$d^+(y) = 2^{-2\beta+1} c^\beta f_z h_s^\beta \mathsf{E}[R^\beta],$$

*where* $\mathsf{E}[R^\beta] = 2^{\beta/2}\Gamma(\beta/2+1)$.

*We next consider the case when $m > 0$ ($m < 0$ can be treated in the similar way). Here, the expression for $d^{\mathrm{rfc}}$ is less explicit and is of the form*

$$d^+(y) = c^\beta f_z h_s^\beta 2^\beta \int_0^{\frac{m}{\sqrt{\omega_0}}} r^{\beta+1} e^{-r^2/2} \mathrm{d}r$$
$$+ c^\beta f_z h_s^\beta \times \int_{\frac{m}{\sqrt{\omega_0}}}^{\infty} \left[ \left( r + \frac{m}{\sqrt{\omega_0}} \right)^\beta + \left( r - \frac{m}{\sqrt{\omega_0}} \right)^\beta \right] r e^{-r^2/2}\, \mathrm{d}r.$$

*Evaluation in closed form of the above integral is not possible but an approximation can be obtained by adopting numerical computing techniques.*

### 10.3.2 MULTIVARIATE LOADS: FATIGUE DAMAGE DUE TO WIND

In this section we take a closer look at the damage intensity and wind load variability.

**Example 10.4.** *We consider the wind transformation* $h(x(t)) = c(x(t)+m)^2$. *Again, for* $m = 0$, *the upper bound for damage intensity can be computed explicitly and shown to be of the form* $d^+(y) = 2c^\beta f_z h_s^\beta \mathsf{E}[R^{2\beta}]$, *where* $h_s = 4\sqrt{\omega_0}$. *When* $m > 0$, *the bound can be written as a one-dimensional integral of the form*

$$d^+(y) = c^\beta f_z h_s^\beta \int_0^{\frac{m}{\sqrt{\omega_0}}} \left[\left(r+\frac{m}{\sqrt{\omega_0}}\right)^2 - \left(r-\frac{m}{\sqrt{\omega_0}}\right)^2\right]^\beta re^{-r^2/2}\,\mathrm{d}r$$
$$+ c^\beta f_z h_s^\beta \int_{\frac{m}{\sqrt{\omega_0}}}^{\infty} \left[\left(r+\frac{m}{\sqrt{\omega_0}}\right)^{2\beta} + \left(r-\frac{m}{\sqrt{\omega_0}}\right)^{2\beta}\right] re^{-r^2/2}\,\mathrm{d}r.$$

The following more detailed analysis of expected rainflow fatigue damage due to wind loadings is based on Rychlik and Gupta (2007) [181]. In this, we will compare the narrow-band bound with a more elaborated method to estimate of damage rate for the wind loads using the Markov chain approximation (MCTP) and numerical simulations. It will demonstrate that in this case, the narrow-band bound is very conservative. For illustrating the proposed method, we ignore the effect of structural vibrations and assume that the fatigue damage is due to the load $F(t)$ presented in Section 10.3.1.1.

#### 10.3.2.1 The Parameters

The force due to wind loadings is given by $F(t) = 0.5C_d\rho lb(V(t)+m_v)^2$, where $V(t)$ is a zero-mean Gaussian random process denoting the wind velocity, $m_v$ is the mean velocity of the wind, $C_d$ is the drag coefficient, $\rho$ is the air density and $l$ and $b$ are the dimensions of a structure. The numerical values for the above parameters have been taken to be $m = 10$ m/s, $C_d = 2\times 10^{-3}$, $\rho = 1.25$ kg/m$^3$, $l = 20$ m and $b = 1$ m.

We assume that $V(t)$ follows the Davenport spectrum from Example 9.8, given by $S_V(f) = \frac{2}{3}\frac{\sigma^2}{f}x^2/((1+x^2)^{4/3})$. Here, $f$ is frequency in Hz, $x = 1200f/V_r$, $V_r$ is the reference wind velocity taken to be 40 m/s and $\sigma = 3.54$. The duration of the storm is assumed to be one hour. The damage function is $f(u,v) = \alpha(v-u)^2$, where $\alpha = 6.94\times 10^{-6}$.

#### 10.3.2.2 Estimation of Damage Rate

The computation of the functions $g(u,v)$ and $G(u,v)$ follows the expressions in Eqs. (10.13–10.14). The contour plots for the new damage functions $g(u,v)$ and $G(u,v)$ are shown in Figures 10.2 top and bottom-plot, respectively. The fact that these functions have non-zero values in one half of the $u-v$ space can be attributed to the fact that the damage function $f(u,v) = 0$ for $v \le u$.

We compute the expected rainflow fatigue damage due to wind loadings of a duration of 200 seconds. An estimate of the expected rainflow fatigue damage can

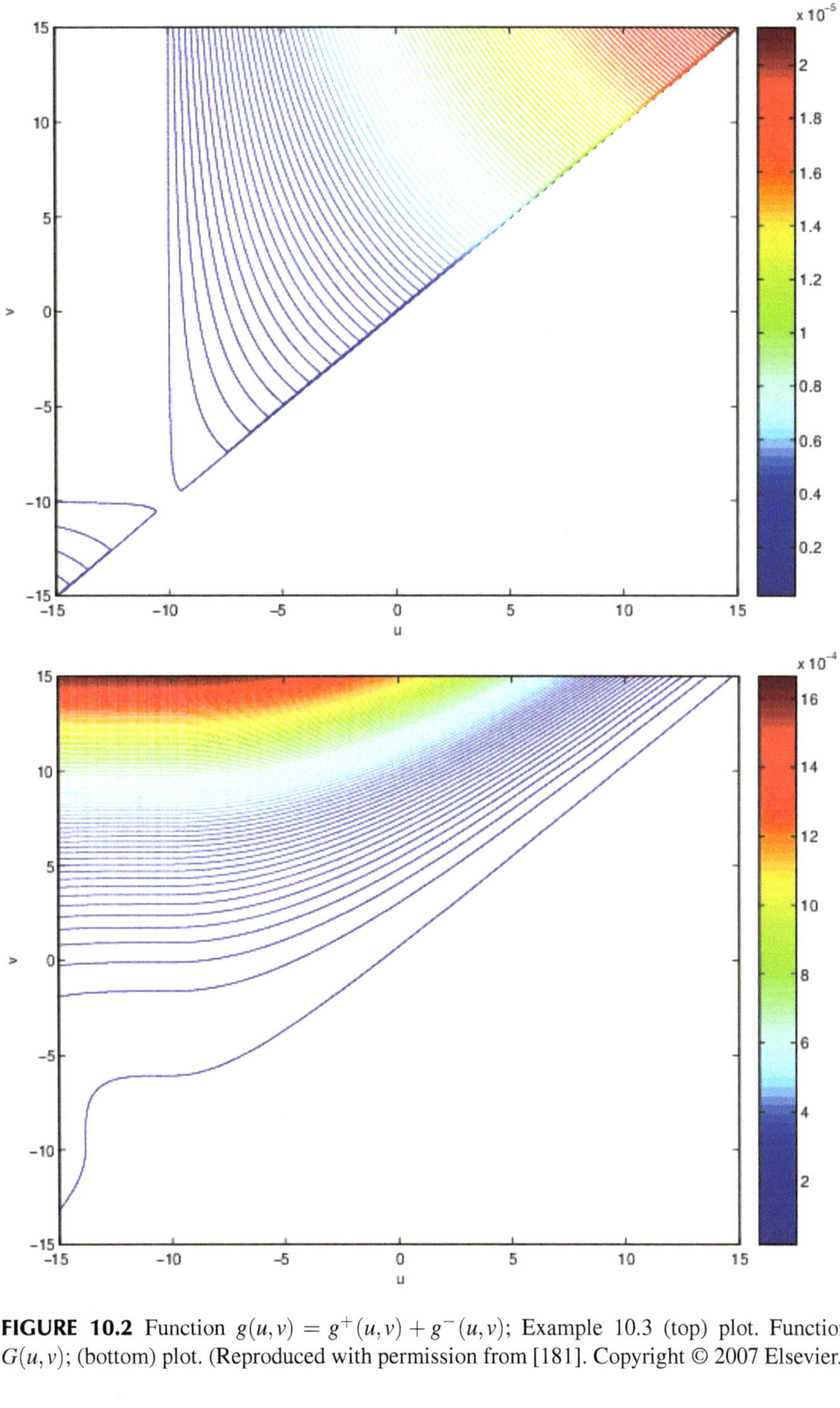


**FIGURE 10.2** Function $g(u,v) = g^{+}(u,v) + g^{-}(u,v)$; Example 10.3 (top) plot. Function $G(u,v)$; (bottom) plot. (Reproduced with permission from [181]. Copyright © 2007 Elsevier.)

be obtained from (4.31). This however requires the knowledge of the joint interval crossing rate $\mu^{osc}(u,v)$. Since this is not easy to compute, we obtain an upper bound for $\mu^{osc}(u,v)$, from Eq. (7.25).

Figure 10.3, top-plot, illustrates the contour plots for $k^+(u,v)$, computed using Eq. (7.25). This leads to a numerical value of 0.636 for the upper bound for the expected rainflow fatigue damage. (Note it is the same value as one would derive by means of the formulas given in Example 10.4.) The observed mean fatigue damage in $Y(t)$, obtained from Monte Carlo simulations on a sample size of 2000 time histories of $Y(t)$, is obtained as 0.125. So, the method is extremely conservative. Consequently, a better approximation of $\mu^{osc}(u,v)$ is needed.

The MCTP method from Section 8.3.2 is then used to approximate $\mu^{osc}(u,v)$. Let $\widehat{\mu}^{osc}(u,v)$ be the derived approximation; see Figure 10.3, bottom-plot. Now, using (4.31) with $\mu^{osc}(u,v)$ replaced by $\widehat{\mu}^{osc}(u,v)$ gives a numerical value for the expected damage to be 0.132 which is very close to the simulated values.

#### 10.3.2.3 A Dynamical Model for the Response of a Structure to Wind Load

Neglecting the dynamical responses of a structure subjected to wind loads is usually a too crude simplification, underestimating the damage. Here, we shall sketch the approach that would include the dynamics of the structure. We will use the Gaussian load from Gupta and Rychlik (2007) [74]. The Laplace load was used with the same model in [89]. Crossing intensities in non-linear response processes under LMA loadings were also the subject in [66]. Jith et al. [88] treated local maxima in non-linear LMA wave loads by the same methods.

The general form of the governing equations of motion, when discretised using finite elements, can be written as

$$\mathbf{M}\ddot{\mathbf{Y}}(t)+\mathbf{C}\dot{\mathbf{Y}}(t)+\mathbf{K}\mathbf{Y}(t)=\mathbf{F}(t). \tag{10.15}$$

Here, $\mathbf{M}$, $\mathbf{C}$ and $\mathbf{K}$ are the structure mass, damping and linear stiffness matrices of dimensions $n\times n$, respectively. $\ddot{\mathbf{Y}}(t)$, $\dot{\mathbf{Y}}(t)$ and $\mathbf{Y}(t)$ are, respectively, the vectors of nodal accelerations, velocities and displacements, and $\mathbf{F}(t)$ is the vector for nodal forces, of size $n\times 1$. Let the focus of our attention be the structure response at a particular location of the structure and denote it by $Y(t)$. We assume that the external excitation can be modelled as stationary Gaussian processes $\{X_j(t)\}_{j=1}^n$ and that the force vector $\mathbf{F}(t)$ is obtained as a non-linear function of the external excitations $\{X_j(t)\}_{j=1}^n$.

Now, we consider the special case when $\mathbf{F}(t)$ is the wind load discussed in Example 10.4, and we consider the rainflow fatigue damage at the support of a cantilever beam illustrated in Figure 10.4, subjected to randomly varying wind loads. Now, the dynamical responses of the structure will be included. The beam is modelled as a simple lumped-mass model, i.e. by (10.15).

As before, the wind velocity is modelled as a stationary Gaussian process with specified mean wind speed, $m_v$, and Davenport spectrum $S$. The variation of the wind velocity along the beam span is taken to be deterministic and is assumed to be parabolic. The vibration analysis of this structure gives the stress $Y(t)$, which,

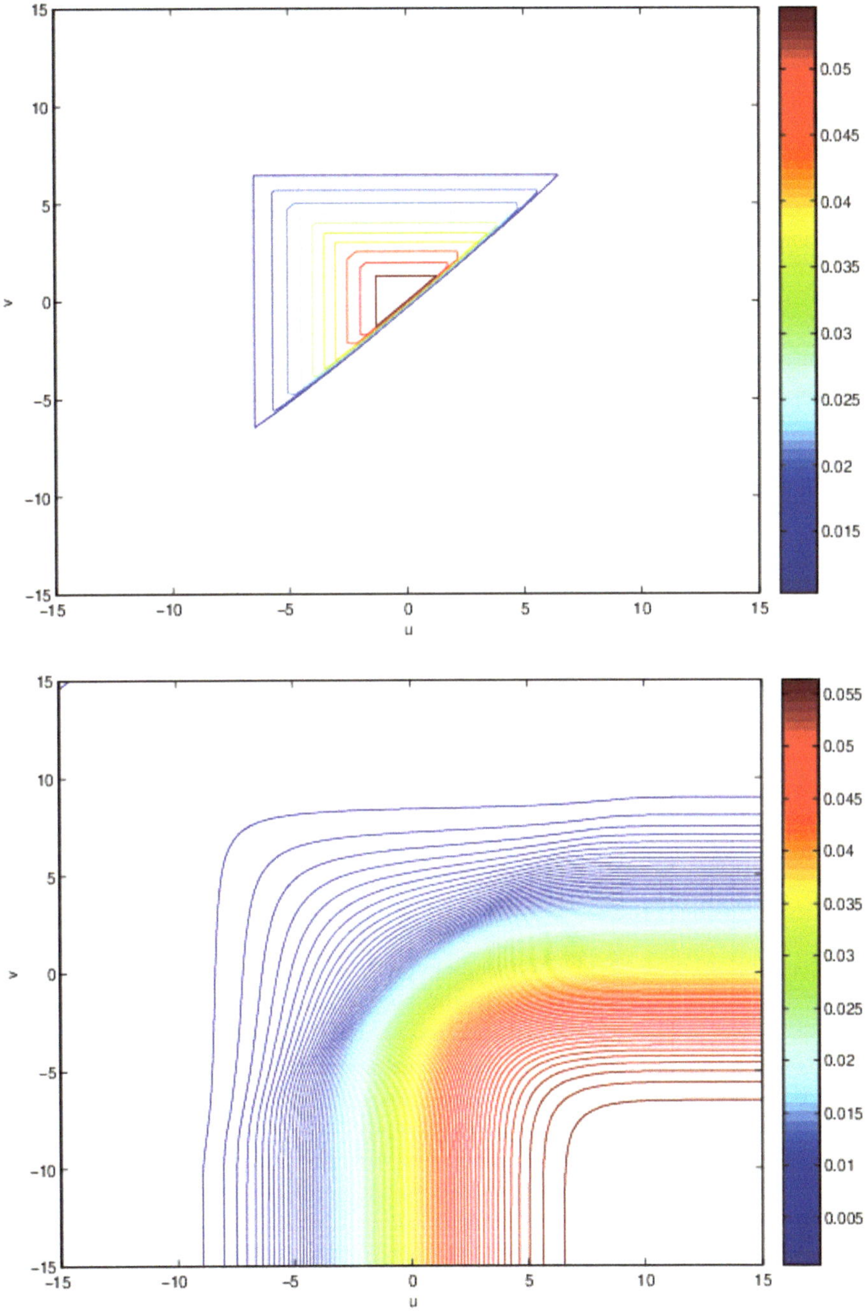


**FIGURE 10.3** Example in Section 10.3.2. Upper bound $k^{+}(u,v)$ for $\mu^{osc}(u,v)$; (top) plot. MCTP approximation for $\widehat{\mu}^{osc}(u,v)$; (bottom) plot. (Reproduced with permission from [181]. Copyright © 2007 Elsevier.)

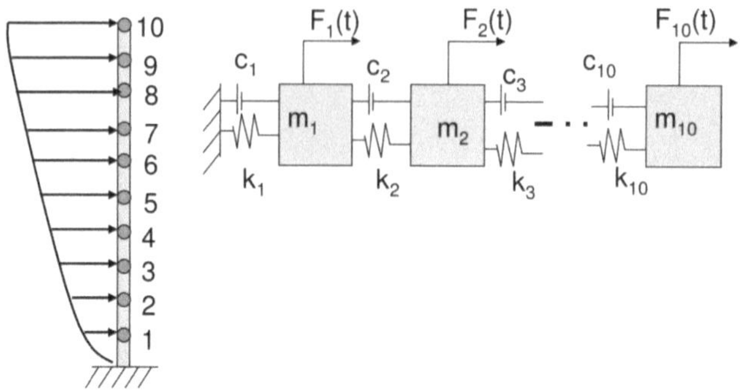


**FIGURE 10.4** Schematic diagram of the 10-dof cantilever beam subjected to random wind excitations. (Reproduced with permission from [74]. Copyright © 2007 Elsevier.)

after some algebra, see Gupta and Rychlik (2007) [74], has the following standard form:

$$Y(t) = \sum_{j=0}^{n} \gamma_j X_j(t) + \sum_{j=0}^{n} \frac{\lambda_j}{2} X_j(t)^2, \tag{10.16}$$

where $\mathbf{X}(t) = (X_1(t), \ldots, X_n(t))$ is a vector-valued stationary Gaussian process, such that for each $t$, $X_j(t) \in N(0,1)$ and the variables $X_j(t), X_k(t)$ are independent. If $\dot{\mathbf{X}}(t) = (\dot{X}_1(t), \ldots, \dot{X}_n(t))$, then $(\mathbf{X}(t), \dot{\mathbf{X}}(t)) \in N(\mathbf{0}, \Sigma)$, where

$$\Sigma = \begin{bmatrix} I & \Sigma_{12} \\ \Sigma_{12}^T & \Sigma_{22} \end{bmatrix}, \tag{10.17}$$

and $I$ is the identity matrix (note that matrices $\Sigma_{12}, \Sigma_{22}$ do not need to be identity matrices). (Many processes can be transformed into this form, e.g. the von Mises stress, envelope processes as in [124], Stokes waves, motion of a moored floating structure, second-order sea-surface and many others.)

Finally, it must be noted that a Gaussian approximation for $Y(t)$ can be obtained by letting all $\lambda_j = 0$ in Eq. (10.16). The motivation for considering this approximation lies in the fact that dealing with Gaussian processes is much simpler, and the quality of this approximation vis-a-vis dealing with processes of the form of Eq. (10.17) needs to be investigated. Thus, we consider the Gaussian approximation $Y_L(t)$, given by

$$Y_L(t) = \sum_{i=1}^{k+1} \gamma_i X_i(t), \tag{10.18}$$

having spectrum shown in Figure 10.5 (top plot).

#### 10.3.2.4 Upper Bound and Markov Approximation for Damage Intensity

Studying the increase in the expected rainflow damage for response $Y(t)$ is a very difficult problem. One obvious possible approach would be to simulate the loads

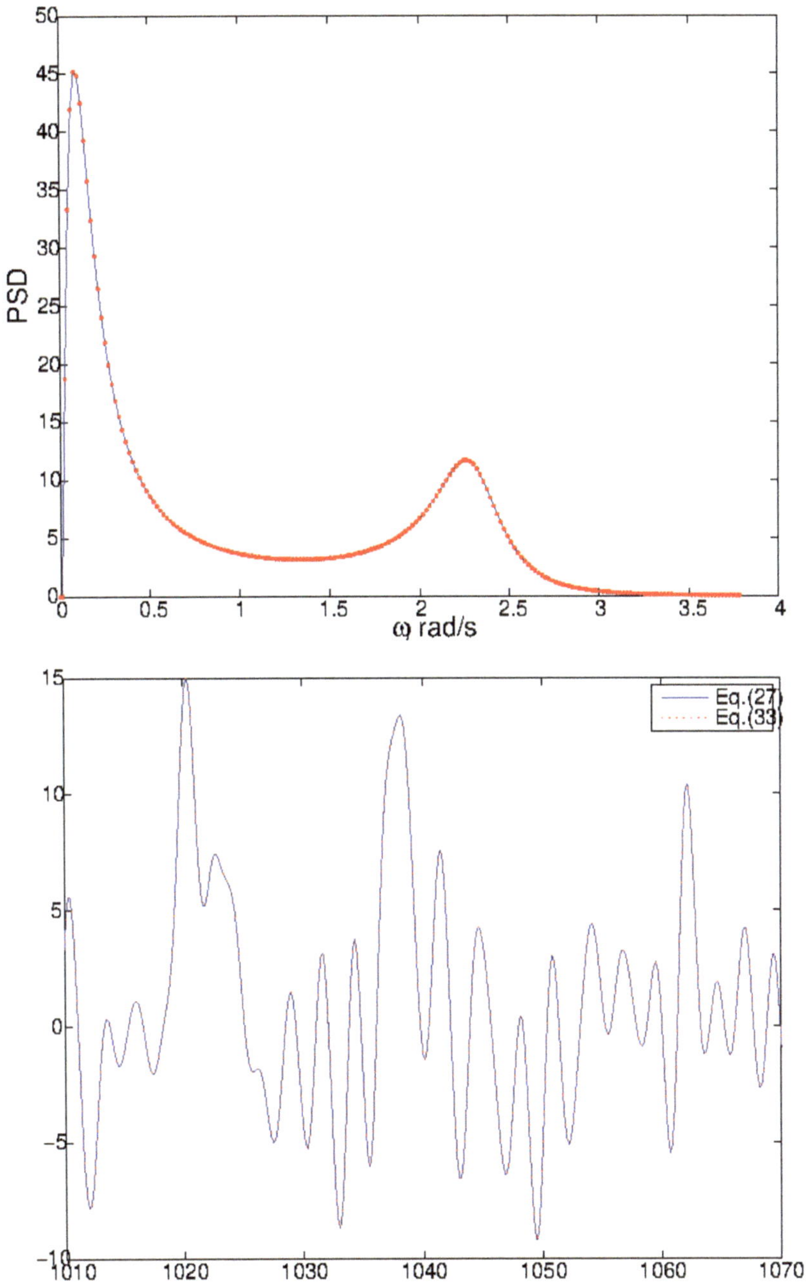


**FIGURE 10.5** The spectral density of the linear approximation $Y_L$ (top plot). Comparison of simulation of $Y(t)$ using $Y(t)$ with $n = 400$ in (10.16) and approximation when only $n = 9$ with the largest absolute values $\lambda_j$ and remaining $\lambda_j$ in (10.16) were set to be zero, (bottom plot). (Reproduced with permission from [74]. Copyright © 2007 Elsevier.)

and to compute the corresponding damage due to the response. However, in this approach, one needs to simulate an ensemble of long sequences of the response to get reliable estimates of the damage. This can be quite time-consuming and expensive, especially at the design stage which includes repetitive iterations. Thus, there is a need for alternative methods which lead to approximate estimates of the damage in a fast and reliable manner.

Two approaches will be taken to estimate the damage intensity. The first one is to compute the upper bound $k^+(u,v)$ for the $\mu^{osc}(u,v;Y)$ and then employ master-formula (4.31) to get an upper bound for the rainflow damage intensity. Here, we need to estimate the upcrossing intensity of $\mu^+(u)$ for the process $Y$, defined in (10.16).

The alternative approach is to use $\mu^+(u)$ to define a transformation $h$ and approximate $Y(t)$ by means of $h(Y_L(t))$. Then, the damage intensity of $h(Y_L(t))$ could be computed if the oscillation intensity $\mu^{osc}(u,v;Y_L)$ were known. Here, we shall approximate $\mu^{osc}(u,v;Y_L)$ by means of an MCTP model and approximate $Y(t)$ by a Markov load.

In the following, the asymptotic SORM method proposed by Breitung (1988) [26], see also Appendix 10.A, is used to approximate $\mu^+(u)$. The derived value is denoted by $\mu^a(u)$.

First, using $\mu^a(u)$, the upper bound $k^+(u,v)$ is approximated by means of $\min_{u\le z\le v}\mu^a(z)$. The results, as reported in [74], are that for $\beta = 1$, the approximate bound is close to the observed damage. For higher values of $\beta$, the bounds are progressively higher and with $\beta = 6$, the approximate bound is about 60% higher than the damage estimated from 100 realisations of response $Y(t)$ of duration 1500 s. However, because 1500 seconds is a relatively short period of time, the observed damage intensity is biased (too small). The bias was estimated to be about 20%, so the true bias is about 40%.

The second approach is to use the crossing spectrum-based method described in Section 10.3.1.2. It transforms a Gaussian load $Y_L$ to $h(Y_L(t))$, a load that has approximately the same crossing spectrum as $Y(t)$, and then estimate the rainflow damage of $h(Y_L(t))$. Following Rychlik et al. (1997) [182], the inverse $h^{-1}(u)$ transformation is given by

$$h^{-1}(u) = \begin{cases} \sqrt{-2\mathrm{Var}[Y_L(0)]\ln(\mu^a(u)/\mu^a(0))} & \text{if } u > 0, \\ -\sqrt{-2\mathrm{Var}[Y_L(0)]\ln(\mu^a(u)/\mu^a(0))} & \text{if } u < 0. \end{cases} \tag{10.19}$$

The process $h(Y_L)(t)$ crosses the level $u$ with intensity $\mu^a(u)$. This is different from the crossing intensity of the response $\mu^+(u;Y)$ for low $u$-levels. In this example, the difference is small (less than 5%) when checked by Monte Carlo simulations. This discrepancy in the computation of crossing intensities for low $u$-levels is more than offset by the gain in the simplicity of the approach, which is a consequence of the fact that the zero upcrossing intensity of $h(Y_L(t))$ process is equal to $\mu^a(0)$; see Appendix 10.A.

As before, the intensity $\mu^{osc}(u,v;Y_L)$ can be approximated by MCTP which gives a Markov load approximation to $Y(t)$. The steps of this approximations are illustrated in Figure 10.6. The damage intensity using transformed MCTP agrees well with the simulated ones.

## 10.4 LOAD WITH VARIABLE SIGNIFICANT AMPLITUDE

In Section 10.3.2, the focus was on the normal variation and its linear and non-linear effects on a complex structure. In this section, the focus is on the statistical character of the load variation, in particular the variation of significant amplitude and frequency. The distribution of $h_s, f_z$ may depend on the type of application. In this section, we shall present two typical cases. The first one is an environmental load when its properties change seasonally (periodically); see Figure 10.7 with measured $h_s(\tau)$. The model is a non-stationary Gaussian model. In the second example, the variability is a consequence of planned missions which define the load variability, e.g. by selection of the type of roads a heavy truck will experience when transporting goods, or planning routes for ships. Here, the planning process often makes it necessary to use a non-Gaussian model.

### 10.4.1 THE NARROW-BAND APPROXIMATION

If the stress $y(\tau)$, $0 \leq \tau \leq t$, is a non stationary process, then the intensity of upcrossings of level $u$ by $y$ may depend on time. Hence, by means of a generalisation of Rice's formula, the time-dependent upcrossing intensity is given by

$$\mu_\tau^+(u) = \int_0^\infty z f_{\dot{y}(\tau),y(\tau)}(z,u)\mathrm{d}z. \tag{10.20}$$

The expected number of times the stress $y$ passes $u$ in upward direction in the interval $[0,T]$ is given by

$$\mu_T^+(u) = \int_0^T \mu_\tau^+(u)\,\mathrm{d}\tau. \tag{10.21}$$

Clearly, using $\mu_T^+(u)$, the upper bound $d^+$ for the rainflow damage can be computed by means of (7.26). However, this is not very convenient since, when time $T$ increases, $\mu_T^+(u)$ has to be updated by means of (10.21) and the double integral in (7.26) recomputed. This can be avoided if the local narrow-band approximation can be employed as will be sketched next.

If, for any $\tau$, $\mu_\tau^+(u) = \mu_\tau^+(-u)$ and $\mu_\tau^+(|u|) > \mu_\tau^+(|v|)$ whenever $|u| < |v|$ then, following Bendat's approach, see (4.44), for any $\tau$ one can define a local amplitude $S_\tau$ having the distribution

$$\mathsf{P}(S_\tau \leq u) = 1 - \frac{\mu_\tau^+(u)}{\mu_\tau^+(0)}, \tag{10.22}$$

and then introduce the local narrow-band damage intensity, for $f(u,v) = (v-u)^\beta$, $d^{\mathrm{nb}}(\tau)$ as follows:

$$d^{\mathrm{nb}}(\tau) = \mu_\tau^+(0)\mathsf{E}[(2\cdot S_\tau)^\beta] = 2^\beta\,\beta \int_0^\infty u^{\beta-1}\mu_\tau^+(u)\,\mathrm{d}u. \tag{10.23}$$

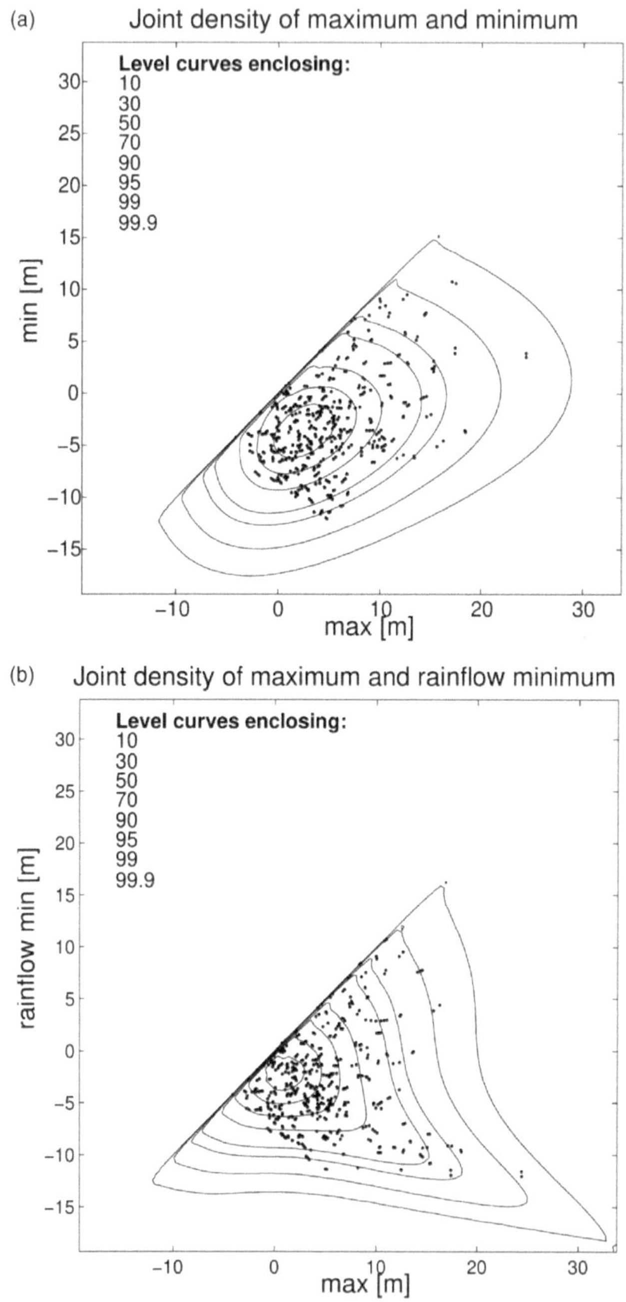


**FIGURE 10.6** (a) Contour lines of the long-run probability density in the transformed Gaussian process $h(Y_L(\tau))$, where $Y_L(\tau)$ is defined by (10.18), while $h(\cdot)$ is defined by (10.19). Dots represent observed min-max cycles in simulated realisations of the response $Y(\tau)$. (b) Contour lines of the expected rainflow matrix for the MCTP load defined by the transition matrices $P^{mM}$, $P^{Mm}$ computed using the density from the left plot. Dots are observed rainflow cycles in the response $Y(\tau)$. (Reproduced with permission from [74]. Copyright © 2007 Elsevier.)

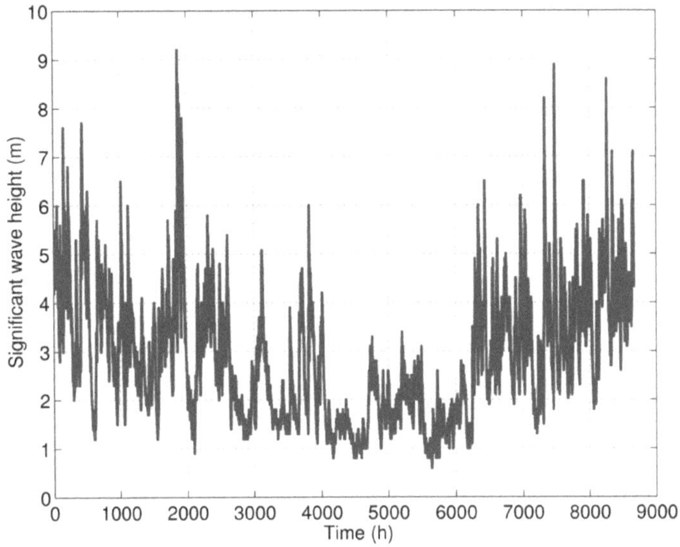


**FIGURE 10.7** One-year measurements of $h_s(\tau)$.

**Lemma 10.1.** *If, for any* $\tau$, $\mu_\tau^+(u) = \mu_\tau^+(-u)$ *and* $\mu_\tau^+(|u|) > \mu_\tau^+(|v|)$ *whenever* $|u| < |v|$, *then*

$$\mathsf{E}[D^{rfc}(t)] \leq \int_0^t d^{nb}(\tau)\,\mathrm{d}\tau. \tag{10.24}$$

For a zero-mean Gaussian load $y(\tau)$ denote the time-dependent variances of $y(\tau), \dot{y}(\tau)$, by $\sigma_y^2(\tau)$, $\sigma_{\dot{y}}^2(\tau)$, respectively, and the correlation between $y(\tau)$ and $\dot{y}(\tau)$ by $\rho(\tau)$. Using (10.20), the upcrossing intensity $\mu_\tau^+(u)$ can be computed explicitly. However, formula (10.57) is quite complex, and we give it in Appendix 10.B. Now, if $\rho(\tau) = 0$, then assumptions of the last lemma are satisfied, and by (7.14), the time-dependent narrow-band approximation of the damage intensity is

$$d^{nb}(\tau) = f_z(\tau)\, h_s(\tau)^\beta\, 2^{-\beta/2}\Gamma(1+\beta/2), \tag{10.25}$$

where $h_s(\tau) = 4\sigma_y(\tau)$ and $f_z(\tau) = (2\pi)^{-1}\sigma_{\dot{y}}(\tau)/\sigma_y(\tau)$, and, by (7.15),

$$d^{nb}(\tau) = \frac{1}{\pi}\,\sigma_{\dot{y}}\sigma_y^{\beta-1}\, 2^{3\beta/2-1}\Gamma(1+\beta/2). \tag{10.26}$$

The expected total damage gets the narrow-band bound

$$\mathsf{E}[D^{rfc}(T)] \leq \int_0^T d^{nb}(\tau)\,\mathrm{d}\tau \equiv D^{nb}. \tag{10.27}$$

As mentioned before, for values of the parameter $\beta \in [2,4]$, which are the most often used values in offshore applications, $0.469 < 2^{-\beta/2}\Gamma(1+\beta/2) \leq 0.5$ and hence

$$\mathsf{E}[D^{rfc}(t)] \leq 0.5 \cdot \int_0^t f_z(\tau)\, h_s(\tau)^\beta\,\mathrm{d}\tau. \tag{10.28}$$

This is again an indication that significant amplitude is the relevant parameter when studying Gaussian loads. In the following, we shall demonstrate two applications of (10.28). The first one is for the offshore applications and the second for vehicle driving on a road with randomly located potholes.

## 10.4.2 GAUSSIAN LOAD IN A VARIABLE ENVIRONMENT

At the design stage, one is often interested in fast estimates of the expected damage and the coefficient of variation. Estimation of the coefficient of variation will be discussed in Chapter 11. Here, we will consider the evaluation of the expected damage. However, we shall use the same method to estimate the expected damage as we will use in the estimation of the coefficient of variation.

### 10.4.2.1 Non-Stationary Holm-de Mare Damage

Adaptation to non-stationary conditions can be achieved by approximating the rainflow damage rate by replacing in (4.15) $H^-(t) = Y(t) - Y^-(t)$ by $|2Y(t)|$, which gives the derivative of Holm-de Maré damage, discussed in Section 7.5, viz.

$$\dot{D}^{HM}(t) = \alpha\beta\,|2Y(t)|^{\beta-1}\dot{Y}(t)^+, \qquad x^+ = \max(0,x). \tag{10.29}$$

It was shown that for any load, the accumulated damage $D^{rfc}(T) \le D^{HM}(T)$. Except for the zero-mean symmetrical load, i.e. a load having symmetrical unimodal crossing intensity, the Holm-de Maré is often a very conservative bound and hence seldom used. Furthermore, in the case of stationary symmetrical loads, $\mathsf{E}[D^{HM}(T)]$ coincides with the narrow-band damage. However, it was successfuly used in the estimation of the coefficient of variation. In many cases, employing the Holm-de Maré damage rate (10.29) gives approximately the same results as when the rainflow damage rate was used.

Let us consider a stationary Gaussian process $Y(t)$, and the damage function $f(u,v) = \alpha(v-u)^\beta$, and the corresponding H-M damage function $G(z) = \alpha\,\mathrm{sign}(z)0.5|2z|^\beta$ , ($\dot{G}(z) = \alpha\beta|2z|^{\beta-1}$). Then, the H-M damage rate is

$$\dot{D}^{HM}(t) = \dot{G}(Y(t))\dot{Y}(t)^+ = \alpha\,\beta|2Y(t)|^{\beta-1}\dot{Y}(t)^+. \tag{10.30}$$

The Holm-de Maré damage intensity is

$$d^{HM} = \mathsf{E}[\dot{D}^{HM}(t)] = \widetilde{d}\,f_z\,h_s^\beta, \tag{10.31}$$

where $\widetilde{d} = \alpha\,2^{-\beta/2}\Gamma(1+\beta/2)$ is the damage for a standardised load with $f_z = 1$ and $h_s = 1$, given in (7.24).

Studying historical data, one however often observes that values of significant amplitude $h_s$ and apparent frequency $f_z$ change depending on time $t$. Then, one may also model the variability of $h_s(t), f_z(t)$ by means of a deterministic relation or, if the variation is rather unpredictable, by random processes. Then, damage intensity also becomes a function of time, $d^{HM}(t) = \mathsf{E}[\dot{D}^{HM}(t)]$. In the following example, we will use both deterministic and random modelling of the parameters.

However, if significant amplitude and frequency vary fast, e.g. changing values after a few zero crossings, then the distribution of $Y(t)$ is likely non-Gaussian and hence formula (11.13) is not applicable and estimation of $f_z, h_s$ is very difficult, if possible. Here, we will assume that $Y$ is Gaussian and locally stationary for a couple of hundred cycles, i.e. zero-level upcrossings. The parameters $h_s, f_z$ may change but in much slower rate. This will be expressed in the notation by introducing $h_s(\tau)$ and $f_z(\tau)$. In the following example, $t$ will be in seconds, while $\tau$ in days.

#### 10.4.2.2 Annual Variation of Wave State

In the first example, we shall estimate the expected wave damage at a fixed location in the North Atlantic. The load, denoted by $W(t)$, is the water elevation. We consider the variation over a one-year period but also shorter seasons: The two months in spring, and in the fall.

Suppose that the stress at some location in an offshore structure is proportional to sea surface elevation $W(t)$, usually $t$ has units seconds. If the wave conditions are stationary for a period $T$, which may vary between 20 minutes up to 3 hours, then the accumulated Holm-de Maré damage is – with the $\alpha$ in the example set to 1 –

$$D(T) = \int_0^T \beta\, |2Y(t)|^{\beta-1} \dot{Y}(t)^+ \, \mathrm{d}t, \tag{10.32}$$

with expectation $\mathsf{E}[D(T)] = T\, d^{HM}$, and $d^{HM} = \widetilde{d} f_z h_z^{\beta}$.

However, for longer periods $T$, the variability of significant wave heights cannot be neglected; see Figure 10.7. Since damage rate $\dot{D}^{HM}(\tau)$ depends on the randomly varying significant wave amplitude, we need to use the formula for averaging conditional expectation (2.29),

$$d^{HM}(\tau) = \mathsf{E}[\mathsf{E}[\dot{D}^{HM}(\tau) \mid h_s(\tau), f_z(\tau)]], \tag{10.33}$$

$\tau$ in days. Consequently, we have to model variability of Holm-de Maré damage rate in random environments, i.e. apparent frequency $f_z(\tau)$ and significant wave height $h_s(\tau)$. We turn next to presentation of a model proposed in Baxevani and Rychlik (2007) [7] .

#### 10.4.2.3 Modelling $h_s(\tau)$ and $f_z(\tau)$ Variability

**Modelling significant waveheight $h_s(\tau)$:** Based on extensive studies of $h_s$, estimated using satellite measurements, it was proposed that variability of $h_s$ at a fixed location in the North Atlantic, can be approximated by a log normal distribution with seasonally varying median, viz.

$$\begin{aligned} \log h_s(\tau) &= m_0 + a_1 \sin(\phi\tau) + a_2 \cos(\phi\tau) + e(\tau) \\ &= m_0 + m(\tau) + e(\tau), \quad \phi = \frac{2\pi}{365.2}, \end{aligned} \tag{10.34}$$

with $\tau$ in days, and where residual $e(\tau)$ is a zero-mean, variance $\sigma_H^2$, stationary Gaussian process. The parameters $a_1, a_2$ and $m_0, \sigma_H^2$ for locations along a specific North

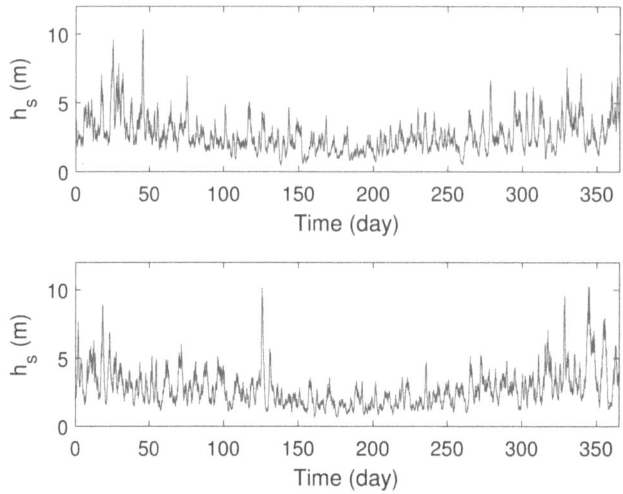


**FIGURE 10.8** Two replicates of simulated one year of daily $h_s$.

Atlantic route were estimated using satellite observations, see [7], where values of parameters can be found. For example, for location –53 W and 50 N, it was estimated $(a_1, a_2, m_0, \sigma_H^2) = (0.39, 0.064, 0.87, 0.16)$. The parameters define the annual variability of $h_s$ (10.34). In the model, $m_0 + m(\tau)$ is the regression fitted to the logarithm of observed $h_s(\tau)$, while $e(\tau)$ is a residual. Often, in regression analysis, one is assuming that the residual is iid Gaussian noise. However, for significant wave data, this assumption is too crude, and one has to model $e(\tau)$ as a stationary Gaussian process. The measured $h_s$ by the buoys shows large scatter around the seasonal mean which is included in the model by means of the residual process $e(\tau)$. The covariance function of $e(\tau)$ was estimated using buoy measurements. Here, we used US NODC Buoy 46005, and the fitted correlation was

$$\mathsf{Cov}[e(\tau_1), e(\tau_2)] = e^{-0.576(\tau_2-\tau_1)^2-0.24|\tau_2-\tau_1|}, \tag{10.35}$$

where $\tau_i$ have units days. Based on this model, we conclude that the $e(\tau)$ are almost uncorrelated after three days. In this time interval, the median $m(\tau)$ is practically constant.

The model seems to replicate nature in Figure 10.7 quite well; in Figure 10.8, one can see the seasonal variation in $h_s(\tau)$ both in mean level and in variability, which is quite large.

**Modelling apparent frequency $f_z$:** Using data coming from radar distance measurements made from the QP structure in the Frigg Field, North Sea, situated 230 km northwest of Stavanger, Labeyrie (1990) [113] studied joint variability of $f_z, h_s$ and proposed the following regression relation between $f_z^{-1}$ and $h_s$,

$$f_z^{-1} = 10^{-3}\sqrt{0.617\,h_s + 1.62} \approx 7.9\,10^{-4}\sqrt{h_s}. \tag{10.36}$$

Units of $f_z^{-1}$ are hours, and a relative error was less than 1%. Since the error is small, we use this deterministic model. This model agrees with a general knowledge that fully developed sea often has a Pierson-Moskowitz spectrum, Appendix 7.B, for which $2.5\sqrt{h_s} < f_z^{-1} < 3.5\sqrt{h_s}$. Note that neglecting the constant 1.62 in (10.36) and changing units to seconds, one obtains that $f_z^{-1} = 2.83\sqrt{h_s}$. Consequently, for simplicity, we shall use the deterministic relation $f_z = \frac{1}{3\sqrt{h_z}}$ per second,

$$f_z = 28800/\sqrt{h_s}, \quad \text{per day.} \tag{10.37}$$

#### 10.4.2.4 The Holm-de Mare Damage Intensity

We compute the expected Holm-de Maré damage rate $d^{HM}(\tau) = \mathsf{E}[\dot{D}^{HM}(\tau)]$ as a function of time $\tau$. Since $\dot{D}^{HM}(\tau)$ depends on the randomly varying significant wave amplitude, we use formula (2.29),

$$d^{HM}(\tau) = \mathsf{E}[\mathsf{E}[\dot{D}^{HM}(\tau) \mid h_s(\tau), f_z(\tau)]].$$

Under the model (10.34), we get[1]

$$\begin{aligned} E[\dot{D}^{HM}(\tau) \mid h_s, f_z] &= 28800\,\widetilde{d}\,h_s(\tau)^{\beta-1/2} \\ &= c\,\widetilde{d}\,\exp((\beta-0.5)(m(\tau)+\sigma_H\, e(\tau))), \end{aligned} \tag{10.38}$$

where $c = 28,800\exp(m_0\,(\beta-0.5))$, $c = 2.5\,10^5$ when $\beta = 3$. Further, $\tau$ is in days, and $e(\tau)$ is a standard Gaussian variable. From now, $c$ is a universal constant, whose value may change between the equations.

Taking expectation once more,[2] we get the damage intensity as a function of time,

$$d^{HM}(\tau) = \mathsf{E}[c\cdot\widetilde{d}\,e^{(\beta-0.5)\cdot(m(\tau)+e(\tau))}] = c\widetilde{d}e^{(\beta-0.5)\cdot m(\tau)},\ c = 4.18\cdot 10^5. \tag{10.39}$$

Finally, the expected accumulated damage in the interval $[\tau_1, \tau_2]$ is bounded by

$$\mathsf{E}[D^{rfc}(\tau_1,\tau_2)] \le \mathsf{E}[D^{HM}(\tau_1,\tau_2)] = c\widetilde{d}\int_{\tau_1}^{\tau_2} e^{2.5\cdot m(\tau)}\,\mathrm{d}\tau, \tag{10.40}$$

where $\widetilde{d} = \alpha\,2^{-\beta/2}\Gamma(1+\beta/2)$ is the standardised H-M damage intensity for a load with $h_s = f_z = 1$; see (11.13). For example, at the chosen location and $\alpha = 10^{-8}$, $\beta = 3$, the expected accumulated rfc damage in months April, May is bounded by 0.098, while in October, November by 0.181.

For the considered location, the expected rainflow damage is given in Table 10.1. (Note that $\alpha$ in the damage function and in the Holm-de Maré $G$-function is a small constant derived from the S-N curve. Here, for convenience only, we choose $\alpha = 10^{-8}$ so that the expected damage for $\beta = 3$ is approximately equal one.)

#### 10.4.2.5 New Challenges for Marine Safety

Ocean crossing ship structures are continuously suffering from wave loads when sailing at sea. The wave loads cause large variation of structural stresses, leading to

**Table 10.1**
**Expected Damage Accumulated during One Year at a Location at Northern Atlantic**

| $\beta = 2$ | $\beta = 3$ | $\beta = 4$ | $\beta = 5$ | $\beta = 7$ |
|---|---|---|---|---|
| 0.25 | 0.92 | 4.47 | 30.2 | 305.1 |

fatigue accumulation. For the fatigue life prediction of ship structures, it is important to obtain both the long-term distribution and the time history of wave-induced loads, see [141, 143]. An essential step is to get reliable wave statistics and an accurate description of the stochastic nature of sea state along a ship's sailing routes during its service time.

Those distributions can be obtained from models of storm movement, strength, and size, combined with planned ship routes and speed. Such models were discussed in several papers; here, we refer to, as examples, [157] and [47].

These types of models have several applications in maritime operation. One important application is the use of natural wind power to reduce dependence on fossil fuel in industrial activities. For example, the maritime industry, which is responsible for more than 90% of the world trade transport, has already started to look for solutions to use wind power as auxiliary propulsion for ships.

The practical installation of the wind facilities often requires large amount of investment, while uncertainties for the corresponding energy gains are large. Therefore, a reliable model to describe the variability of wind speeds is needed to estimate the expected available wind power, coefficient of the variation of the power and other statistics of interest, e.g. expected length of the wind conditions favourable for the wind-energy harvesting. In [184], wind speed was modelled by means of a spatio-temporal transformed Gaussian field. Its dependence structure is localised by the introduction of time and space-dependent parameters in the field. The model has the advantage of having a relatively small number of parameters. These parameters have natural physical interpretation and are statistically fitted to represent variability of observed wind speeds in ERA Interim reanalysis data set.

Another challenge is in planning of exploration of regions with sensitive marine environment, e.g. maritime operations in Arctics that were prone to accident should be minimised. Strong winds over the oceans can create dangerous metocean conditions for ships and offshore engineering activities, such as oil and gas rigs or renewable energy installations. As sea ice declines, human activities in the Arctic Ocean are increasing, the Arctic shipping is attracting an increasing number of sailing ships and Arctic offshore development is also increasing. Hence, investigations of environmental loads in this regions are important.

Statistical properties of wind speed in Arctic vary in terms of geographic location and season. A spatio-temporal transformed Gaussian field has been proposed to model wind variability in the northern North Atlantic. In the Arctic area, in most locations, this model is sufficiently accurate. However, in some regions, e.g., the Laptev and Beaufort Seas, this model severely underestimates the frequencies of extreme wind speeds, while along the east coast of Greenland, the frequencies of high wind speeds are severely overestimated. In [142], the hybrid model, similar to the one proposed in the chapter on Laplace loads, has been proposed to improve the accuracy.

### 10.4.3 NON-GAUSSIAN LOAD IN A VARIABLE ENVIRONMENT

#### 10.4.3.1 A Mixture Model for Road Irregularity

Following Bogsjö and Rychlik [22], we shall present a qualitative study of fatigue damage accumulation in vehicle components based on a Gaussian mixture model. For an alternative, based on the Laplace model, see [93]. Clearly, road roughness is a major source of fatigue-inducing vehicle loads. Properties of roads' surfaces, in scale of kilometres, are often modelled as stationary processes, often Gaussian, with a standardised type of spectrum, which may change with geographical region, age or type of road.

As was shown in Bogsjö (2007) [20], most of the fatigue damage accumulated in a vehicle is a consequence of short-duration oscillations caused by a local, higher degree, of road roughness. Those irregularities are sometimes called potholes, and in [20], a road model was proposed that includes such irregularities.

The variable stresses that may cause fatigue failure of a component are functions of road surface variability. As a first approximation, the vehicle responses (stresses) are modelled by means of linear filters having road surface profiles as input. Although the true stresses acting on a component in a vehicle will differ from the one computed using linear filters (due to the complicated and non-linear interaction of tires and the road surface), the results of this approximate analysis can be used qualitatively to compare different design concepts at an early stage.

**Remark 10.1.** *The problem of damage processes for Gaussian loads with superimposed transients is quite often met in applications. For example, responses of a sailing vessel can be modelled as a Gaussian process (responses to sea surface variability – waves) with added transients (whippings) due to "slams" which occur when a ship proceeds at a certain speed in rough seas and the front part of the hull bottom sustains large forces as the result of impact with the sea surface.*

Suppose that the external load $x(\tau)$ can be considered a sum of a stationary Gaussian process $x_0(t)$ having mean zero and spectral density $S(\omega)$ and non-stationary Gaussian loads $x_i(\tau)$ such that $x_i(t) = 0$ for all $t$ outside the interval $[t_i, t_i + \Delta_i]$, the derivatives $\dot{x}_i(t_i) = \dot{x}_i(t_i + \Delta_i) = 0$ and such that $\mathsf{E}[x_i(\tau)] = 0$. The total load is then

$$x(\tau) = x_0(\tau) + \sum_{i>0} x_i(\tau).$$

Obviously, the exact location and duration of an irregularity is unknown, and hence we consider random versions $(t_i, \Delta_i)$, independent of the process $x_0(t)$. Further, we suppose that the stress $y(t)$ acting on a component, caused by the load $x(t)$, can be adequately modelled by means of a linear filter,

$$y(\tau) = \int_{-\infty}^{\tau} h(t-s)x(s)\,\mathrm{d}s = y_0(\tau) + \sum_{i>0} y_i(\tau), \tag{10.41}$$

where $h(t)$ is the impulse response. Obviously, given position and length of irregularities, $y_i(\tau) = \int_{-\infty}^{\tau} h(\tau - s)x_i(s)\,\mathrm{d}s$ are all Gaussian processes, where only $y_0(t)$ is stationary.

#### 10.4.3.2 Fatigue Accumulation in a Vehicle

A vehicle travelling at velocity $v$ m/s will experience the road as a function of time, $\tau = x/v$. Thus, the vehicle load $y(\tau)$ is induced by the road profile $z(x)$, viz. $y(t) = z(vt)$. If the "tire" is in contact with the road, then the relationship between $y(t)$ and $z(vt)$ is linear, and the load is given by

$$y(\tau) = \int_{-\infty}^{\tau} h(\tau - s)z(vs)\,\mathrm{d}s, \tag{10.42}$$

where $h(t)$ is the vehicle's impulse response. The peculiarity with this example is that it is a road surface irregularity that causes the fatigue load.

The main variability in the road is described by the Gaussian process, $z_0(x)$. Irregularities of random shape, length and location are superimposed to $z_0(x)$. The $i$:th irregularity is denoted by $z_i(x)$, $i > 0$, and the road with superimposed irregularities is denoted by $z(x)$; see Figure 10.9 for illustration. The irregularities are modelled as (non-stationary) conditional Gaussian processes. The locations and lengths of the sections with added roughness are random. More precisely, the distance between the end of an irregularity and the start of the next is exponentially distributed with mean $\theta_1$. The lengths of the irregularities are also exponential, but with mean $\theta_2$.

Thus, the road is described as a sum of a stationary (smooth) part and a non-stationary (rough) part, $z(vt) = z_s(vt) + z_r(vt)$. Using the linearity assumption, the

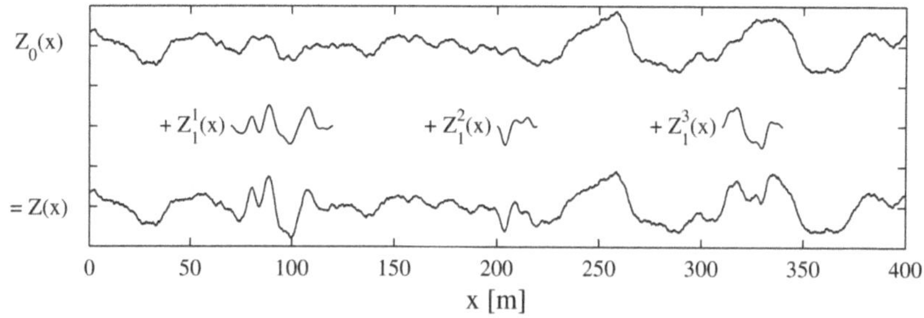


**FIGURE 10.9** Simulation of road model. $z_0(x)$ (stationary Gaussian) the top line, $z_j(x)$ the superimposed non-stationary Gaussian irregularities in the middle, and the road surface $z(x) = z_0(x) + \sum_j z_j(x)$ in the lowest graph. (Reproduced with permission from [22]. Copyright © 2007 Wiley.)

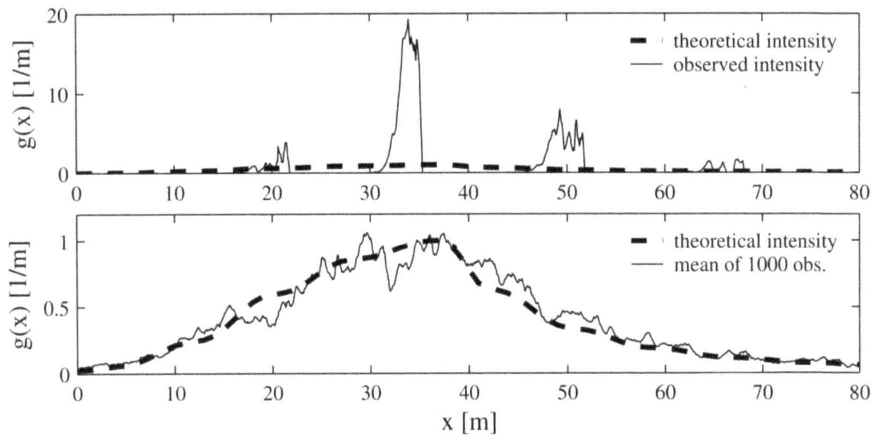


**FIGURE 10.10** Comparison of the observed and expected damage intensities as functions of the distance defined using (10.43). Top: The observed $\dot{D}^{rfc}(x)$ (solid line) compared with $d^{nb}(x)$ computed by means of (10.23) (dashed line). Bottom: The estimated $\mathsf{E}[\dot{D}^{rfc}(x)]$ by averaging 1000 simulated values of $\dot{D}^{rfc}(x)$ (solid line) compared with the approximation (10.23) (dashed line). (Reproduced with permission from [22]. Copyright © 2007 Wiley).

load $y(t)$ is also divided into a stationary and non-stationary part,

$$y(t) = \int_{-\infty}^{t} h(t-s)z_s(vs)\,\mathrm{d}s + \int_{-\infty}^{t} h(t-s)z_r(vs)\,\mathrm{d}s \equiv y_s(t) + y_r(t),$$

$$\dot{y}(t) = \dot{y}_s(t) + \dot{y}_r(t) = \dot{y}_s(t) + vh(0)\dot{z}_r(vt) + \int_{-\infty}^{t} \dot{h}(t-s)z_r(vs)\,\mathrm{d}s.$$

The processes $y(t)$ and $\dot{y}(t)$ are non-stationary Gaussian. The computations of the variances $\sigma_y^2(t)$ and $\sigma_{\dot{y}}^2(t)$ and the correlation between $y(t)$ and $\dot{y}(t)$, $\rho(t)$, are rather technical and are given in Appendix 10.B. In addition, it is easy to see that $\mathsf{E}[y(t)] = 0$. In the example studied in [22], it was demonstrated that for a particular choice of vehicle model, $\rho(t)$ is small and can be set to zero.

In Figure 10.10, the observed damage intensities D and d are compared with approximation (10.23). The damage intensity is expressed as a function of distance using the change of variable $x = vt$,

$$\int d^{nb}(t)\,\mathrm{d}t = \int \underbrace{d^{nb}(x/v)1/v}_{=g(x)}\,\mathrm{d}x. \tag{10.43}$$

In the example, the velocity is 20 m/s and the irregularity length is 40 m. In the above plot, in Figure 10.10, the observed damage intensities $\dot{D}(x)$ and $d^{nb}(x)$ is compared. As illustrated, the observed damage due to a particular irregularity may differ significantly from $d^{nb}(x)$. However, in the lower plot, 1000 observed intensities are averaged, and the result resembles $d^{nb}(x)$.

#### 10.4.3.3 Expected Damage Caused by a Road with Several Irregularities

Conditionally, on the position and length of irregularities, the load $y(\tau)$ is Gaussian, and the expected damage is given by the integral (10.43). Suppose that the transient in the load is dying out (can be neglected) after a distance $x_1$ or time $t_1$. Further, let irregularities be separated by more than $x_1$, i.e. the transients are isolated in time. Let also transfer function (and velocity of the vehicle) be constant, while irregularities be defined as a conditional Gaussian process $z(\tau)$ given $z(0) = z(L) = 0$ and $\dot{z}(0) = \dot{z}(L) = 0$. Then, the damage caused by an irregularity depends only on its length $L$. Suppose the damage starts at time zero and have length $L$. Then, the damage accumulated in time $L/v + t_1$ will be denoted by $D^{tr}(L/v + t_1)$. For load between irregularities, when conditions are stationary, the damage intensity will be denoted by $d^{st}$. Then, conditionally that there are $n$ irregularities with length $L_i$, the total damage is

$$D(t) = \left( t - n \cdot t_1 - \sum_{i=1}^{n} L_i/v \right) \cdot d^{st} + \sum_{i=1}^{n} D^{tr}(L_i/v + t_1).$$

Let us further suppose that the number of irregularities and their lengths are statistically independent. Then, $\mathsf{E}[n] = \theta_1 t$ and hence the expected damage is

$$\begin{aligned} \mathsf{E}[D(t)] &= \left( t - \frac{t}{\theta_1}(t_1 + \mathsf{E}[L]/v) \right) + \frac{t}{\theta_1}\mathsf{E}[D^{tr}(L/v + t_1)] \\ &= t\left( 1 - \frac{t_1 + \theta_2/v}{\theta_1} \right) + \frac{t}{\theta_1}\mathsf{E}[D^{tr}(L/v + t_1)]. \end{aligned}$$

Now, $\mathsf{E}[D^{tr}(L/v + t_1)]$ is computed using

$$\begin{aligned} \mathsf{E}[D^{tr}(L/v + t_1)] &= \int_0^\infty \mathsf{E}[D(l/v + t_1) \mid \text{irregularity length} = l] f(l)\,\mathrm{d}l \\ &= \int_0^\infty \left( \int_0^{L/v+t_1} d^{nb}(\tau)\,\mathrm{d}\tau \right) f(l)\,\mathrm{d}l \end{aligned} \tag{10.44}$$

where $d^{nb}(\tau)$ is computed using (10.25). The probability density function is the exponential $f(l) = \theta_2^{-1} e^{-l/\theta_2}$.

## 10.5 LOADS WITH NON-CONSTANT MEAN

An important class of loads for which a simple transformed Gaussian process $Y(\tau) = h(X(\tau))$ may give poor damage predictions is the case when load is a sum of a slowly varying component and a fast varying "noise".

For example, let the mean load $m(\tau)$ be piecewise constant, i.e. it changes abruptly and then stays constant for a relatively long period of time. If the time instances of jumps vary in an unpredictable manner, then one often needs to model $m(\tau)$ as a random process, e.g. a Markov chain. The property of the added noise may depend on the mean. Here, we consider a simple example of such an analysis which can be easily extended to other situations.

### 10.5.1 LATERAL FORCES

Suppose that lateral loads in a car are of interest. (This problem was studied in Karlsson (2007) [101].) The load has three regimes, driving straight, turn to the left and turn to the right. We assume that when driving straight, the average stress $m$ is zero, when turning to the right, the mean stress is positive, while to the left, it is negative.

**Specifying the mean $m(\tau)$:** Suppose that, at the beginning, a car drives straight for a random period of time $T_0$, say. Let curves be numbered by $i = 1, 2, \ldots, n$ and denote by $Z_i$ the mean stress when turning on a curve $i$. Suppose $R_i$ are random independent variables describing how severe the curve is. Their distributions depend on velocity, and curve radius, driver, etc. For simplicity, let $|R_i|$ be equally distributed.

Next, suppose that the length of times it takes to pass a curve $i$ is a random variable $S_i$, say. Again, for simplicity assume that $S_i$ are iid, e.g. exponential with mean $\theta_1$. Finally, let the times $T_i$, $i = 0, 1, 2, \ldots$, one is driving straight between the curves $i-1$ and $i$ are also independent and equally distributed random variables, e.g. exponential with mean $\theta_2$.

Finally, in order to fully define the mean load, one also needs to model the order of curves. This could be a Markov chain $Z_i$ taking two values 1 and $-1$ to indicate the right, left curve, respectively. Here we shall use the simplest MC in which the direction of a curve is switched each time, i.e. left curve is always followed by the right curve. Let $\mathsf{P}(Z_1 = 1) = 1/2$.

**Specifying the noise $X(\tau)$:** Here, for simplicity only, let $X$ be a stationary Gaussian load with apparent frequency $f_z$ much higher than the frequency of curves, i.e. $f_z >> \max(\theta_1^{-1}, \theta_2^{-1})$, and the significant amplitude much smaller than $\mathsf{E}[|R_i|]$.

**Rainflow analysis of $Y(\tau) = m(\tau) + X(\tau)$:** Often when the load is a sum of two functions, then one would like also to add damages caused by the components. However, this is not an appropriate procedure due to a non-linear character of rainflow cycle procedure. (Note that even crossing intensity $\mu^+(u; X+Y) \neq \mu^+(u; X) + \mu^+(u; Y)$ as long as both processes are not constant.) However, for the load specified in this section, one can assume that the damage caused by $Y$ is, approximately, a sum of damages caused by $X(\tau)$ and a time series $X_n = R_n + e_n$ where

$$e_n = \begin{cases} \max_{t_n \le \tau \le t_n + S_n} X(\tau) & \text{if } Z_n = 1, \\ \min_{t_n \le \tau \le t_n + S_n} X(\tau) & \text{if } Z_n = -1. \end{cases}$$

Here, $t_n$ is a time when car enters curve $n$. In other words, where $e_n$ is the maximum value of $X$ during the $n$th curve if it is to the right and minimum of $X$ if it is the left curve. For ergodic Gaussian loads, $e_n$ are approximately independent.

Consequently, we model $X_n$ as an MCTP with turning points $M_i > 0 > m_i$, and heights of turning points are independent. Obviously, maximum $M$ has distribution $M = |R_1| + e_1$ while minimum $m = -|R_1| + e_1$. For this MCTP, the oscillation intensity is given by (5.3), viz.

$$\mu^{osc}(u, v; X_n) = \frac{\mathsf{P}(m < u)\mathsf{P}(M > v)}{1 - P(M \le v)P(m \ge u)}.$$

The frequency of curves is $f_c = (2(\theta_1 + \theta_2))^{-1}$.

Now, the oscillation intensity in $X(\tau)$ can be bounded by $k^+(u,v;X)$ which gives the damage intensity about $0.5\, f_z h_s^\beta$ if $2 \leq \beta \leq 4$ and hence we approximate the oscillation intensity of $Y$ by means of

$$\begin{aligned}\mu^{osc}(u,v;Y) &\approx f_c\mu^{osc}(u,v;X_n) + \mu^{osc}(u,v;X)\\ &\leq f_c\mu^{osc}(u,v;X_n) + k^+(u,v;X).\end{aligned} \tag{10.45}$$

It remains to specify the distributions of $M$ and $m$. Here, we will use the simple Rice's method: For any continuously differentiable process $Z(\tau)$, the tail distribution is bounded by

$$\mathsf{P}(\max_{0\leq\tau\leq t} Z(\tau) > u) \leq \mathsf{P}(Z(0) > u) + \mathsf{E}[N_t^+(u;Z)], \tag{10.46}$$

where $N_t^+(u;Z)$ is the number of $u$-upcrossings by the process $Z(\tau)$ in the time interval $[0,t]$. For large $t$, $P(Z(0) > u)$ is usually an order of magnitude smaller than $\mathsf{E}[N_t^+(u;Z)]$ and hence can be neglected.

Suppose that for a given curve, the severity $R_i$ and duration $S_i$ are independent. Then

$$\mathsf{E}[N_S^+(u) \mid R = r, S = s] = s\, f_z e^{-8\frac{(u-r)^2}{h_S^2}}$$

and hence, for $u < 0 < v$,

$$\begin{aligned}\mathsf{P}(M > v) &\leq \theta_1\, f_z \mathsf{E}\left[e^{-8(v-|R_1|)^2/h_s^2}\right],\\ \mathsf{P}(m < u) &\leq \theta_1\, f_z \mathsf{E}\left[e^{-8(u+|R_1|)^2/h_s^2}\right].\end{aligned} \tag{10.47}$$

# APPENDICES

## 10.A SORM METHOD TO EVALUATE RICE'S FORMULA

The following generalisation of Breitung's (1988) [26] approximation is based on results of Hagberg (2005) [75]; see also Lindgren 1989 [124] and Baxevani et al. (2005) [5]. The formulae presented are, in principle, explicit, although they require finding the design point.

**Theorem 10.2.** *Let $g : \mathbb{R}^n \to \mathbb{R}$ be a function such that the surface*

$$S = \{\mathbf{x} = (x_1, \ldots, x_n); g(\mathbf{x}) = 0\}$$

*has a point $\mathbf{x}_0$ such that $||\mathbf{x}_0|| = 1$ and $||\mathbf{x}|| > 1$ for all other $\mathbf{x} \in S$. By $\mathbf{x}$, we denote both the vector $(x_1, \ldots, x_n)$ and the $n \times 1$ column matrix. Suppose $\mathbf{Z}(t)$ is an n-dimensional, stationary, differentiable, Gaussian vector process with mean zero, and let $\dot{\mathbf{Z}}(t)$ denote its derivative. The covariance matrix of $(\mathbf{Z}(t), \dot{\mathbf{Z}}(t))$ is denoted by $\Sigma$,*

$$\Sigma = \begin{bmatrix} I & \Sigma_{12} \\ \Sigma_{21} & \Sigma_{22} \end{bmatrix}. \tag{10.48}$$

*For a family of processes $g(\mathbf{Z}(t)/\beta)$, $\beta > 0$, under some mild technical assumptions, the intensity of zero upcrossings is given by*

$$\mu_\beta(0) = \frac{e^{-\beta^2/2}}{2\pi}(c + \mathcal{O}(\beta^{-2})), \quad c = \sqrt{\frac{\mathbf{x}_0^T(\Sigma_{22} - \Sigma_{21}G_0\Sigma_{12})\mathbf{x}_0}{\det(I + P_0G_0P_0)}}, \tag{10.49}$$

*as $\beta$ tends to infinity, where $G_0 := \frac{1}{|\nabla g(\mathbf{x}_0)|}\left[\frac{\partial^2 g}{\partial x_i \partial x_j}(\mathbf{x}_0)\right]_{i,j=1,2,\ldots,n}$ and $P_0 := I - \mathbf{x}_0\mathbf{x}_0^T$. Since $\mathbf{Z}(t)$ is stationary, the same formula is valid for downcrossings.*

Based on the above theorem, the following remarks can be made:

1. Formula (10.49) is in a sense finitely additive, i.e. if there are a finite number of points with minimal distance to the origin, the asymptotic formula for the upcrossing intensity of the process $g(\mathbf{Z}(t)/\beta)$ is the sum of the upcrossing intensities estimated using Eq. (10.49) for each point separately. Breitung's asymptotic approximation fails in the case of an infinite number of such points.
2. Theorem 10.2 lends itself to a geometric interpretation. Note that $g(\mathbf{Z}(t)/\beta)$ crosses the zero level, if and only if, the vector process $\mathbf{Z}(t)$ crosses the surface $\beta S$. Hence, instead of saying that the formula is asymptotic "as $\beta$ tends to infinity", we may say "as the surface $S$ is inflated".
3. The simpler FORM approximation is derived by assuming that locally at the design point $\mathbf{x}_0$, the curvature of the surface is zero, i.e. $G_0$ is a matrix containing only zeros giving the following approximation

$$\mu_\beta(0) \approx \frac{e^{-\beta^2/2}}{2\pi}\sqrt{\mathbf{x}_0^T\Sigma_{22}\mathbf{x}_0}. \tag{10.50}$$

Here, we are interested in quadratic processes of type

$$Y(t) = \sum_{j=0}^{n} \gamma_j X_j(t) + \sum_{j=0}^{n} \frac{\lambda_j}{2} X_j(t)^2, \tag{10.51}$$

where $\mathbf{X}(t) = (X_0(t), \ldots, X_n(t))$ is a vector of stationary Gaussian process, such that, for each $t$, $X_j(t) \in N(0,1)$ and the variables $X_j(t)$ and $X_k(t)$ are uncorrelated. Denote $\dot{\mathbf{X}}(t) = (\dot{X}_0(t), \ldots, \dot{X}_n(t))$. Let the correlation structure of $\mathbf{X}(t), \dot{\mathbf{X}}(t)$ be given by Eq. (10.48). We shall now demonstrate how the theorem can be used to approximate $\mu(u)$, the upcrossing intensity of the level $u$ by the process $Y(t)$ defined in Eq. (10.51).

### 10.A.1 GAUSSIAN CASE, ALL $\lambda_i = 0$

Let us consider two special cases when the stress is a Gaussian process i.e. the quadratic correction term in Eq. (10.16) may be ignored. This implies that $\lambda_j = 0$. Thus, Eq. (10.16) simplifies to $Y(t) = \gamma^T\mathbf{X}(t)$, where $\gamma = (\gamma_1, \ldots, \gamma_n)$. Define

$$g(\mathbf{x}) = 1 - \frac{1}{||\gamma||}\gamma^T\mathbf{x},$$

and note that, if $\beta = u/||\gamma|| := \beta_u$, the process $g(\mathbf{X}(t)/\beta)$ downcrosses the zero level exactly when $Y(t)$ upcrosses the level $u$. (Here $||\cdot||$ denotes norm in $\mathbb{R}^n$, e.g., $||\gamma||^2 = \sum \gamma_j^2 = \mathsf{Var}[X(0)]$.) On the surface $g(\mathbf{x}) = 0$, the point closest to the origin is $\mathbf{x}_0 = \gamma/||\gamma||$, and since all second-order derivatives of $g$ are zero, Breitung's approximation gives

$$\mu_\beta(0) = \frac{1}{2\pi}\frac{\sqrt{\gamma^T\Sigma_{22}\gamma}}{||\gamma||}\exp\left(\frac{-\beta_u^2}{2}\right), \tag{10.52}$$

since $\mathbf{x}_0 = \gamma/||\gamma||$. As expected, the approximation is exact, as given by Rice's formula for a Gaussian process, since with $\mathsf{Var}[X(0)] = ||\gamma||^2$ and $\mathsf{Var}[\dot{X}(0)] = \gamma^T\Sigma_{22}\gamma$.

## 10.A.2 PURE QUADRATIC CASE, ALL $\gamma_j = 0$

Suppose next that the linear term in Eq. (10.16) is negligible. In this case, we may write $Y(t) = \frac{1}{2}\mathbf{X}(t)^T\Lambda\mathbf{X}(t)$, where $\Lambda = \operatorname{diag}([\lambda_1,\ldots,\lambda_n])$ is the diagonal matrix, with $\prod_{i=1}^n \lambda_i \neq 0$. We also assume that $\lambda_1 \le \lambda_2 \le \cdots \le \lambda_n$. Additionally, for $u > 0$, we need to assume that $\lambda_n > 0$, since otherwise $Y(t) \le 0$ for all $t$. By defining

$$g(\mathbf{x}) = 1 - \frac{1}{\lambda_n}\mathbf{x}^T\Lambda\mathbf{x},$$

it can be seen that for $\beta = \sqrt{2u/\lambda_n} := \beta_u$, the zero downcrossing intensity of the process $g(\mathbf{X}(t)/\beta)$ equals the upcrossing intensity of the level $u$ by the process $Y(t)$. Consequently, $\mu_\beta(0) = \mu(u)$ and Theorem 10.2 may be used to compute $\mu_\beta(0)$. Note, however, that there are two points

$$\mathbf{x}_\pm := \pm\begin{bmatrix} 0 & 0 & \ldots & 1 \end{bmatrix}^T$$

of minimal distance to the origin. Hence, one has to compute formula (10.49) for each one of these points separately ($G_0 = -\frac{1}{\lambda_n}\Lambda$ for both points), and add the results. Consequently, by Theorem 10.2,

$$\begin{aligned}\mu(u) &= \frac{1}{\pi}\exp\left(-u/\lambda_n\right)\\ &\times\left(\sqrt{\frac{\mathsf{Var}[\dot{X}_n(0)] - \sum_{i=1}^{n-1}(\lambda_i/\lambda_n)\mathsf{Cov}[X_i(0),\dot{X}_n(0)]^2}{(1-\lambda_1/\lambda_n)\ldots(1-\lambda_{n-1}/\lambda_n)}} + \mathcal{O}(u^{-1})\right),\end{aligned} \tag{10.53}$$

since we have two design points. Note that for $\mu(v)$ with $v < 0$, one can use Eq. (10.53) with $u = -v$ by changing signs of $\lambda_i$, which then have to be again ordered into increasing order.

## 10.A.3 GENERAL QUADRATIC LOADS

Before proceeding further, it should be emphasised that Theorem 10.2 for the two cases of purely linear/quadratic loads gave asymptotic formulas *as the level tends to infinity*. The general case is more complicated. This is not surprising since we have the mix of two different limiting cases. We have to construct a somewhat artificial asymptotic. The idea is as follows. Fix the level $u$, assumed to be large, and let

$$p(\mathbf{x}) := \gamma^T \mathbf{x} + \frac{1}{2}\mathbf{x}^T \Lambda \mathbf{x}. \tag{10.54}$$

Assume that there is only one point $\mathbf{x}_u$ on the surface

$$\{\mathbf{x} \in \mathbb{R}^n; p(\mathbf{x}) = u\}$$

of minimal distance to the origin, and define $\beta_u := ||\mathbf{x}_u||$. Let

$$g(\mathbf{x}) := 1 - \frac{1}{u}\left(\beta_u \gamma^T \mathbf{x} + \beta_u^2 \frac{1}{2}\mathbf{x}^T \Lambda \mathbf{x}\right).$$

As before, the process $g(\mathbf{X}(t)/\beta_u)$ crosses the zero level when the process $p(\mathbf{X}(t))$ crosses the level $u$; hence, using Breitung's method, with $\mathbf{x}_0 = \mathbf{x}_u/\beta_u$, for each level $u$ separately, we have $\mu_{\beta_u}(0)$ equal to the $u$-upcrossing intensity for the process $p(\mathbf{Z}(t))$. Therefore, it is reasonable to believe that if $\beta_u$ is large, then the term $O(\beta^{-2})$ for $\beta = \beta_u$ is small. Hence, the approximation is good. However, the problem is that for each value $u$, one has to find the design point $\mathbf{x}_u$ and define a new function $g(\cdot)$. This implies that we cannot use the theorem, which is valid for a fixed $g$, to motivate that the error term decreases to zero as $h$ tends to infinity.

However, there are good reasons to use Breitung's approximation in this way instead of a formula which is truly asymptotic as the level $u$ tends to infinity, especially if the linear terms dominate over the quadratic for the levels of interest, $u$. This is because, for $u$ large enough, the $u$-crossings of the process will almost exclusively depend on the quadratic part. In contrast for $u$-level close to zero ($\beta_u \approx 0$), the quadratic term in Eq. (10.54) is negligible, and one has the pure Gaussian case. Note that the method approximates the zero crossing intensity by zero crossing intensity in the pure Gaussian process. It is suggested that the use of Breitung's method will give a proper balance between the linear and the quadratic terms.

### 10.A.4 COMPUTATION OF BREITUNG'S APPROXIMATION

The form

$$\mu_{\beta_u}(0) = \frac{\exp(-\beta_u^2/2)}{2\pi} c(\beta_u),$$

was given in Theorem 10.2. Note that opposed to the formulation in Theorem 10.2, we indicate $c'$s dependence of $\beta_u$. Evaluating the terms in (10.49) gives

$$\nabla g(\mathbf{x})^T|_{\mathbf{x}=\mathbf{x}_u/\beta_u} = -\left(\frac{\beta_u}{u}\gamma + \frac{\beta_u^2}{u}\Lambda\mathbf{x}\right)|_{\mathbf{x}=\mathbf{x}_u/\beta_u} = -\frac{\beta_u}{u}(\gamma + \Lambda\mathbf{x}_u),$$

$$G_0 = \frac{-\beta_u \Lambda}{||(\gamma + \Lambda\mathbf{x}_u)||}\Lambda, \quad P_0 = I - \frac{1}{\beta_u^2}\mathbf{x}_u\mathbf{x}_u^T.$$

Now, we can use the following approximation

$$\mu(u) \approx \frac{\mathrm{e}^{-\beta_u^2/2}}{2\pi} c(\beta_u), \tag{10.55}$$

where

$$c(\beta_u) = \sqrt{\frac{\mathbf{x}_u^T\left(\Sigma_{22} - \Sigma_{21} G_0 \Sigma_{21}\right)\mathbf{x}_u}{\beta_u^2 \det\left(I + P_0 G_0 P_0\right)}}.$$

The simpler FORM approximation is, by Eq.(10.50), given by

$$\mu(u) \approx \mathrm{e}^{-\beta_u^2/2} \frac{\sqrt{\mathbf{x}_u^T \Sigma_{22} \mathbf{x}_u}}{2\pi\beta_u}. \tag{10.56}$$

It must be noted that the point of minimum norm, $\mathbf{x}_u$, can be found by standard optimisation methods, see Ditlevsen and Madsen (1996) [55] .

## 10.B COMPUTATION OF $\mu_t^+(u)$

In order to compute $\mu_t^+(u)$ with Rice's formula, one needs the joint density of $y(t)$ and $\dot{y}(t)$. However, since $y(t)$ and $\dot{y}(t)$ are multivariate Gaussian, it is sufficient to compute their mean, variance and covariance as a function of time. Their mean values are zero: $\mathsf{E}[y(t)] = \int_{-\infty}^{t} h(t-s)\mathsf{E}[z(vs)]\,\mathrm{d}s = 0$ and $\mathsf{E}[\dot{y}(t)] = 0$. The variances and covariance are written as

$$\begin{aligned}
\mathsf{Var}[y(t)] &= \mathsf{Var}[y_s(t)] + \mathsf{Var}[y_r(t)] \\
&\equiv \sigma_{ys}^2 + \sigma_{yr}^2(t) \equiv \sigma_y^2(t), \\
\mathsf{Var}[\dot{y}(t)] &= \mathsf{Var}[\dot{y}_s(t)] + \mathsf{Var}[\dot{y}_r(t)] \\
&\equiv \sigma_{\dot{y}s}^2 + \sigma_{\dot{y}r}^2(t) \equiv \sigma_{\dot{y}}^2(t), \\
\mathsf{Cov}(\dot{y}(t), y(t)) &= \mathsf{Cov}(\dot{y}_s(t) + \dot{y}_r(t), y_s(t) + y_r(t)) \\
&= \mathsf{E}[\dot{y}_r(t) y_r(t)] = c(t).
\end{aligned}$$

The time varying parameters $\sigma_{yr}^2(t), \sigma_{\dot{y}r}^2(t)$ and $c(t)$ parameters are computed by

$$\begin{aligned}
\sigma_{yr}^2(t) &= \mathsf{E}[y_r^2(t)] \\
&= \int_{-\infty}^{t}\int_{-\infty}^{t} h(t-u)h(t-s)\mathsf{E}[z_r(vs)z_r(vu)]\,\mathrm{d}s\,\mathrm{d}u, \\
\sigma_{\dot{y}r}^2(t) &= \mathsf{E}[\dot{y}_r^2(t)] \\
&= \int_{-\infty}^{t}\int_{-\infty}^{t} \widetilde{h}(t-u)\widetilde{h}(t-s)\mathsf{E}[z_r(vs)z_r(vu)]\,\mathrm{d}s\,\mathrm{d}u, \\
c(t) &= \mathsf{E}[\dot{y}_r(t) y_r(t)] \\
&= \int_{-\infty}^{t}\int_{-\infty}^{t} \widetilde{h}(t-u)h(t-s)\mathsf{E}[z_r(vs)z_r(vu)]\,\mathrm{d}s\,\mathrm{d}u.
\end{aligned}$$

The constant parameters $\sigma_{y_S}^2$ and $\sigma_{\dot{y}_S}^2$ are given by the frequency response function,

$$\sigma_{y_S}^2 = \int_0^\infty |H(\omega)|^2 \widetilde{R}_v(\omega)\, d\omega,$$
$$\sigma_{\dot{y}_S}^2 = \int_0^\infty \omega^2 |H(\omega)|^2 \widetilde{R}_v(\omega)\, d\omega,$$

where $H(\omega) = \int_{-\infty}^{\infty} h(t)e^{i\omega t}\, dt$, $\widetilde{R}_v(\omega) = R(\frac{\omega}{2\pi v})\frac{1}{2\pi v}$ and $R$ is the spectrum of road surface elevation without irregularities. Now, the two-dimensional density is

$$f_{\dot{y}(t),y(t)}(z,u) = f_{\dot{y}(t)|y(t)=u}(z) f_{y(t)}(u)$$
$$= \frac{1}{\sigma(t)\sqrt{2\pi}} e^{\left(\frac{-(z-m(t)u)^2}{2\sigma^2(t)}\right)} \frac{1}{\sigma_y(t)\sqrt{2\pi}} e^{\left(\frac{-u^2}{2\sigma_y^2(t)}\right)},$$

where correlation, mean and standard deviations are $\rho(t) = c(t)/(\sigma_y(t)\sigma_{\dot{y}}(t))$, $m(t) = \rho(t)\sigma_{\dot{y}}(t)/\sigma_y(t)$ and $\sigma(t) = \sigma_{\dot{y}}(t)\sqrt{1-\rho^2(t)}$, respectively. The upcrossing intensity is

$$\mu_t^+(u) = \int_0^\infty z f_{\dot{y}(t)|y(t)=u}(z) f_{y(t)}(u)\, dz$$
$$= \frac{e^{\frac{-u^2}{2\sigma_y^2(t)}}}{\sigma_y(t)\sqrt{2\pi}} \left( m(t)u(1-\Phi(-\frac{m(t)u}{\sigma(t)})) + \frac{\sigma(t)}{\sqrt{2\pi}} e^{-u^2 m^2(t)/2\sigma^2(t)} \right), \quad (10.57)$$

where $\Phi(z) = \int_{-\infty}^{z} \frac{e^{-u^2/2}}{\sqrt{2\pi}}\, du$. Now, if we set $\rho(t)$ to zero, we get

$$\mu_t^+(u) = \frac{1}{2\pi} \frac{\sigma_{\dot{y}}(t)}{\sigma_y(t)} \exp\left(-\frac{u^2}{2\sigma_y^2(t)}\right), \quad (10.58)$$

and inserting (10.58) into (10.23) gives the local narrow-band approximation (10.25).

## Notes

[1]Recall that $\widetilde{d} = \alpha 2^{-\beta/2}\Gamma(1+\beta/2)$ is the damage for a standardised load with $f_z = 1$ and $h_s = 1$, given in (7.24).

[2]The factor $c$ is evaluated using that for $X \in N(m,\sigma^2)$, $E[exp(\alpha X)] = exp(\alpha m + \alpha^2\sigma^2/2)$ and some algebra gives

$$c = 2.5 \cdot 10^5 exp((\beta - 1/2)^2 \sigma_H^2/2) = 2.1 \cdot 10^5.$$

# 11 Variance of Damage Estimate

## 11.1 INTRODUCTION

Often, in statistical analysis of fatigue damage $D(t)$, only the expected value $\mathsf{E}[D(t)]$ is considered. However, in many practical situations also, the variability is of interest. For example, decisions of safe service intervals of vehicles may be based on damage variability; see [78]. To decide if a stochastic load model gives the same fatigue damage as actual load measurements, one needs to know the uncertainty of the fatigue damage estimates; see [19]. Computations of safety index for fatigue failure include the computation of the coefficient of variation $\mathsf{CoV}[D(t)] = \sqrt{\mathsf{Var}[D(t)]}/\mathsf{E}[D(t)]$; see [15].

The chapter is focused on computation of the coefficient of variation CoV. First, we consider stationary and ergodic loads. The ergodicity assumption is needed in order to estimate the expected damage and covariance of damage intensity from one sample path, i.e. a single measurement of a load. Here, both Gaussian and non-Gaussian loads will be considered. We concentrate on symmetric loads for which the local narrow-band approximation proposed in [22] can be used to estimate the expected damage. The method will be extended to give approximation of the coefficient of variation. In situations when the asymmetry of the load cannot be neglected, we propose a more complex estimation method of CoV. Note that for stationary random loads, when a probabilistic model is known, one can use Monte Carlo methods to estimate both expected damage and variance, and the assumption of ergodicity is not necessary.

Assuming stationary conditions for a long period, e.g. a year, can be unrealistic in many situations. Here, we consider a locally stationary load, i.e. the stationarity conditions change slowly relative to the variability of the load. As a special case, we analyse the seasonal variation of wave load conditions for a marine structure at a fixed position during one year.

Estimation of the coefficient of variation for stationary loads is discussed in Section 11.2, and in Section 11.3, stationary Gaussian loads will be considered. The results will be illustrated by three processes, two having broad and narrow spectra, respectively, and one being a bimodal spectrum estimated from measured stress on a vessel sailing under stationary conditions for 30 minutes. It will be demonstrated that the measured stress is slightly non-Gaussian which considerably affects the coefficient of variation. We conclude that suitably normalised, the narrow-band Gaussian load has higher coefficient of variation than the broad band Gaussian loads. For the wave loads, we consider a Gaussian model with slowly changing spectrum due to weather conditions and season of the year.


DOI: 10.1201/9781003019947-11

In Section 11.4, we make a quantitative study of the vehicle fatigue from Section 10.4.3. There, we considered a non-Gaussian load for the acceleration of a heavy vehicle travelling with constant speed on an irregular Gaussian road with added extra irregularities representing "potholes". The added irregularities were placed at random, and hence the load was still stationary but no more Gaussian.

We shall use the simple damage function $f(u,v) = (v-u)^{\beta}$. For the rainflow analysis, we shall use the alternative formula for the rainflow damage employing the damage accumulation rate $\dot{D}(t)$ given in (4.15) to estimate the variation coefficient for the damage $D(T)$.

More precisely, as was shown in [173], the rainflow damage can be written as

$$D(T) = \int_0^T \beta(Y(t) - Y^-(t))^{\beta-1}\dot{Y}(t)^+ \,\mathrm{d}t \equiv \int_0^T \dot{D}(t)\,dt, \qquad (11.1)$$

where $Y^-(t)$ is the lowest value in the backward direction between $t$ and the nearest exceedance of $Y(t)$; see Figure 11.1. In (4.17), $H^-(t) = Y(t) - Y^-(t)$ was named rainflow memory at time $t$. Obviously, if $\dot{Y}(t) < 0$, then $H^-(t) = 0$; consequently, if $\beta = 1$, the damage rate simplifies to

$$D(T) = \int_0^T \dot{Y}(t)^+ \,\mathrm{d}t, \qquad (11.2)$$

and $\dot{D}(t) = \dot{Y}(t)^+$. The function $\dot{D}(t)$ is interpreted as the rainflow damage rate.

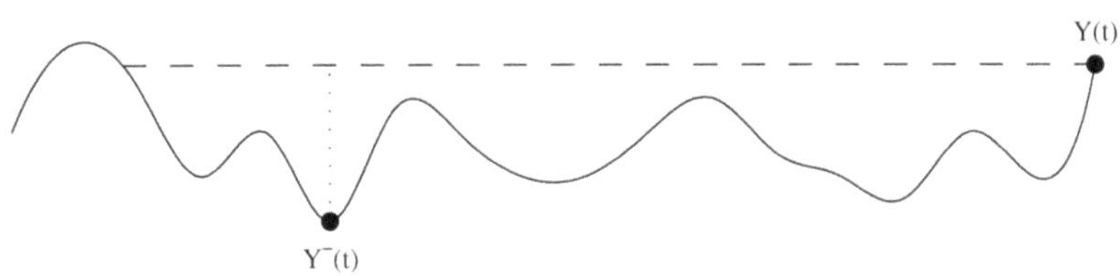


**FIGURE 11.1** Definition of $Y^-(t)$

## 11.2 COMPUTATION OF THE COEFFICIENT OF VARIATION

### 11.2.1 ESTIMATION OF THE COEFFICIENT OF VARIATION

When a model for a random load is fully specified, e.g., one is assuming that the load is a stationary Gaussian process with a known mean and spectrum, then both expected damage and coefficient of variation can be estimated by means of a Monte Carlo method. More precisely, one may simulate a large number of loads with fixed lengths from the model and compute the accumulated damage for each of the loads. Then, expected damage and variance can be estimated using standard statistical estimators.

In the situation, when one has only one measured load, it may still be important to express a statement about how safe any possible conclusion can be. The unpredictable irregularity of data indicates that a random process is a suitable model for the load. If the expected damage and coefficient of variation for the random load are needed, then one could, based on data, fit a stochastic model and then use Monte Carlo methodology to estimate mean and variance. In such an approach, a possibility of modelling error should not be neglected. A non-parametric approach, where no particular model for the measured load is specified, will be presented next.

Suppose that the load is stationary and ergodic; it means that one infinitely long realisation defines the random mechanism that generated the load. Here, we shall use the fact that the damage $D(t)$ is absolutely continuous, i.e. the damage rate $\dot{D}(t)$ exists and $D(T) = \int_0^T \dot{D}(t)\,\mathrm{d}t$. Further, we suppose that also the damage rate is a stationary process. (This is not true for the rainflow damage when $t$ is close to the beginning of the measured load, but we neglect this fact here.)

For stationary damage rate, $d = \mathsf{E}[\dot{D}(t)]$ is referred to as the damage intensity, and the expected total damage is $\mathsf{E}[D(T)] = T\,d$. If the load has been observed for a time period of length $t$, then a natural estimate of $d$ is $\widehat{d} = D(t)/t$. The variation coefficient is obviously a measure of how accurate this estimate is. We turn now to the estimate of the coefficient.

Let the stationary positive process $\dot{D}(t)$ be the damage rate for the load process $Y(t)$. For example, rainflow damage rate

$$\dot{D}(t) = \alpha\beta H^-(t)^{\beta-1}\dot{Y}^+(t), \tag{11.3}$$

where $H^-(t) = Y(t) - Y^-(t)$. Suppose that $Y$ is stationary and ergodic and let $r_{\dot{D}}(s) = \mathsf{Cov}[\dot{D}(0), \dot{D}(s)]$ be the covariance function of $\dot{D}(s)$. Then, with $d = \mathsf{E}[\dot{D}(t)]$, $\mathsf{E}[D(T)] = T\,d$, the variance of the accumulated damage $D(T) = \int_0^T \dot{D}(s)\,\mathrm{d}s$ is, according to (2.54),

$$\mathsf{Var}\,[D(T)] = \int_0^T \int_0^T \mathsf{Cov}[\dot{D}(s), \dot{D}(t)]\,\mathrm{d}s\,\mathrm{d}t = T\int_0^T 2(1 - s/T)r_{\dot{D}}(s)\,\mathrm{d}s.$$

If $r_{\dot{D}}(s)$ converges fast to zero, then asymptotically

$$\mathsf{Var}\,[D(T)] \approx T\,2\int_0^\infty r_{\dot{D}}(s)\,\mathrm{d}s = T\,\sigma^2, \quad \text{say.} \tag{11.4}$$

Let $\widehat{r}(s)$ be an estimate of the covariance function $r_{\dot{D}}(s)$ based on observations $y(s)$ of $Y(s)$ for $0 \le s \le t_0 < t$. Then

$$\widehat{\sigma}^2 = 2\int_0^{t_0} \widehat{r}(s)\,\mathrm{d}s, \tag{11.5}$$

and hence the coefficient of variation can be estimated as

$$\mathsf{CoV}[D(T)] \approx \frac{\widehat{\sigma}}{\widehat{d}}\frac{1}{\sqrt{T}}. \tag{11.6}$$

### 11.2.2 APPROXIMATION OF CoV FOR SYMMETRIC LOADS

At the design stage, one is often interested in fast estimates of the expected damage and the coefficient of variation. As it was proposed in Section 10.4.2, for zero-mean symmetric loads, one can approximate the rainflow damage rate by the Holm-de Maré damage rate, which is obtained by replacing $H^{-}(t)$ in (11.3) by $|2Y(t)|$, leading to the approximate time-dependent damage rate

$$\dot{D}(t) = \beta\,(|2Y(t)|)^{\beta-1}\dot{Y}(t)^{+}. \tag{11.7}$$

For stationary Gaussian loads, the Holm-de Maré damage intensity is, (7.24),

$$\mathsf{E}[\dot{D}(t)] = f_z h_s^{\beta}\, 2^{-\beta/2}\Gamma(1+\beta/2). \tag{11.8}$$

The Holm-de Maré damage has the same damage intensity as the narrow-band bound since $Y(t)$ is a symmetric, zero-mean process. In the text, when we need to distinguish between the two, we shall add the superscript $\cdot^{rfc}$, $\cdot^{HM}$, respectively.

It was shown that for any load, the accumulated damage $D^{rfc}(T) \leq D^{HM}(T)$. Except for zero-mean symmetric loads, i.e. having symmetric unimodal crossing intensity, the Holm-de Maré is a very conservative bound and hence seldom used. Furthermore, in the case of stationary symmetric loads, $\mathsf{E}[D^{HM}(T)]$ coincides with the narrow-band damage by Bendat (1964) [12], and our hypothesis is that

$$\mathsf{CoV}[D^{rfc}(T)] \approx \mathsf{CoV}[D^{HM}(T)]. \tag{11.9}$$

## 11.3 STATIONARY GAUSSIAN LOADS

### 11.3.1 SCALING DAMAGE AND VARIABILITY

In Chapter 9, where we discussed the stationary Gaussian and Laplace models, we often used the normalised spectra, i.e. their spectral moments $\omega_0 = \omega_2 = 1$. For Gaussian models with normalised spectrum, the significant amplitude $h_s = 4$, while

$$f_z = \frac{1}{2\pi}\sqrt{\frac{\mathsf{Var}[\dot{Y}(0)]}{\mathsf{Var}[Y(0)]}} = \frac{1}{2\pi}$$

is the apparent frequency, i.e. the intensity of zero level upcrossings. For the Gaussian load with normalised spectrum, the Holm-de Maré damage intensity is

$$d^{HM} = \frac{2^{3\beta/2}}{2\pi}\Gamma(1+\beta/2). \tag{11.10}$$

Note that $d^{HM} = d^{rfc}$ for $\beta = 1$. However, the function $d^{HM}(\beta)$ is growing very fast with $\beta$ which can be inconvenient when comparing the variation coefficients for different types of load spectra. In addition, the damage exponent $\beta$, often considered as an "uncertain" quantity, may have larger influence on safety measures than other uncertain factors.

Another normalisation was used in Example 7.2, introduced in [22], see also [15]. In that approach, one is scaling the spectrum so that $\omega_0 = 1/16$ and $\omega_2 = 2\pi$. The so normalised spectrum has significant amplitude $h_s = 1$ and apparent frequency $f_z = 1$, too. The corresponding normalised load is then defined as

$$Z(t) = \frac{1}{h_s} Y(t/f_z), \tag{11.11}$$

where $h_s$ and $f_z$ are the significant amplitude and apparent frequency of $Y(\tau)$.

Let us denote the spectrum of $Z$ by $\widetilde{S}$ and let its damage intensity and variance of damage (11.4) be $\widetilde{d}$, $\widetilde{\sigma}^2$. Note that we are not specifying the definition of damage rate $\dot{D}(t)$,

The narrow-band bound of rainflow damage for Gaussian loads employs the two parameters $h_s, f_z$. Both are "natural" estimates of sea harshness, $f_z$ is frequency of waves, mean level upcrosing frequency, while $h_s$ is an estimate of the average of the largest 33% of waves, for stationary Gaussian seas estimated as four times the standard deviation.

Let $d, \widetilde{d}$ be damage intensities $\mathsf{E}[D(T)]/T$ (rainflow, Holm-de Maré) caused by load $Y$ and for the normalised load $Z$, respectively. Furthermore, let $\sigma^2, \widetilde{\sigma}^2$ be variances $\mathsf{Var}[D(T)]/T$ computed using (11.4), for the two loads. Then, the simple scaling rules reads, respectively,

$$d = f_z h_s^{\beta} \widetilde{d}, \quad \sigma^2 = f_z h_s^{2\beta} \widetilde{\sigma}^2. \tag{11.12}$$

At this moment, we do not specify how the damage rate is used, but the same definition is used for both $Y$ and $Z$. Now, it is easy to see that the coefficient of variation scales as

$$\sqrt{f_z} \mathsf{CoV}[D_Y(T)] = \mathsf{CoV}[D_Z(T)].$$

It follows from (7.24) that for the standardised load $Z(t)$, the Holm-de Maré damage intensity is

$$\widetilde{d} = 2^{-\beta/2} \Gamma(1 + \beta/2), \tag{11.13}$$

i.e. it is independent of the load spectrum, and hence we propose to use

$$\frac{\widetilde{\sigma}}{\widetilde{d}} = \sqrt{f_z} \frac{\sigma}{d} \tag{11.14}$$

to compare the variation coefficients for different spectra. Consequently, (11.9) can be motivated. Note that $\widetilde{d} \approx 0.5$ for $\beta \in [2, 4]$, i.e. for most important values of $\beta$, only the variance $\widetilde{\sigma}^2$ needs to be estimated, employing the spectrum $\widetilde{S}$ since it does not depend on $f_z$ or $h_s$.

### 11.3.2 THE VARIATION OF MARINE ENVIRONMENT

Environmental loads are seldom stationary. Studying historical data, one often observes that values of $h_s(t)$ and $f_z(t)$ change in a rather unpredictable manner, and then one may also model their variability as random processes.

For example, in offshore engineering, one often approximates sea conditions by assuming surface elevation is a zero-mean Gaussian process with an adequate spectrum type, e.g., Pierson-Moskowitz, Jonswap, etc. Here, we will use the P-M spectrum defined by the two parameters $h_s$ and $f_z$. Further simplification will be an approximation that $f_z^{-1} = 3\sqrt{h_s}$.

Change of sea conditions is modelled by slowly changing random processes $h_s(\tau)$, $\tau$ in days, since it is generally assumed that $h_s$ remains almost constant from 20 minutes up to three hours. It is convenient since we can separate the variability of environment with $h_s(\tau), f_z(\tau)$ from the local damage growth variability, due to randomness of individual waves, expressed with the parameter $\widetilde{\sigma}^2(\tau)$. In the case when the load $Y$ has Pierson-Moskowitz spectrum, $\widetilde{\sigma}^2$ is constant during seasons.

In this section, we demonstrate the accuracy of the proposed approximation (11.9) using two zero-mean stationary Gaussian loads; the first one has the Pierson-Moskowitz spectrum, the second is narrow-band; see Figure 11.2.

We shall compare the coefficient of variation of the Holm-de Maré damage model with that of observed damage intensity in the rainflow model. To that end, we simulate the Gaussian load and estimate the covariance function of H-M damage

$$\dot{D}^{HM}(t) = \beta|2Y(t)|^{\beta-1}\dot{Y}(t)^+. \tag{11.15}$$

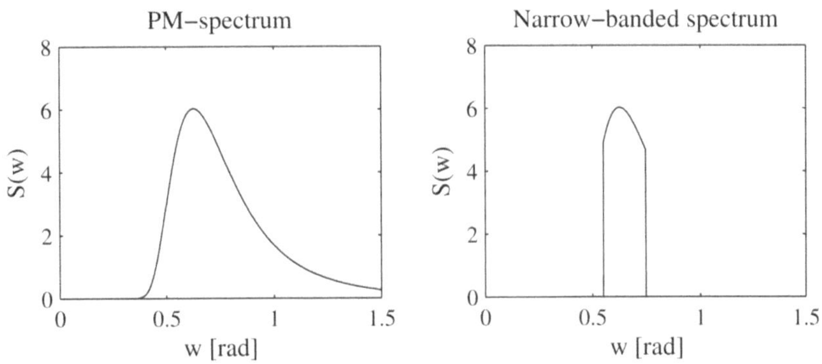


**FIGURE 11.2** P-M spectrum and narrow-banded spectrum for sea example.

#### 11.3.2.1 Gaussian Load, Pierson-Moskowitz Spectrum

As a first example, we consider Gaussian loads $Y(t)$ with a Pierson-Moskowitz (P-M) spectrum. This is a suitable model for fully developed sea. The P-M spectrum is shown in the left plot in Figure 11.2. In the upper plot in Figure 11.4, a simulated Gaussian load with P-M spectrum is presented.

One thousand 30-minute-long realisations of the load are simulated, and the ratio between the empirical standard deviation and the average rainflow damage gives the empirical coefficient of variation. This is compared to the average of the 1000 estimates of the coefficient of variation, using (11.6) with damage rate $\dot{D}^{HM}(t)$ defined by (11.15). The cut-off time $t_0$ in (11.5) is set to 30 s.

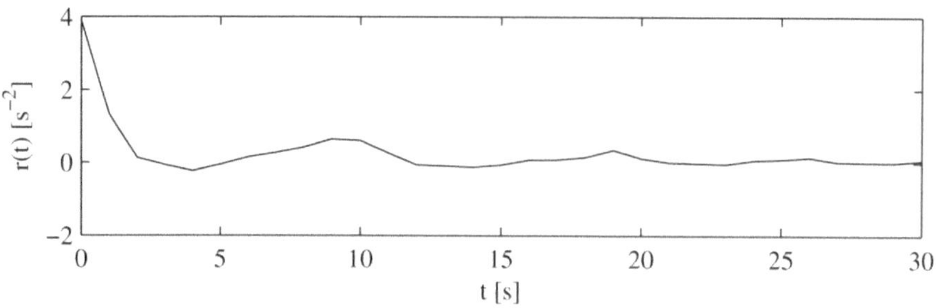


**FIGURE 11.3** Covariance function, $r(t) = \mathsf{Cov}[\dot{D}^{HM}(t+t_0), \dot{D}^{HM}(t_0)]$, estimated from 1000 simulations of 30-minute-long realisations of the Gaussian load with P-M spectrum, for $\beta = 3$.

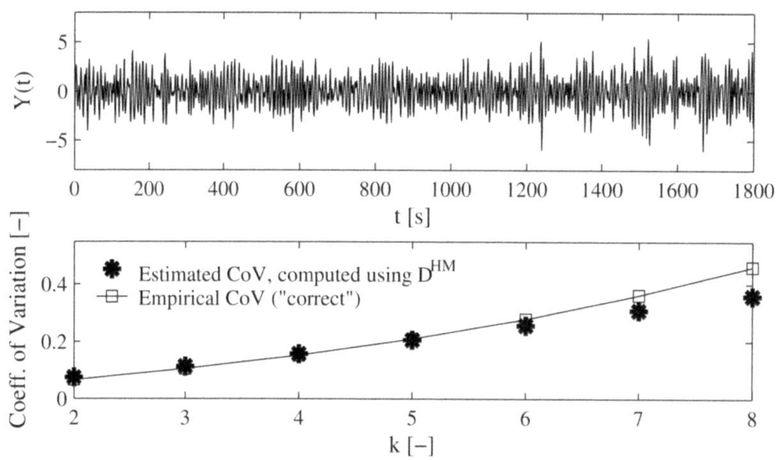


**FIGURE 11.4** Upper plot: Simulation of one 30-minute-long realisation of the Gaussian load with P-M spectrum $f_z = 0.13$ $[s^{-1}]$. Lower plot: Comparison of the coefficients of variation of $\mathsf{CoV}[D^{HM}(T)]$ and $\mathsf{CoV}[D^{rfc}(T)]$, $T = 30$ minutes, for the load. (Note: $k = \beta$).

In Figure 11.3, we present the covariance function $r(s)$ used in (11.4) to compute $\sigma^2$, for damage rate $\dot{D}^{HM}(t)$ with $\beta = 3$. The estimated coefficient of variation $\mathsf{CoV}[D^{HM}(T)]$ for $T = 30$ minutes loads is presented in Figure 11.4, (lower) plot. It is compared with the estimates of $\mathsf{CoV}[D^{rfc}(T)]$, and we can see that agreement is very good showing that the claim (11.9) is correct for the Gaussian load.

#### 11.3.2.2 Gaussian Load, Narrow-Band Spectrum

In fatigue damage applications, it is common to suppose that the Gaussian load is narrow-band since then the so-called narrow-band approximation, see Section 7.4, can be used to estimate the expected damage. Actually, the narrow-band approximation is equal to $\mathsf{E}[D^{HM}(T)]$. When the load is narrow-band, then the large cycles are coming in groups causing a lot of damage. The groups are separated by periods with lower damage rate. This property causes higher uncertainty than in broad band case when large cycles are more evenly spread in time. This will be illustrated in this example.

Here, a narrow-band spectrum is created by simply cutting off low and high frequencies in the P-M spectrum shown in Figure 11.2. The narrow-band spectrum is shown in Figure 11.2 (right) plot, with a simulated sample in Figure 11.6.

As before, 1000 Gaussian load sequences are simulated, but now using the narrow-banded spectrum. From each sequence, an estimate of the coefficient of variation is obtained. The average estimate is compared to the empirical coefficient of variation of the rainflow fatigue damage. The tendency to under-estimate the coefficient of variation for large $\beta$-values is higher in this example. Note also that the coefficient of variation, as expected, is larger for the narrow-band case compared to the broad-band case. As before, the covariance $r(s)$, for $\beta = 3$, is presented in Figure 11.5. We can see that the correlation between $X(0)$ and $X(10)$ is now much higher than in the broad banded case.

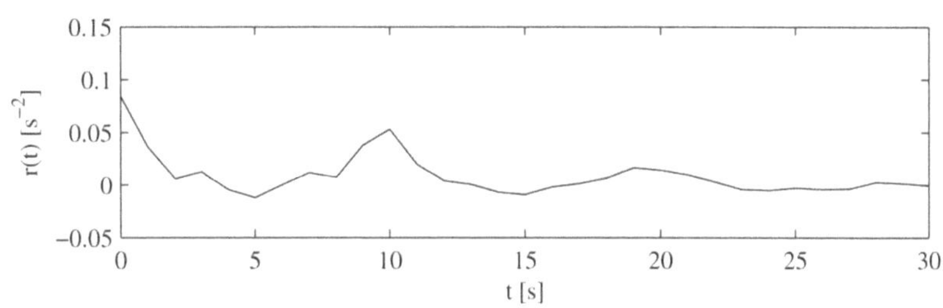


**FIGURE 11.5** Covariance function, $r(t) = \text{Cov}[\dot{D}^{HM}(t+t_0), \dot{D}^{HM}(t_0)]$ estimated from 1000 sample paths of Gaussian load with narrow-banded spectrum, for $\beta = 3$.

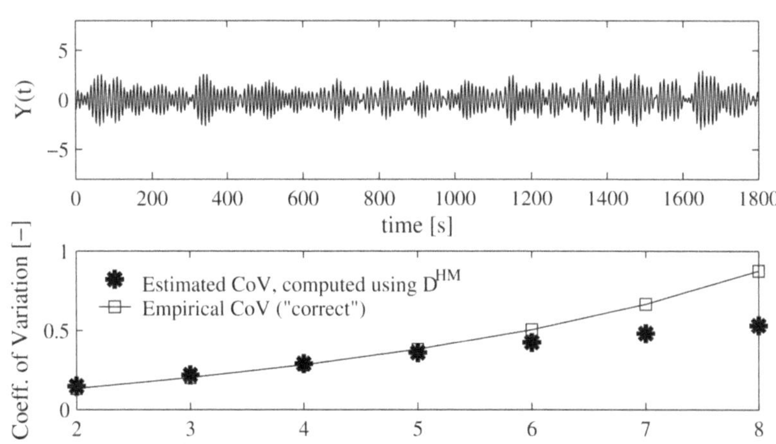


**FIGURE 11.6** A simulation of 30 minutes of the narrow-band Gaussian load with $f_z = 0.102$ $[s^{-1}]$, (upper) plot. The comparison between estimates $\text{CoV}[D^{HM}(T)]$ and $\text{CoV}[D^{rfc}(T)]$, derived from 1000 simulations of 30-minute Gaussian loads with narrow-banded spectrum, (lower) plot.

### 11.3.3 MEASURED WAVE STRESSES IN A VESSEL

We finish this section by analysing the uncertainty in the estimate of the accumulated damage for a measured stress, as in Figure 11.7, at some location in a vessel. The load is a superposition of wave-induced stresses and transients due to whippings. Here, one has only one measurement and hence Equation (11.6) will be used to estimate the variation coefficient. In Figure 11.8 upper right plot, the observed upcrossing intensity (irregular line) is compared with the theoretical one, under the assumption that the measured load is a realisation of a stationary Gaussian process with spectrum estimated from the measured signal. The spectrum is shown in Figure 11.8 upper left plot. The observed crossing intensity is symmetric and resembles well the theoretical one. Consequently, we expect that the observed Holm-de Maré damage rate (11.15) will be close to the observed rainflow damage rate defined by (11.3). This is also the case and for $\beta = 3$ and $\alpha = 1$, the observed rainflow damage is $7.7 \cdot 10^8$, while the observed Holm-de Maré damage is $9.1 \cdot 10^8$.

Next, we will investigate if one can model the load by means of a Gaussian process with spectrum from Figure 11.8. The values of the measured load plotted on normal probability plot, see Figure 11.8 lower left plot, do not contradict an assumption that the load could be Gaussian. In addition, the theoretical narrow-band Gaussian upper bound (7.14) for the expected damage is $9.2 \cdot 10^8$ which is very close to the observed Holm-de Maré damage.

Next, 100 realisations of the Gaussian load with the spectrum were simulated which gave an estimate of the expected rainflow damage equal to $7.1 \cdot 10^8$. We conclude that for computation of expected damage, the Gaussian load model gives sufficiently accurate estimates.

However, the load is non-Gaussian which can be seen in Figure 11.8 lower right plot where the estimated $((Y(t+\Delta t) - Y(t))/\Delta t)$ derivatives are plotted on normal probability paper. One can see that the derivative has heavier tails than the Gaussian

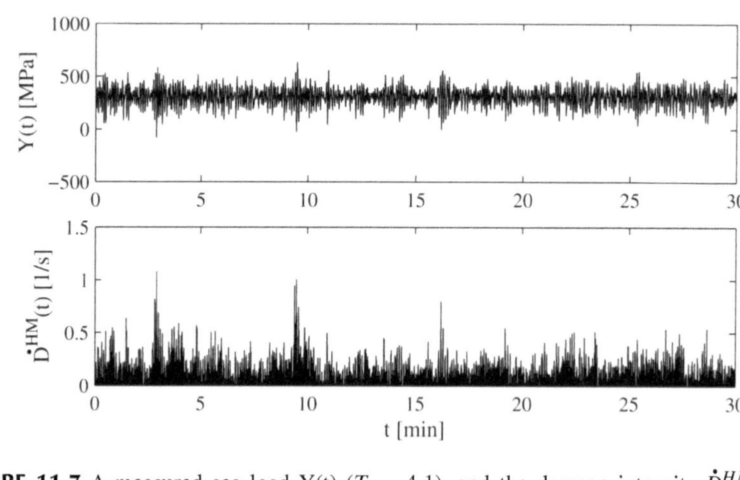


**FIGURE 11.7** A measured sea load Y(t) ($T_z = 4.1$), and the damage intensity $\dot{D}^{HM}(t)$ for $\beta = 2$.

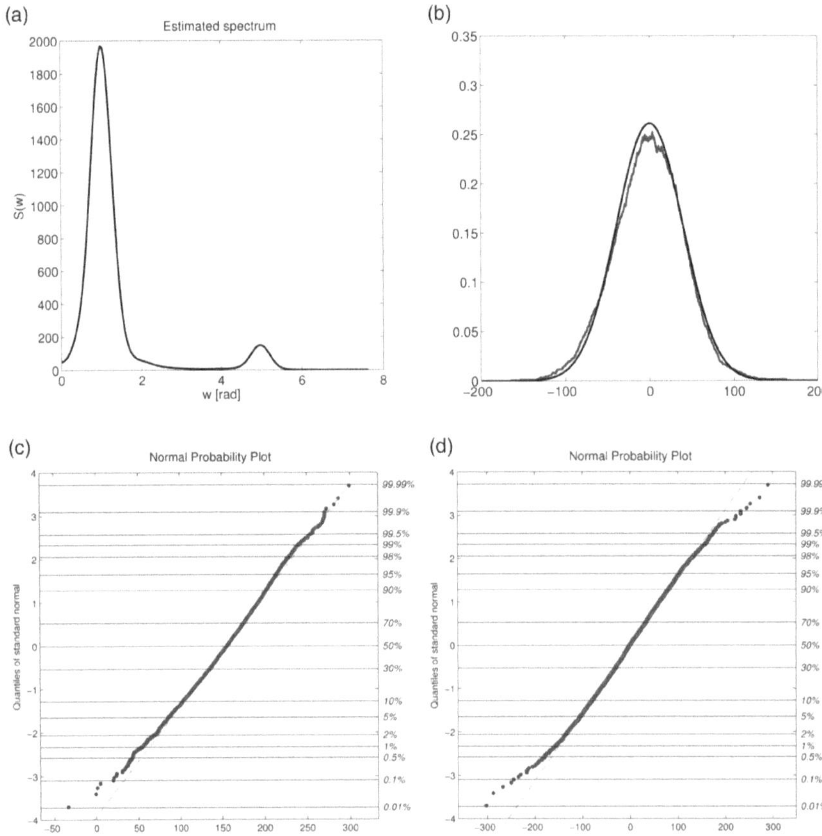


**FIGURE 11.8** (a) Estimated spectrum for the measured sea load; see Figure 11.7 upper plot. (b) The observed upcrossing intensity, irregular line, compared with the theoretical upcrossing intensity for a Gaussian process with the estimated spectrum. (c) Measured load plotted on normal probability paper. (d) Estimates of the derivative of the load plotted on normal probability paper.

load. This will affect the coefficient of variation for the expected damage as can be seen in Table 11.1 two last rows. The coefficient of variation for the Gaussian model is significantly lower than the observed coefficient of variation. Although the spectrum is clearly broad banded, the observed coefficient of variation is closed to the narrow-band Gaussian one, second row in the table.

## 11.4 VEHICLE FATIGUE CAUSED BY ROUGH ROADS

In the previous section, we have considered Gaussian (and close to Gaussian) wave loads. We concluded that the narrow-band loads have higher coefficient of variation of the expected damage than the broad band load has. For the measured load, which

**Table 11.1**
**Table of $\sqrt{f_z}\frac{\sigma}{d}$ for the Holm-de Mare Damage**

| Description | Figures | $\beta = 3$ | $\beta = 5$ | $\beta = 7$ |
|---|---|---|---|---|
| Gaussian broadband | 11.4 | 1.7 | 3.1 | 4.7 |
| Gaussian narrow-band | 11.6 | 3.0 | 4.9 | 6.5 |
| Gaussian two peaks | 11.8 | 1.6 | 3.2 | 5.8 |
| Measured "sea load" | 11.7 | 3.0 | 5.3 | 7.8 |

is a superposition of wave-induced stresses and whippings (added transients) and hence slightly non-Gaussian, the observed variation of coefficient was higher but of the same order as the Gaussian one. In this section, we demonstrate an example of a stationary load which will have much higher coefficient of variation than the loads studied in the previous section.

Commonly, stochastic models are used to describe the randomness of measured road profiles. Vehicle models travelling on road profiles modelled as stationary Gaussian processes have been extensively studied (see e.g., [198] and [148] for recent studies). However, measured profiles are not accurately described by a stationary Gaussian model [20]. In Section 10.4.3, we presented a road model with random non-stationary irregularities; a realisation of the model is shown in Figure 10.9.

Fatigue damage is assessed by studying a quarter-vehicle model travelling at constant velocity on random road profiles; see Figure 11.9. This very simple model cannot be expected to predict loads on a physical vehicle exactly, but it will highlight the most important road characteristics as far as fatigue damage accumulation is concerned. The parameters in the model are set to mimic heavy vehicle dynamics; see Table 11.2. In order to assess vehicle fatigue damage, the total force, $Y(t)$, for $t \in [0, T]$, acting on the sprung mass is rainflow-counted, and the damage is given by (10.26).

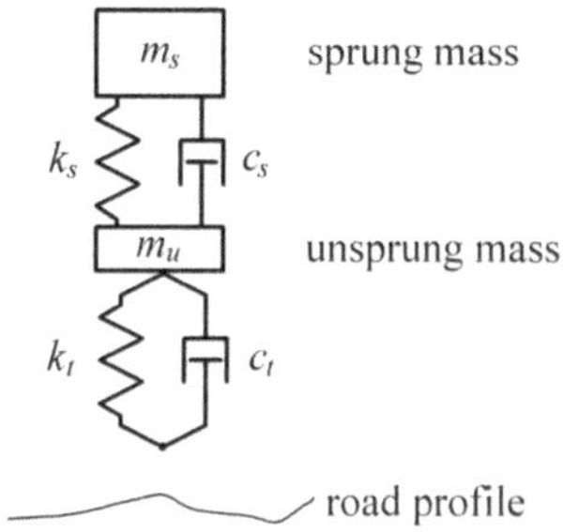


**FIGURE 11.9** Quarter vehicle model. (Reproduced with permission from [14]. Copyright © 2009 Wiley.)

**Table 11.2**
**Quarter Vehicle Parameters**

| Description | Symbol | Value | Unit |
|---|---|---|---|
| Sprung mass | $m_s$ | 3400 | kg |
| Suspension spring stiffness | $k_s$ | 270,000 | N/m |
| Suspension damper stiffness | $c_s$ | 6000 | Ns/m |
| Unsprung mass | $m_u$ | 350 | kg |
| Tire spring stiffness | $k_t$ | 950,000 | N/m |
| Tire damper stiffness | $c_t$ | 300 | Ns/m |

Analysis of measured road tracks shows that actual roads contain short sections with above-average irregularity. Such irregularities are shown to cause most of the vehicle fatigue damage [20]. The main variability in the road is described by the stationary Gaussian process $Z_0(t)$, with spectrum

$$S_0(f) = \begin{cases} 10^{a_0}\left(\frac{f}{f_0}\right)^{-w_1}, & f \in [0.01, 0.20], \\ 10^{a_0}\left(\frac{f}{f_0}\right)^{-w_2}, & f \in [0.20, 10], \\ 0, & \text{otherwise}, \end{cases} \tag{11.16}$$

where $f_0 = 0.2$. Here, $10^{a_0}$ is the basic roughness coefficient. The exponent $w_1$ describes energy distribution for wavelengths between 100 and 5 m. The exponent $w_2$ used for wavelengths between 5 and 0.1 m describes the state of road deterioration.

Irregularities of random shape, length and location are superimposed to $Z_0(x)$. The $j$th irregularity is denoted by $Z_j(x)$, $j > 0$, and the road with superimposed irregularities is denoted by $Z(x)$. To exemplify, a 400-m-long road is simulated and plotted in Figure 11.10.

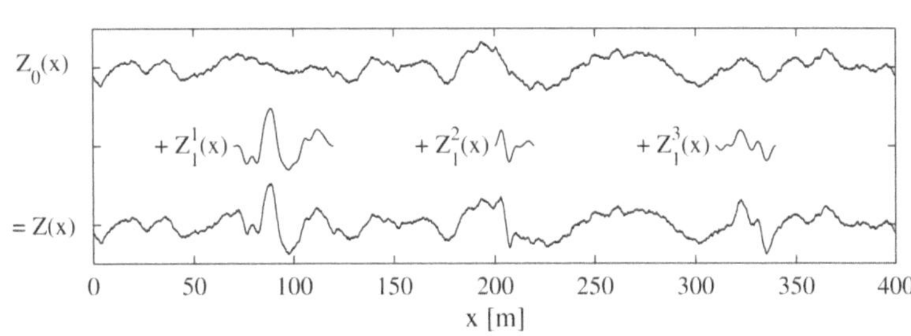


**FIGURE 11.10** A synthetic (computer simulated) road profile ($a_0 = -5.2$, $w_1 = -3.1$, $w_2 = -2.2$, for $Z_0$, and $a_0 = -4.1$, for $Z_1$). (Reproduced with permission from [14]. Copyright © 2009 Wiley.)

The irregularities are modelled as (non-stationary) conditional Gaussian processes. To avoid discontinuities at the start and end of the rough sections, the added irregularities start and end with zero slope and zero level. The irregularities are simulated conditioned on the zero boundary values, see [18] for more details. Furthermore, the location and length of the sections (with added roughness) are random. More precisely, the distance between the end of an irregularity and the start of the next is exponentially distributed. The irregularity length is also exponentially distributed.

The process $Y(t)$ that we study is the load acting on the sprung mass when the quarter vehicle travels at speed $v$ on the random road. The load process is stationary but non-Gaussian. It can also be shown that it has also symmetric crossing spectrum (see Figure 11.11), and hence the proposed method to estimate the coefficient of variation can be used, i.e. we shall approximate the coefficient of variation of the rainflow damage by coefficient of variation of the Holm-de Maré damage.

In Figure 11.12, the rainflow and the Holm-de Maré damage rates are compared. In this particular example, 400-m-long roads are realised from the road model described above, but here the irregularities are placed deterministically at the intervals $[70,120]$, $[200,220]$ and $[310,340]$ (meter). The quarter vehicle is simulated travelling at 70 km/h, so these intervals correspond to the following intervals expressed in seconds: [3.6, 6.2], [10.3, 11.3] and [15.9, 17.5]. We can expect that the intensities will be higher at these locations.

In the upper figure, both rates are computed from a particular $Y(t)$, for $t \in [0,100]$, and in the lower plot, the average from 100 simulations are plotted. This illustrates how the two rates differ for a particular load, but on average, they are quite similar.

The upper plot in Figure 11.13 shows a road profile $Z$, simulated from the model with random irregularities. The middle plot shows the load $Y(t)$ acting on the sprung mass ($v = 80$ km/h). The load peaks correspond to road sections with increased roughness. The lower plot shows the Holm-de Maré damage intensity for $\beta = 3$.

For an ergodic series, the covariance function $r(t)$ can be estimated from a single time-series $X(t)$. In this case, $r(t)$ is estimated from the damage intensity in Figure 11.13. The result is shown in Figure 11.14. Then, the variance and coefficient of variation (CoV ) is estimated using (11.4) and (11.6).

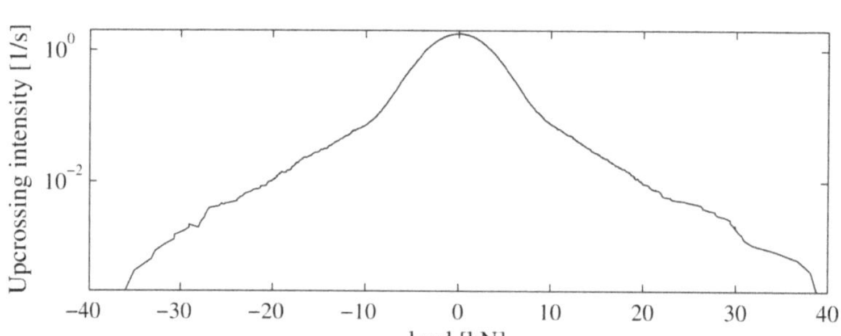


**FIGURE 11.11** Observed upcrossing intensity in the load $Y(t)$ in Figure 11.13. (Reproduced with permission from [14]. Copyright © 2009 Wiley.)

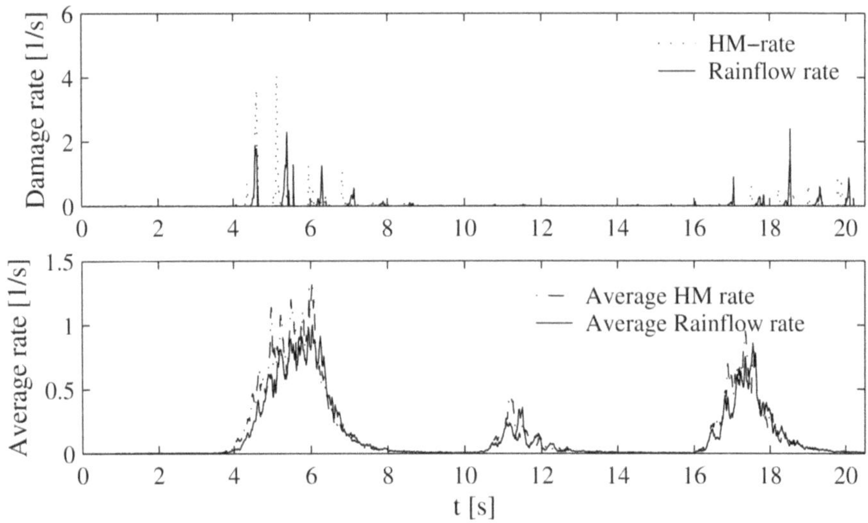


**FIGURE 11.12** Comparison between the rainflow and Holm-de Maré damage rate. (Reproduced with permission from [14]. Copyright © 2009 Wiley.)

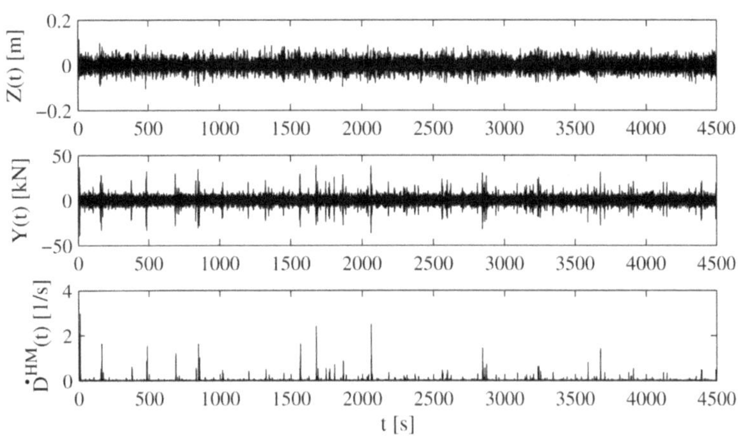


**FIGURE 11.13** A road profile Z(t) and corresponding vehicle load Y(t) ($f_z = 1.79$ [$s^{-1}$]) when $v = 80$ km/h, and the damage intensity $\dot{D}^{HM}(t)$ for $\beta = 3$. (Reproduced with permission from [14]. Copyright.© 2009 Wiley.)

A Monte Carlo study will show the accuracy of the CoV estimation. One hundred roads of length 100 km are realised with expected distance between irregularities 400 m and the expected length of an irregularity $\tau = 32$ m. The quarter vehicle is simulated travelling at 80 km/h, and the vehicle damage caused by each road is computed according to (10.26).

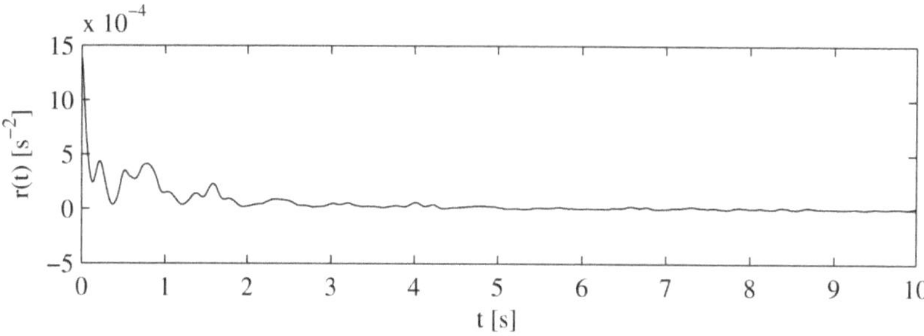


**FIGURE 11.14** Covariance function, $r(t) = \mathrm{Cov}[\dot{D}^{HM}(t+t_0), \dot{D}^{HM}(t_0)]$ estimated from the sample path $\dot{D}^{HM}(t)$ in Figure 11.13. (Reproduced with permission from [14]. Copyright © 2009 Wiley.)

Then, we compute the empirical standard deviation and the average damage using these 100 outcomes of $D(T)$. The empirical standard deviation divided by the average damage gives an estimate of the coefficient of variation of $D(T)$, which is referred to as the empirical coefficient of variation.

For comparison, from each load sequence, we also compute the Holm-de Maré intensity and estimate the variance using (11.6). The average result is close to the empirical coefficient of variation, computed using the standard rainflow approach, as illustrated by Figure 11.15. Thus, on average, we will get a coefficient of variation close to the true value. As expected, when $\beta$ increases, the CoV also increases. So, for $\beta = 8$, the coefficient of variation is as high as 0.5, despite the length of the road ($L$ = 100 km, with $T$ = 1 h 15 min). This has practical consequences, when vehicle loads have been measured on a 100-km-long road, with these properties, it is not enough to consider only the expected damage, since the variability is still quite high.

It is also interesting to study the uncertainty of the estimate. In Figure 11.16, a histogram is shown of all 100 estimates, compared to the empirical coefficient of variation. We conclude the uncertainty in the estimated value of the variation coefficient is quit large, even in so long as 100-km-long measurement.

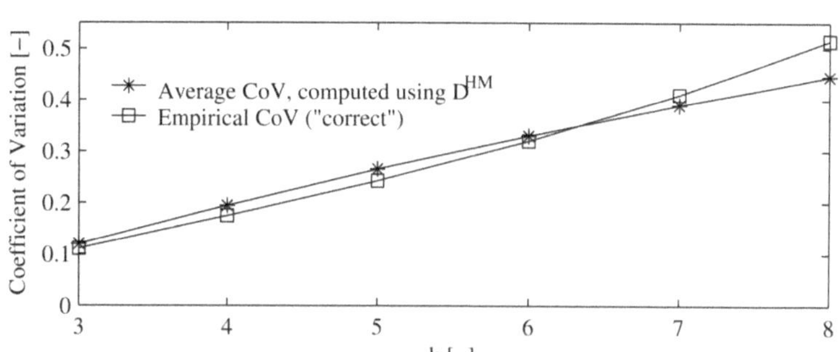


**FIGURE 11.15** Average CoV computed using (11.6), compared to empirical CoV. (Reproduced with permission from [14]. Copyright © 2009 Wiley.)

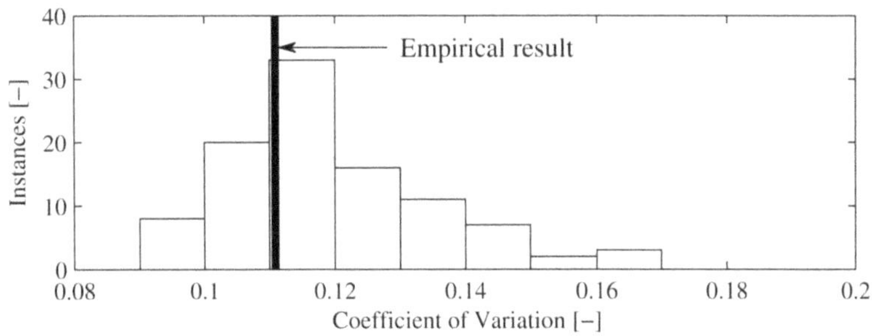


**FIGURE 11.16** Uncertainty of CoV-estimate, for $\beta = 3$. (Reproduced with permission from [14]. Copyright © 2009 Wiley.)

**Table 11.3**

**Table of $\sqrt{f_z}\frac{\sigma}{d}$ for the Holm-de Mare Vehicle Damage**

| Description | Figures | $\beta = 3$ | $\beta = 5$ | $\beta = 7$ |
|---|---|---|---|---|
| Vehicle load | 11.13, 11.15 | 10.7 | 23.7 | 34.9 |

Finally, in order to compare the uncertainties in the estimates of the expected damage for the studied vehicle load and the wave loads presented in the previous section, see Table 11.5. In Table 11.3, we give the normalised coefficient of variation $\sqrt{f_z}\frac{\sigma}{d}$ for the vehicle load. We can see that this is about six times higher.

We finish this section with some comments on alternative modelling of vehicles responses, presented and studied in [21]. Let us again consider the load $Y(t)$ presented in Figure 11.13. Its upcrossing frequencies, shown in Figure 11.11, look similar to the crossing spectra of the hybrid models presented in Chapter 9. Actually, the Gaussian model with added irregularities, discussed here, was an inspiration for the hybrid model presented first in [21].

In that work, five models were compared: The first was a homogeneous Gaussian model, the second was a non-homogeneous Gaussian model, the third was a non-homogeneous Laplace model, the fourth was a Laplace moving average (LMA) and, finally, the fifth was a combination of the Gaussian and the LMA model named the hybrid model.

The comparisons facilitated thorough analysis of the actual road records comprising eight measured roads of total length about 230 km and of varying quality, ranging from rough gravel and asphalt roads to smooth asphalt roads. These signals had different lengths and were scaled to have variance one and zero-mean.

**Conclusions:** The conclusions of the investigations were quite clear. The non-homogeneous Gaussian model seems to work relatively well when enough data are available for estimation although some tendency for underestimation of the rainflow damage intensity for lower speed has been observed. The hybrid model worked very well in all the considered cases, except in one, where it was slightly conservative, but still better than the other models. The LMA model worked very well for shorter road segments, which are more likely to be homogeneous. Finally, the homogeneous Gaussian model severely underestimates the expected rainflow damage.

## 11.5 GAUSSIAN LOADS IN A VARIABLE ENVIRONMENT

Often, marine loads can be well modelled as a zero-mean stationary Gaussian processes for quite long periods of time $t$, say, when the significant wave height $h_s$ and apparent frequency $f_z$ are almost constant. A typical example is loads changing with seasons, e.g., wave loads. Let us consider a wave load $W(t)$ at a fixed locations, height defined as the sea surface elevation around still water level. In North Atlantic, the load can be approximated by a zero-mean, stationary process for a period of 30 minutes to three hours. Following Section 10.4.2.3, we let $W$ have Pierson-Moskowitz spectrum, a quite broad-banded spectrum for which loads become independent after a few cycles. (During a stationarity period, we may have several hundreds of cycles.)

Consequently, by (11.9), we assume that coefficient of variation for rainflow damage can be approximated by coefficient of variation for Holm-de Maré damage with the damage increase rate

$$\dot{D}(t) = \alpha\beta\,|2W(t)|^{\beta-1}\dot{W}(t)^+,$$

with $t$ in seconds. Here, we will estimate the coefficient of variation $\mathsf{CoV}$ for the damage accumulated in longer periods of time with $T$ a season, or year, viz $D(T) = \int_0^T \dot{D}(\tau)\,d\tau$, $\tau$ in days.

### 11.5.1 COEFFICIENT OF VARIATION OF YEARLY DAMAGE

The coefficient of variation is a measure of uncertainty in prediction of expected damage and is defined by

$$\mathsf{CoV}(D(T)) = \sqrt{\mathsf{Var}[D(T)]}/\mathsf{E}[D(T)].$$

The method is based on prediction of damage rate $\dot{D}(\tau)$ using the significant wave hight $h_s(\tau)$ varying seasonally in a random manner. The damage intensity $d(\tau) = \mathsf{E}[\dot{D}(\tau)] = \mathsf{E}[\mathsf{E}[\dot{D}(\tau) \mid h_s, f_z]]$, where

$$\mathsf{E}[\dot{D}(\tau) \mid h_s, f_z] = \tilde{d}(\beta) f_z h_z^{\beta}, \quad \tilde{d}(\beta) = \alpha 2^{-\beta/2}\Gamma(1+\beta/2),$$

with the damage function $f(u,v) = \alpha(v-u)^{\beta}$.

In Section 10.4.2.4, we computed the expected yearly damage $\mathsf{E}[D(T)]$; see Table 10.1. The method employed a deterministic (regression) model for $f_z$ as a function of $h_s$, see (10.37) which is

$$f_z(\tau) = 28800/\sqrt{h_s(\tau)},$$

per day, which results in

$$\mathsf{E}[\dot{D}(\tau) \mid h_s, f_z] = \widetilde{d}(\beta)\, h_s(\tau)^{\beta-0.5}. \tag{11.17}$$

Further, the $h_s(\tau)$ is a lognormal process with median $m_0 + m(\tau) + e(\tau)$; see (10.34). Here, $m_0$ is a location-dependent constant, while $m(\tau)$ is modelling the seasonal variability of the $\mathsf{E}[\ln(h_s(\tau)]$. The residual $e(\tau)$ is a stationary Gaussian process having variance $\sigma_H^2$, and the covariance function given in (10.35), viz

$$\mathsf{Cov}[e(t), e(s)] = e^{-0.576(s-t)^2-0.24|s-t|} = r_e(s-t) \tag{11.18}$$

where $t, s$ have unit days. We can see that the $e(\tau)$ becomes almost independent if separated by three days. This is the combined result of the shape of storm systems and their velocity when passing the observing station in the North Atlantic.

Since the coefficient of variation does not depend on multiplicative constants, we move all non-random factors into $\widetilde{d}(\beta)$, e.g. $\alpha$, 28800 and $e^{m_0(\beta-0.5)}$, i.e.

$$\widetilde{d}(\beta) = 2.88 \cdot 10^{-4} \cdot e^{m_0(\beta-0.5)}\, 2^{-\beta/2}\Gamma(1+\beta/2), \tag{11.19}$$

see also (10.38). Now, the model for conditional damage rate can be written

$$\mathsf{E}[\dot{D}(\tau) \mid h_s, f_z] = g(\tau) R(\tau), \tag{11.20}$$

where $g(\tau)$ is a periodic function, while $\ln R(\tau)$ is a stationary Gaussian process describing variability of significant wave height, defined by

$$g(\tau) = \widetilde{d}(\beta) e^{(\beta-0.5)m(\tau)},$$
$$R(\tau) = e^{(\beta-0.5)\sigma_h e(\tau)}.$$

The covariance function of the random factor $R(\tau)$ is given by

$$\begin{aligned} \mathsf{Cov}[R(t), R(s)] &= \mathsf{Cov}[e^{(\beta-1/2)\sigma_H e(t)}, e^{(\beta-0.5)\sigma_H e(s)}] \\ &= e^{(\beta-0.5)^2\sigma_H^2}\left(e^{(\beta-0.5)^2 r_e(s-t)} - 1\right) = r_R(s-t), \end{aligned} \tag{11.21}$$

with $t, s$ in days.

We turn next to computation of the variance of yearly damage increase. As can be seen in Figure 10.7, it is the variability of $h_s$ that contributes most to the accumulated damage variance. However, there is also another source of uncertainty to the yearly damage, and it is randomness of cycle amplitudes. We will compare those two sources of uncertainty in the prediction of the accumulated damage $\mathsf{Var}[\int_0^T \dot{D}(\tau)\, \mathrm{d}\tau]$. This is done by employing the total variance formula which splits the variability of $D(T)$ into two parts, the expected conditional variance of $X = D(T)$ given $Y = h_s(\cdot)$ and the variance of the conditional expectation of $X = D(T)$ given $Y = h_s(\cdot)$, viz.

$$\mathsf{Var}[X] = \mathsf{Var}[\mathsf{E}[X \mid Y]] + \mathsf{E}[\mathsf{Var}[X \mid Y]]. \tag{11.22}$$

The resulting variance $\mathsf{Var}[D(T)] = V_E + E_V$ can be used to compute the coefficient of variation of accumulated yearly damage $\sqrt{\mathsf{Var}[D(T)]}/\mathsf{E}[D(T)]$, which can be used to compute the reliability index; see [15].

#### 11.5.1.1 Variance of the Conditional Expectation $V_E$

We turn first to the variance of the conditional expectation $\mathsf{Var}[\mathsf{E}[X \mid Y]]$. Since, by (11.20),

$$\mathsf{E}[D(T)] \mid h_s(\cdot)] = \int_0^T g(\tau)R(\tau)\,\mathrm{d}\tau,$$

the variance is

$$\mathsf{Var}[\mathsf{E}[D(T)] \mid h_s(\cdot)] = \int_0^T \int_0^T g(t)g(s)r_R(s-t)\,\mathrm{d}t\,\mathrm{d}s, \tag{11.23}$$

where $r_R(s-t)$ is defined in (11.21) and hence

$$g(t)g(s)r_R(s-t) = \widetilde{d}(\beta)^2 e^{(\beta-0.5)\,(m(t)+m(s))} e^{(\beta-0.5)^2} r_R(s-t). \tag{11.24}$$

Combining (11.23) and (11.24), the $\mathsf{Var}[\mathsf{E}[X|Y]]$ can be evaluated by means of numerical integration.

#### 11.5.1.2 Expectation of the Conditional Variance $E_V$

In [15], it was shown that for the studied location in the North Atlantic, the term $\mathsf{E}[\mathsf{Var}[D(T) \mid h_s(\cdot)]]$ is negligible relative to the term $\mathsf{Var}[\mathsf{E}[D(T) \mid h_s(\cdot)]]$, except for very high damage exponents, e.g. $\beta = 7$. However, for completeness, we present means to compute the variance.

The term $\mathsf{E}[\mathsf{Var}[D(T) \mid h_s(\cdot)]]$ describes uncertainty of the yearly damage originated in variation of individual waves. In Section 11.3.2.1, the variation coefficient was computed for a zero-mean Gaussian process with Pierson-Moscovitz spectrum with parameters $h_s = 6.5$ m, and $f_z = (3\sqrt{h_s})^{-1} = 0.13$ were studied. It has been demonstrated that the coefficient of variation for the accumulated damages (rfc and H-M) during half of hour is very close even for so high damage exponents as $\beta = 7$.

The sea level $W(t)$ is modelled by a zero-mean stationary Gaussian process having Pierson-Moscovitz spectrum which rests for about half an hour, i.e. $\Delta\tau = 0.02$ of a day, say. The load $W(t)$ has a very short memory and $\mathsf{Cov}[W(0), W(s)]$ is practically zero for $s$ greater than 30 seconds. Consequently, the damage increments in different stationary periods are independent, and the conditional variance can be found by integration,

$$V[D(T) \mid h_s(\cdot)] = \int_0^T V[\dot{D}(\tau) \mid h_s(\tau)]\,\mathrm{d}\tau.$$

Employing the normalised process $Z(t)$ defined in (11.11) and (11.12)

$$\begin{aligned} V[D(T) \mid h_s(\cdot)] &= \widetilde{\sigma}^2 \int_0^T f_z(\tau)h_s(\tau)^{2\beta}\,\mathrm{d}\tau \\ &= \frac{\widetilde{\sigma}^2}{3} \int_0^T h_s(\tau)^{2\beta-0.5}\,\mathrm{d}\tau, \end{aligned} \tag{11.25}$$

where $\widetilde{\sigma}^2$ is the variance of the H-M damage for $Y(t)$ having P-M spectrum with $h_s = 1$ and $f_z = 1$.

Using (11.25), we obtain

$$\mathsf{E}[\mathsf{Var}[D(T) \mid h_s(\cdot)]] = T\,\frac{\widetilde{\sigma}^2}{3}\,\mathsf{E}[H_s^{2\beta-0.5}],$$

where $H_s$ has the "yearly" distribution of significant wave heights, i.e. value of $h_s$ at times taken at random during a year, e.g. a histogram of observed values. Similarly, the expected yearly damage increase, studied in Section 10.4.2.4, (10.39), can be written as

$$\mathsf{E}[D(T)] = T\,c\widetilde{d}\,\mathsf{E}[H_s^{\beta-0.5}],$$

and hence this parts contribution to the coefficient of variation is proportional to

$$\frac{\widetilde{\sigma}(\beta)}{\widetilde{d}(\beta)} \cdot \frac{\sqrt{\mathsf{E}[H_s^{2\beta-0.5}]}}{\mathsf{E}[H_s^{\beta-0.5}]},$$

where $\widetilde{d} = 2^{-\beta/2}\Gamma(1+\beta/2)$.

Note that $\mathsf{Var}[\mathsf{E}[D(T) \mid h_s(\cdot)]]$ depends on the process $H_s(\tau)$, e.g. on mean and correlation of the $\log H_s(\tau)$, as the result of shape and duration of passing storms.

For completeness, we present the dependence of $\widetilde{\sigma}^2$ on $\beta$ (Table 11.4). The values have been estimated using MC methods, with 30 minutes as a period of constant $h_s$. Note that log-values are basically a linear function of $\beta$.

**Table 11.4**
**Simulated $\widetilde{\sigma}^2$ Related to $\beta$**

| $\beta$ | 1 | 2 | 3 | 4 | 5 | 6 | 7 | 8 |
|---|---|---|---|---|---|---|---|---|
| $10^{-3}\,\widetilde{\sigma}^2$ | 0.129 | 0.404 | 0.915 | 2.128 | 5.544 | 16.21 | 58.08 | 209.95 |

Note that the damage increment distribution depends only on $h_s$. The second parameter $f_z$ is modelled deterministically by means of the approximation $f_z^{-1} = 3\sqrt{h_s}$, valid for fully developed sea with P-M spectrum, which is an ergodic process. Then, an efficient procedure is to simulate a long sample of $W(t), \dot{W}(t)$ with $h_s = 1$ and $f_z = 1$, then evaluate the H-M damage (11.7) and estimate the covariance function $\widehat{r}(s)$. Finally, an estimate $\widetilde{\sigma}^2$ is given by (11.5) with $t_0$ equal half or one hour, dependent of how fast $h_s(\tau)$ changes.

The conditional variance of the yearly damage increment is given by (11.25). The expected conditional variance is then evaluated in a similar way as we computed the expected damage increase during a year, viz. by means of formula (11.20). The resulting coefficients of variation for the considered location and time period $T$ equal to one year were reported in [14] and are given in Table 11.5.

**Remark 11.1.** *More details on spatio-temporal models of sea surface can be found in [7] where means to estimate the expected damage and its coefficient of variation*

**Table 11.5**
**Estimated Coefficients of Variations** $\mathsf{CoV}(D(T)) = \sqrt{\mathsf{Var}[D(T)]}/\mathsf{E}[D(T)]$ **for Accumulated Damage during One Year at a Location at Northern Atlantic**

| $\beta = 2$ | $\beta = 3$ | $\beta = 4$ | $\beta = 5$ | $\beta = 7$ |
|---|---|---|---|---|
| 0.04 | 0.09 | 0.17 | 0.34 | 1.82 |

*accumulated in a ship detail on a route in Northern Atlantic were presented. This type of estimates can be used to plan the route which minimises the accumulated damage; see Mao et al. (2010) [141].*

*The variation coefficient* $\sqrt{\mathsf{Var}[D(T)]}/\mathsf{E}[D(T)]$ *can also be used to compute the reliability index, see [15].*

## 11.6 MARKOV LOADS

Markov loads were introduced in Chapter 8. These are functions of discrete valued stationary Markov chain and constitute a large class of models for stresses. The loads are non-differentiable, and hence we cannot use formulas (7.14) or (7.15), both based on the load derivative, to approximate the damage intensity. However, the Holm-de Maré damage can be computed using the exhaustion function. So, let $X_n$ be a Markov load. Then, the damage increase between time $n$ and $n+1$ is $\Delta D_{n+1} = G(X_{n+1}) - G(X_n)$ if $X_{n+1} > X_n$ and zero otherwise. Consequently, the accumulated damage at time $n$ is

$$D_n = \sum_{i=1}^{n} \Delta D_i = \sum_{i=1}^{n} \left(G(X_{n+1}) - G(X_n)\right)^+ .$$

Now, $Y_n = G(X_n)$ is a function of a Markov chain, and statistically, it has the same properties as a Markov load. Hence, in order to find the variance of the Holm-de Maré damage, one needs to find the covariance $r_{n-1}$ between $(Y_n - Y_{n-1})^+$ and $(Y_1 - Y_0)^+$, and then use

$$\mathsf{Var}[D_n] = 2\sum_{i=1}^{n-1} (n-i)r_i + nr_0,$$

which for large $n$ gives the *damage variance factor*

$$\mathsf{Var}[D_n]/n \approx \sigma_D^2 := \sum_{-\infty}^{\infty} r_i. \tag{11.26}$$

Warning: These formulas are of less use when the damage distribution is very skewed.

## 11.7 UNCERTAINTY OF FATIGUE LIFE PREDICTION

### 11.7.1 EMPIRICAL OR THEORETICAL ERROR ESTIMATE?

Due to the additive character of the fatigue process, postulated in this book, the variability and uncertainty of the estimated fatigue damage can be calculated for every load model where the variance and covaricances between damage cycles can be observed or calculated. We will here give an example of how the technique is used to make a statement about the predicted fatigue life. The example is inspired by the work by Rychlik, Lindgren and Lin (1993) [183].

**Example 11.1** (Sea wave variability as a Markov load). *We consider a Gaussian sea with Pierson-Moskowitz spectrum with significant wave height* $s = 7\,m$ *and peak period* $T_p = 11\,s$*, with spectrum truncated at* $\omega_o = 3$*. It has peak frequency* $f_p = 2\pi/11 = 0.57$ *[rad/s], and there are on average eight seconds between zero upcrossings and* 5.6 *seconds between local maxima.*

*Since there are a finite number of local extremes in finite time (due to the truncation of the spectrum), we can allow us to use a Markov chain of turning points (MCTP) to compute fatigue damage. Our study is designed as follows.*

**Simulate load** *Simulate continuous load*
**Find turning points** *Get TP*
**Find Mm cycles** $(m_i^{mM}, M_i^{mM})$
**Find RF cycles** $(m_i^{rfc}, M_i^{rfc})$
**Get total damage** $D = \alpha \sum |M_i - m_i|^\beta$
**Repeat** $n$ **times to get mean and std** *Average total damage* $D_{emp} = \overline{D_n}$ *and standard deviation* $\sigma_{emp} = \sqrt{V(D_n)}$
**Compare with theoretical std** $\sigma_{teo} \approx \sqrt{N_{cycles}} \times \sqrt{r(0) + 2\sum_1^\infty r_i}$

*The cycle matrices are shown in Figure 11.17 for the min-Max and rainflow cycles, while Figure 11.18 shows the estimated covariance functions for the sequences.*

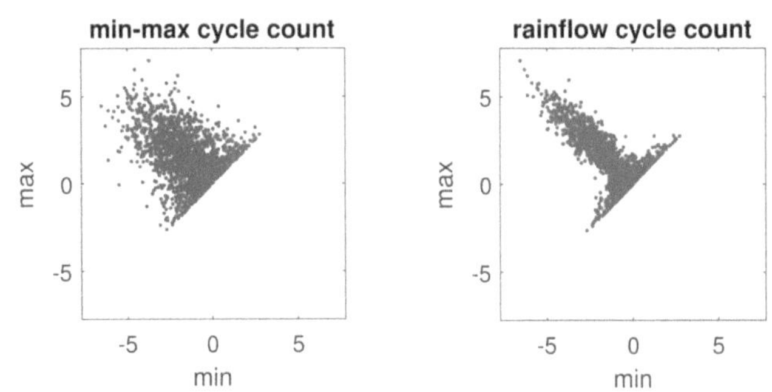


**FIGURE 11.17** min-Max and rainflow cycle counts for Pierson-Moskowitz waves.

*We simulated the Gaussian load over a long time span,* $T = 10{,}000$ *seconds,* $\Delta t = 0.1$ *and generated 100 samples. With damage function* $\alpha|M_i - m_i|^\beta$, $\beta = 3,4,5,$

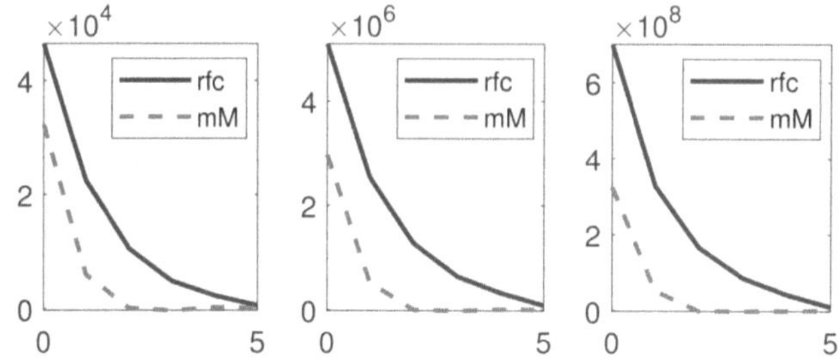


**FIGURE 11.18** Estimated covariance functions for the sequence of min-Max and rainflow cycles for $\beta = 3,4,5$.

**Table 11.6**
**Empirical Damage, Standard Deviations and CoV for Example 11.1**

| $\alpha = 1$ | $D_{emp}$ $\times 10^{-6}$ | $\sigma_{emp}$ $\times 10^{-5}$ | $\sigma_{theory}$ $\times 10^{-5}$ | $CoV_{emp}$ |
|---|---|---|---|---|
| Mm $\beta = 3$ | 0.167 | 0.075 | 0.088 | 0.045 |
| $\beta = 4$ | 1.153 | 0.741 | 0.827 | 0.064 |
| $\beta = 5$ | 8.771 | 7.659 | 8.423 | 0.087 |
| Rfc $\beta = 3$ | 0.193 | 0.080 | 0.135 | 0.041 |
| $\beta = 4$ | 1.412 | 0.837 | 1.449 | 0.059 |
| $\beta = 5$ | 11.392 | 9.286 | 16.374 | 0.082 |

*the empirical total damage together with empirical and theoretical approximated standard deviations are found in Table 11.6 together with the empirical coefficient of variation. As seen, the theoretical approximation* (11.26) *grossly underestimates the variability for the rainflow damage, which has a very skewed distribution.*

### 11.7.2 ERROR IN FATIGUE LIFE ESTIMATION

The theme in the paper by Rychlik et al. [183] was to make a statement about the reliability of fatigue life prediction based on the uncertainty in fatigue damage. The cycle model was a bivariate Markov load process of peaks and troughs with transition sequence $(m_{k-1}, M_k) \mapsto (M_k, m_k) \mapsto (m_k, M_{k+1})$ governed by the sets of transition probabilities

$$p_{ij} = \mathsf{P}(m_k = u_j \mid M_k = u_i),$$

$$\widehat{p}_{ij} = \mathsf{P}(M_{k+1} = u_j \mid m_k = u_i).$$

For any damage function $f(u,v)$, let $(m_0, M_1)$ have the stationary distribution of the bivariate Markov chain and define $d = \mathsf{E}[f(m_0, M_1)]$ as the average damage increment with covariance $r_f(k) = \mathsf{Cov}[f(m_0, M_1), f(m_k, M_{k+1}]$, with the damage variance factor $\sigma_D^2 = \sum_{-\infty}^{\infty} r_f(k)$.

Assuming there is a deterministic critical level $d_{crit}$ such that fatigue failure occurs as soon as $D_n > d_{crit}$, we define $n_0 = d_{crit}/d$ as a mean for an approximate distribution for the number och cycles to failure. Then, the following theorem holds.

**Theorem 11.1** ([183]). *For small $\alpha$, the number of cycles to failure $N_f$ has an approximate normal distribution*

$$N_f \stackrel{\mathscr{L}}{=} n_0 + \sqrt{n_0}\,\frac{\sigma_D}{d}\,\widetilde{X},$$

*where $\widetilde{X}$ is a standard normal variable.*

To get the *time* to failure, one has to model cycle length $\tau_k$ as the time interval between the completion of cycles $k-1$ and $k$, and also have an idea about its relation to cycle damage. (Such information can be obtained experimentaly or theoretically, e.g. in a Gaussian load model.) Define the covariances

$$r_\tau(k) = \mathsf{Cov}[\tau_0, \tau_k], \quad r_{f\tau}(k) = \mathsf{Cov}[f(m_0, M_0), \tau_k],$$

and the parameters

$$\tau = \mathsf{E}[\tau_k], \quad \sigma_\tau^2 = \sum_{-\infty}^{\infty} r_\tau(k), \quad \sigma_{D\tau} = \sum_{-\infty}^{\infty} r_{f\tau}(k).$$

The reduced variance

$$\sigma_{T\cdot D}^2 = \frac{\sigma_{\tau^2}}{\tau^2} - 2\frac{\sigma_{D\tau}}{d\tau} + \frac{\sigma_D^2}{d^2}$$

takes care of the correlation between cycle length and cycle damage.

**Theorem 11.2** ([183]). *Under the assumption that the triple $(M_k, m_k, \tau_k)$ forms a stationary Markov chain, the time to failure can be approximated as*

$$T_f \stackrel{\mathscr{L}}{=} \tau n_0 + \tau\sqrt{n_0}\,\sigma_{T\cdot D}\widetilde{Y},$$

*where $\widetilde{Y}$ has a standard normal distribution.*

Theorems 11.1 and 11.2 can easily be extended to models where also the material quality is random and independent of the loading. The simplest way to achieve this is to replace the critical level $d_{crit}$ by a random level $K$ and the parameter $n_0$ by the random $K/d$.

## 11.8 NUMERICAL TOOLS – TOOLBOX WAFO

WAFO (Wave Analysis for Fatigue and Oceanography) is a toolbox of Matlab routines for statistical analysis and simulation of random waves and random loads. It is freely redistributable software; see WAFO licence, cf. the GNU General Public License (GPL). It is included in the wafo-project on GitHub [208], where also support for Octave is available, as well as a Python version.

WAFO contains tools for empirical and theoretical analysis of fatigue and wave loads, i.a.

- *Fatigue analysis*: Fatigue life prediction for random loads; computation of theoretical and approximative distributions of rainflow cycles
- *Sea modelling*: Simulation of Gaussian waves and wave-fields; Stokes and Lagrange waves and wave-fields; estimation of seawave models (wave spectra); exact computation of wave distributions for joint wave height and period and for wave steepness.
- *Statistics*: Extreme value analysis; Kernel density estimation; hidden Markov models

WAFO was used in all examples presented in this book, those original and for the figures taken from published works. In some sense, this toolbox is the numerical version of this book. It is recommended to use example in WAFO for practical illustrations of the material presented in the book.

### 11.8.1 THE AUDIENCE

There are three major audiences to which this book and the WAFO toolbox can have a great deal of appeal. First, ocean engineers will find a comprehensive set of computational tools for statistical analysis of random waves and ship's responses to them. Second, the toolbox contains a number of procedures of prime importance for mechanical engineers working on random loads or damage and fatigue analysis. Finally, any researcher who is interested in statistical analysis of random processes will find an extensive and up-to-date set of computational and graphical tools, including simulation from spectrum of Gaussian and non-Gaussian processes and fields.

### 11.8.2 TWO SIMPLE EXAMPLES

The WAFO toolbox consists of powerful modules for simulation stationary processes and fields in time and space, empirical analysis of wave and load data, numerical computation of exact and approximate wave and load distributions, statistical analysis and plotting.

We illustrate how these can be easily combined with two examples, first to define and simulate a Gaussian sea load as a random Fourier series, and extract turning points and rainflow cycles and ranges, and then to compute the theoretical rainflow cycle distribution by the Markov method.

```
%%WAFO-example 1
%Simple Matlab code to simulate rainflow cycles
%from Gaussian load with PM-spectrum

% Define parameters and generate spectrum
hs = 7; tp = 11; wc = 3;
PM = pmspec(wc,[hs,tp])
% Define simulation parameters and simulate
ns = 5000; dt = 0.1; Nsim = 100;
[Range,Cycles] = spec2rfc(PM,ns,dt,Nsim);
figure
subplot(221), histogram(Range,'normalization','pdf')
title('Histogram, rfc ranges')
subplot(222), ccplot(Cycles,6)
title('Rfc cycles')
-------------------------------------------
function [H,RFC] = spec2rfc(S,ns,dt,Nsim)
%SPEC2RFC Simulates RFC from spectrum
%
%   [H,RFC] = spec2rfc(S,ns,dt,Nsim)
%       H       rainflow ranges M - m^(rfc)
%       RFC     rainflow cycles (m^(rfc), M)
%                       size = (ns*Nsim,2)
%       S       Spectrum structure
%       ns      # data in each replicate
%       dt      sampling interval
%       Nsim    # replicates

H = [];RFC = [];
for n = 1:Nsim
    L = spec2sdat(S,ns,dt);
    tp = dat2tp(L);
    rfc_ = tp2rfc(tp);
    H_ = rfc_(:,2)-rfc_(:,1);
    H = [H; H_];
    RFC = [RFC; rfc_];
end
```

In WAFO-example 1, the commands `spec2sdat`, `dat2tp` and `tp2rfc` refer to basic modules which natural command structure. The result of the example is shown in Figure 11.19.

To compute the theoretical rainflow cycle distribution, one can use the module `spec2cmat` as in WAFO-example 2; the results are shown in Figure 11.20.

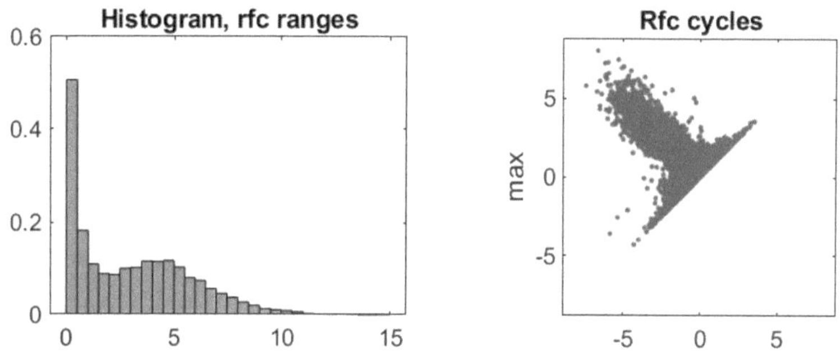


**FIGURE 11.19** Result of WAFO-example 1.

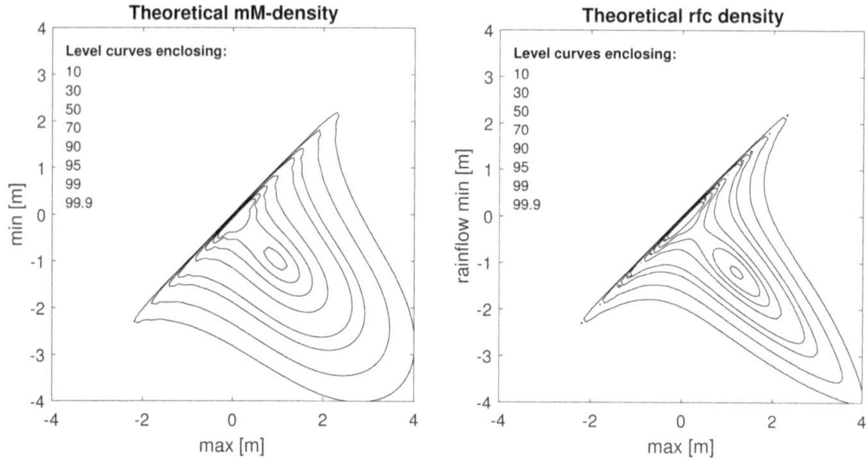


**FIGURE 11.20** Result of WAFO-example 2.

```
%%WAFO-example 2
%Matlab module to compute minMax and rainflow cycle
%distribution for Gaussian load with PM-spectrum

S = normspec(PM); %Normalise the PM-spec
param = [-4 4 81]; %Define grid
NIT = 4; %Define integration depth
mm = spec2cmat(S,[],'mM',[],param,NIT);
rfc = spec2cmat(S,[],'rfc',[],param,NIT);
figure, pdfplot(mm)
title('Theoretical mM-density')
figure, pdfplot(rfc)
title('Theoretical rfc density')
```

### 11.8.3 WHY IT WAS CREATED

A common problem with research involving complex scientific (numerical) computations is that when researchers try to advance and leverage their colleagues work, they often spend a considerable amount of time just reproducing it. Often, after few months since the completion of their own work, authors are not capable of reproducing it without a great deal of agony, due to various circumstances such as the loss of the original input data or/and parameter values, etc. Thus, many scientific articles are reproducible in principle, but not in practice. To deal with this and to organise computational scientific research and hence to conveniently transfer our technology, we impose a simple filing discipline on the authors contributing to the Wafo-toolbox. (A positive side effect of this discipline is a reduced amount of errors which are prone to occur in computational science.) This philosophy was inspired by the article by Matthias Schwab et al. "Making scientific computations reproducible", [79]. The idea is to develop reproducible knowledge about the results of the computational experiments (research) done and to make it available to other researchers for their inspection, modification, re-use and criticism.

#### 11.8.3.1 Planned Development

The most recently developed models, Laplace and hybrid, uni- and multiaxial-loads, are not yet included in the toolbox. A temporary solution has been to give, in the original papers, the codes used to evaluate results presented in those works. For example: [2, 21, 93, 100, 111, 126].

# References


1. S. Åberg and K. Podgórski. A class of non-Gaussian second order random fields. *Extrems*, 14:187–222, 2010.
2. S. Åberg, K. Podgórski, and I. Rychlik. Fatigue damage assessment for a spectral model of non-Gaussian random loads. *Probab. Eng. Mech.*, 24:608–617, 2009.
3. H. Agerskov. Fatigue in steel structures under random loading. *J. Construct. Steel Res.*, 53:283–305, 2000.
4. T.L. Anderson. *Fracture mechanics: Fundamentals and applications.* CRC Press, 1995.
5. A. Baxevani, O. Hagberg, and I. Rychlik. Note on the distribution of extreme waves crests. In *Proceeding of 24th International Conference on Ocean, Offshore and Arctic Engineering*, 2005.
6. A. Baxevani, K. Podgórski, and J. Wegener. Sample path asymmetries in non-Gaussian random processes. *Scand. J. Statist.*, 41:1102–1123, 2014.
7. A. Baxevani and I. Rychlik. Fatigue life prediction for a vessel sailing the North Atlantic route. *Probab. Eng. Mech.*, 22:159–169, 2007.
8. J. Beirlant, Y. Goegebeur, J. Teugels, and J. Segers. *Statistics of extremes: Theory and applications.* Wiley, 2004.
9. D. Benasciutti. *Fatigue analysis of random loadings.* PhD thesis, University of Ferrara, Department of Engineering, 2004.
10. D. Benasciutti and R. Tovo. Spectral methods for lifetime prediction under wide-band stationary random processes. *Int. J. Fatigue*, 27(8):867–877, 2005.
11. D. Benasciutti and R. Tovo. On fatigue damage assessment in bimodal random processes. *Int. J. Fatigue*, 29:232–244, 2007.
12. J.S. Bendat. Probability functions for random responses: Prediction of peaks, fatigue damage and catastrophic failures. Technical report, NASA, 1964.
13. S. Bendat, J. *Principles and applications of random noise theory.* John Wiley and Sons, 1958.
14. A. Bengtsson, K. Bogsjö, and I. Rychlik. Fatigue damage uncertainty. In B. Bergman, J. de Maré, S. Lorén, and T. Svensson, editors, *Robust design methodology for reliability: Exploring the effects of variation and uncertainty.* Wiley, 2009.
15. A. Bengtsson and I. Rychlik. Uncertainty in fatigue life prediction of structures subject to Gaussian loads. *Probab. Eng. Mech.*, 24:224–235, 2009.
16. A. Beste, K. Dreßler, H. Kötzle, W. Krüger, B. Mainer, and J. Petersen. Multiaxial rainflow a consequent continuation of professor Tatsuo Endo's work. In Y. Murakami, editor, *The rainflow method in fatigue.* Butterworth-Heinemann, 1992.
17. N.W.M. Bishop and F. Sherratt. A theoretical solution for the estimation of rainflow ranges from power spectral density data. *Fatigue Fract. Eng. Mater. Struct.*, 13:311–326, 1990.
18. K. Bogsjö. Development of analysis tools and stochastic models of road profiles regarding their influence on heavy vehicle fatigue. *Suppl. Veh. Syst. Dyn.*, 44:780–790, 2006.
19. K. Bogsjö. Evaluation of stochastic models of parallel road tracks. *Probab. Eng. Mech.*, 22:362–370, 2007.

20. K. Bogsjö. *Road profile statistics relevant for vehicle fatigue*. PhD thesis, Mathematical Statistics, Lund: Lund University, 2007.
21. K. Bogsjö, K. Podgórski, and I. Rychlik. Models for road surface roughness. *Veh. Syst. Dyn.*, 50:725–747, 2012.
22. K. Bogsjö and I. Rychlik. Vehicle fatigue damage caused by road irregularities. *Fatigue Fract. Eng. Mater. Struct.*, 32:391–402, 2009.
23. V.V. Bolotin. *Prediction of service life for machines and structures*. New York: ASME Press, 1989.
24. L. Bondesson. On simulation from infinitely divisible distributions. *Adv. Appl. Probab.*, 14:855–869, 1982.
25. K. Breitung. Asymptotic approximations for multinormal integrals. *J. Eng. Mech.*, 110(3):357–366, 1984.
26. K. Breitung. Asymptotic crossing rates for stationary Gaussian vector processes. *Stoch. Proc. Appl.*, 29:195–207, 1988.
27. K. Breitung. Asymptotic approximations for probability integrals. *Probab. Eng. Mech.*, 4(4):187–190, 1989.
28. K. Breitung. Probability approximations by log likelihood maximization. *J. Eng. Mech.*, 117(3):457–477, 1991.
29. K. Breitung. New approximations for reliability integrals. *J. Eng. Mech.*, 126(2):207, 2001.
30. P.A. Brodtkorb. *The probability of occurrence of dangerous wave situations at sea*. PhD thesis, Norwegian University of Science and Technology, 2004.
31. P.A. Brodtkorb. Evaluating nearly singular multinormal expectations with application to wave distributions. *Method. Comput. Appl. Probab.*, 8:65–91, 2006.
32. P.A. Brodtkorb, P. Johannesson, G. Lindgren, I. Rychlik, J. Rydén, and E. Sjö. WAFO – A Matlab toolbox for analysis of random waves and loads. In *Proceedings of 10th International Offshore Polar Engineering Conference*, volume 3, pages 343–350, 2000.
33. E.V. Bulinskaya. On the mean number of crossings of a level by a stationary Gaussian process. *Theory Probab. Appl.*, 6:435–438, 1961.
34. R.W. Butler, U.B. Machado, and I. Rychlik. Distribution of wave crests in a non-Gaussian sea. *Appl. Ocean Res.*, 31:57–64, 2009.
35. G.Q. Cai and I. Elishakoff. Refined second-order reliability analysis. *Struct. Saf.*, 14:267–276, 1994.
36. G.Q. Cai, J.S. Yu, and Y.K. Lin. Fatigue life and reliability of randomly excited structures. *J. Ship Res.*, 39(1):62–69, 1995.
37. E. Castillo. *Extreme value theory in engineering*. Elsevier Science, 1988.
38. X. Chen and N.C. Lind. Fast probability integration by three parameter normal tail approximation. *Struct. Saf.*, 1:269–276, 1983.
39. A.B. Clarke and S.H. Coverman. *Structural steelwork: Limit state design*. CRC Press, 1987.
40. S. Coles. *An Introduction to statistical modeling of extreme values*. Springer, 2001.
41. C.A. Cornell. A probability-based structural code. *ACI Journal Proceedings*, 66:974–985, 1969.
42. H.T. Corten and T.J. Dolan. Cumulative fatigue damage. In *Proceedings of International Conference on Fatigue of Metals, ASME and IME*, pages 235–246, London, 1956. Instititution of Mechanical Engineers.
43. H. Cramér and M.R. Leadbetter. *Stationary and related stochastic processes*. New York: Wiley, 1967.

44. H. Cramér. On the intersections between the trajectories of a normal stationary stochastic process and a high level. *Ark. Mat.*, 6:337–349, 1966.
45. H.E. Daniels. Saddelpoint approximations in statistics. *Ann. Math. Statist.*, 25:631–650, 1954.
46. A.G. Davenport. The spectrum of horizontal gustiness near the ground in high winds. *Q. J. Roy. Meteor. Soc.*, 87:194–211, 1961.
47. L. De Gracia, H. Wang, W. Mao, N. Osawa, I. Rychlik, and G. Storhaug. Comparison of two statistical wave models for fatigue and fracture analysis of ship structures. *Ocean Eng.*, 187:106161, 2019.
48. L. de Haan and A. Ferreira. *Extreme value theory: An introduction.* New York: Springer, 2006.
49. L. de Haan, C. Mercadier, and C. Zhou. Adapting extreme value statistics to financial time series: Dealing with bias and serial dependence. *Financ. Stat.*, 20(2):321–354, 2016.
50. J.B. de Jonge. *The analysis of load-time histories by means of counting cethods.* National Aerospace Laboratory, 1982.
51. A. Der Kiureghian, H.Z. Lin, and S.J. Hwang. Second-order reliability approximations. *J. Eng. Mech.*, 113(8):1208–1225, 1987.
52. A. Der Kiureghian and M.D. Stefano. Efficient algorithm for second order reliability analysis. *J. Eng. Mech.*, 117(12):2904–2923, 1991.
53. T. Dirlik. *Application of computer in fatigue analysis.* PhD thesis, School of Engineering, University of Warwick, 1985.
54. O. Ditlevsen. Random fatigue crack growth – A first passage problem. *Eng. Fract. Mech.*, 23:467–477, 1986.
55. O. Ditlevsen and H.O. Madsen. *Structural reliability methods.* New York: Wiley, 1996.
56. N.E. Dowling. Fatigue failure predictions for complicated stress strain histories. *J. Mater.*, 7:71–87, 1972.
57. N.E. Dowling. *Mechanical behaviour of materials.* Pearson Education, 2017.
58. K. Dreßler, M. Hack, and W. Krüger. Stochastic reconstruction of loading histories from a rainflow matrix. *Z. angew. Math. Mech.*, 77:217–226, 1997.
59. R.A. Fisher and L.H.C. Tippett. Limiting forms of the frequency distribution of the largest or smallest member of a sample. *Proc. Cambridge Phil. Soc.*, 24:180–190, 1928.
60. J.-Y. Fortin and M. Clusel. Applications of extreme value statistics in physics. *J. Phys. A: Math. Theor.*, 48:183001, 2015.
61. M. Fréchet. Sur la loi de probabilité de l'écart maximum. *Ann. Soc. Polon. Math.*, 6:93–116, 1927.
62. M. Frendahl and I. Rychlik. Rainflow analysis: Markov method. *Int. J. Fatigue*, 15:265–272, 1993.
63. H.O. Fuchs and R.I. Stephens. *Metal fatigue in engineering.* John Wiley and Sons, 1980.
64. J. Galambos. *The asymptotic theory of extreme order statistics.* Florida: Kreiger, 1987.
65. T. Galtier. Note on the estimation of crossing intensity for Laplace moving average. *Extremes*, 14:157–166, 2011.
66. T. Galtier, S. Gupta, and I. Rychlik. Crossings and second-order response processes subjected to LMA loadings. *J. Probab. Stat.*, 2010:1–22, 2010.
67. E. Gaßner. Betriebsfestigkeit eine Bemessungsgrundlage für Konstruktionsteile mit statisch wechselnden Betriebsbeanspruchungen. *Konstruktion*, 6:97–104, 1954.

68. E. Gaßner. Effect of variable load and cumulative damage of fatigue in vehicle and airplane structures. In *Proceedings of International Conference on Fatigue of Metals, ASME and IME*, pages 304–309. Instititution of Mechanical Engineers, 1956.
69. W. Gawronski. Three models of wind.gust disturbanes for the analysis of antenna pointing accuracy. Technical Report IPN Progress Report 42–149, NASA, 2002.
70. R. Ghanem and P.D. Spanos. *Stochastic finite elements: A spectral approach*. Berlin: Springer, 1991.
71. B.V. Gnedenko. Sur la distribution limite du terme maximum d'une série aléatoire. *Ann. Math.*, 44:423–453, 1943.
72. E. J. Gumbel. *Statistics of extremes*. New York: Columbia University Press, 1958.
73. S. Gupta and D. Ghosh. Uncertainty quantification in structural engineering: Current status and future directions. In S. Sarkar and J. Witteveen, editors, *Uncertainty Quantification in computational science: Theory and application in fluids and structural mechanics*, pages 119–149. World Scientific, 2016.
74. S. Gupta and I. Rychlik. Rain-flow fatigue damage due to nonlinear combination of vector Gaussian loads. *Probab. Eng. Mech.*, 22:231–249, 2007.
75. O. Hagberg. *Asymptotic expansions of crossing rates of stationary random processes*. PhD thesis, Mathematical statistics, Lund: Lund University, 2005.
76. A. Haldar, A. Guran, and B.M. Ayyub. *Uncertainty modeling in finite element, fatigue and stability of systems*. World Scientific, Singapore, 1997.
77. A. Haldar and S. Mahadevan. *Reliability assessment using stochastic finite element analysis*. New York: John Wiley and Sons, 2000.
78. N. Hammarström. Required distance for vehicular service load measurements. Master's thesis, Royal Institute of Technology, Stockholm, 2007.
79. A. Hansen. The three extreme value distributions: An introductory review. *Front. Phys.*, 8:604053, 2020.
80. A.M. Hasofer and N.C. Lind. Exact and invariant second-moment code-format. *J. Eng. Mech. Div.*, 100:111–121, 1974.
81. K. Hasselmann, T.P. Barnett, E. Bouws, H. Carlson, D.E. Cartwright, K. Enke, J.A. Ewing, H. Gienapp, D.E. Hasselmann, P. Kruseman, A. Meerburg, P. Müller, D.J. Olbers, K. Richter, W. Sell, and H. Walden. Measurements of wind-wave growth and swell decay during the joint north sea wave project (JONSWAP). *Dtsch. Hydrogr. Z. Suppl. A*, 8, 1973.
82. D.L. Henry. Theory of fatigue damage accumulation in steel. *ASME Transactions*, 77:913–918, 1955.
83. M. Hohenbichler and R. Rackwitz. Improvement of second-order reliability estimates by importance sampling. *J. Eng. Mech.*, 14(12):2195–2199, 1988.
84. S. Holm and J. de Maré. A simple model for fatigue life. *IEEE Trans. Reliab.*, 37:314–322, 1988.
85. H.P. Hong. Simple approximations for improving second-order reliability estimates. *J. Eng. Mech.*, 125(5):592–595, 1999.
86. ISO-8608. Mechanical vibration – Road surface profiles. Technical report, International Organization for Standardization, 2016.
87. G. Jiao and T. Moan. Probabilistic analysis of fatigue due to Gaussian load processes. *Probab. Eng. Mech.*, 5:76–83, 1990.
88. J. Jith, S. Gupta, and I. Rychlik. Local maxima in quadratic LMA processes. In *Proc. 11th ASCE Joint Spec. Conf.*, pages 1–10, 2012.

89. J. Jith, S. Gupta, and I. Rychlik. Estimating crossing rate statistics of second-order response of structures subjected to LMA loadings. In S. Chakraborty and G. Battacharya, editors, *Proc. ISEUSAM-2012*. Springer India, 2013.
90. H. Joe. *Multivariate models and dependence concepts.* London: Chapman & Hall, 997.
91. H. Joe. *Dependence modeling with copulas.* London: Chapman & Hall/CRC Press, 2014.
92. P. Johannesson. *Rainflow analysis of switching markov loads.* PhD thesis, Mathematical Statistics, Lund: Lund University, 1999.
93. P. Johannesson, K. Podgórski, and I. Rychlik. Laplace distribution models for road topography and roughness. *Int. J. Veh. Perfor.*, 3:224–258, 2017.
94. P. Johannesson and M. Speckert, editors. *Guide to load analysis for durability in vehicle engineering.* Chichester: Wiley, 2013.
95. P. Johannesson, T. Svensson, and de Maré J. Fatigue life prediction on variable amplitude tests – Methodology. *Int. J. Fatigue*, 27:954–965, 2005.
96. P. Johannesson and J.-J. Thomas. Extrapolation of rainflow matrices. *Extremes*, 4:241–262, 2001.
97. M. Kac. On the average number of real roots of a random algebraic equation. *Bull. Am. Math. Soc.*, 49:314–320, 1942.
98. M. Kac. Random walk and the theory of brownian motion. *Am. Math. Mon.*, 54:369–391, 1947.
99. M. Kac and D. Slepian. Large excursions of Gaussian processes. *Ann. Math. Statist.*, 30:1215–1228, 1959.
100. J. Karlsson, K. Podgórski, and I. Rychlik. The Laplace multi-axial response model for fatigue analysis. *Int. J. Fatigue*, 85:11–17, 2016.
101. M. Karlsson. *Load modelling for fatigue assessment of vehicles – A statistical approach.* PhD thesis, Mathematical statistics, Göteborg: Chalmers, 2007.
102. A.Y. Khinchin. Sur la loi des grands nombres. *Comptes Rendu*, 188:477, 1929.
103. M. Kleiber and T.D. Hien. *The stochastic finite element method.* John Wiley, 1992.
104. S. Kotz, T.J. Kozubowski, and K. Podgórski. *The Laplace distribution and generalizations.* Boston: Birkhäuser, 2001.
105. S. Kotz and S. Nadarajah. *Extreme value Distributions: Theory and applications.* London: Imperial College Press, 2000.
106. J. Kowalewski. On the relationship between component life under irregularly fluctuating and ordered load sequences. Part 1. *MIRA Translations*, 43, 1966.
107. J. Kowalewski. On the relationship between component life under irregularly fluctuating and ordered load sequences. Part 2. *MIRA Translations*, 60, 1966.
108. H.U. Koyluoglu and S.R.K. Nielsen. New approximations for SORM integrals. *Struct. Saf.*, 13:235–246, 1994.
109. T.J. Kozubowski, K. Podgórski, and I. Rychlik. Multivariate generalized laplace distribution and related random fields. *J. of Multiv. Anal.*, 113(C):59–72, 2013.
110. S. Krenk and H. Gluver. A markov matrix for fatigue load simulation and rainflow range evaluation. *Struct. Saf.*, 6:247–258, 1989.
111. M. Kvarnström, K. Podgórski, and I. Rychlik. Laplace moving average model for multi-axial response in fatigue analysis of a cultivator. *Probab. Eng. Mech.*, 34:12–25, 2013.
112. R.P. LaBarre, R.T. Forbes, and S. Andrews. *The measurement and analysis of road surface roughness.* Motor Industry Research Association (MIRA), 1969.

113. J. Labeyrie. Time scales and statistical uncertainties in the prediction of extreme environmental conditions. *Reliab. Eng. Syst. Saf.*, 32:243–266, 1990.
114. C.E. Larsen and L.D. Lutes. Predicting the fatigue life of offshore structures by the single-moment spectral method. *Probab. Eng. Mech.*, 6:96–108, 1991.
115. M.R. Leadbetter, G. Lindgren, and H. Rootzén. *Extremes and related properties of random sequences and processes.* New York: Springer-Verlag, 1983.
116. Y. Lee, M. Lu, R.C. Segar, C.D. Welch, and R.J. Rudy. Reliability based cumulative fatigue damage assessment in crack initiation. *Int. J. Mater. Prod. Technol.*, 14(1):1–16, 1999.
117. Y.-L. Lee. Fatigue damage theories. In Y.-L. Lee, J. Pan, R.B. Hathaway, and M.E. Barkey, editors, *Fatigue testing and analysis: Theory and practice.* Butterworth-Heinmann: Elsevier, 2005.
118. E. Leporati. *The assessment of structural safety.* Oregon: Research Studies Press, 1979.
119. Y.K. Lin and J.N. Yang. On statistical moments of fatigue crack propagation. *Eng. Fract. Mech.*, 18:243–256, 1983.
120. Y.K. Lin and J.N. Yang. A stochastic theory of fatigue crack propagation. *AIAA Journal*, 23(1):117–124, 1985.
121. G. Lindgren. Some properties of a normal process near a local maximum. *Ann. Math. Stat.*, 41:1870–1883, 1970.
122. G. Lindgren. Wave-length and amplitude in Gaussian noise. *Adv. Appl. Probab.*, 4:81–108, 1972.
123. G. Lindgren. Functional limits of empirical distributions in crossing theory. *Stoch. Proc. Appl.*, 5:143–149, 1977.
124. G. Lindgren. Slepian models for $\chi^2$-processes with dependent components with application to envelope upcrossings. *J. Appl. Probab.*, 26:36–49, 1989.
125. G. Lindgren. *Stationary stochastic processes – Theory and applications.* Boca Raton: CRC Press, 2013.
126. G. Lindgren. Gaussian integrals and Rice series in crossing distributions – To compute the distribution of maxima and other features of Gaussian processes. *Stat. Sci.*, 34:100–128, 2019.
127. G. Lindgren and K. B. Broberg. Cycle range distributions for Gaussian processes – Exact and approximative results. *Extremes*, 7:69–89, 2004.
128. G. Lindgren and K. Holmström. Wamp – A computer program for wave-length and amplitude analysis of Gaussian waves. Technical report, Mathematical Statistics, Umeå University, 1978.
129. G. Lindgren and M. Prevosto. The relation between wave asymmetry and particle orbits analysed by Slepian models. *J. Fluid. Mech.*, 924: A12, 607–633, 2021. doi:10.1017/jfm.2021.607
130. G. Lindgren, H. Rootzén, and M. Sandsten. *Stationary stochastic processes for scientists and engineers.* Boca Raton: CRC Press, 2014.
131. G. Lindgren and I. Rychlik. Rain flow cycle distributions for fatigue life prediction under Gaussian load processes. *Fatigue Fract. Engng. Mater. Struct.*, 10:251–260, 1987.
132. G. Lindgren and I. Rychlik. Slepian models and regression approximations in crossing and extreme value theory. *Int. Stat. Rev.*, 59:195–225, 1991.
133. P.L. Liu and A. Der Kiureghian. Finite element reliability of geometrically nonlinear uncertain structures. *J. Eng. Mech.*, 117(8):1806–1825, 1991.

134. L.D. Lutes and C.E. Larsen. Improved spectral method for variable amplitude fatigue prediction. *J. Struct. Eng.*, 116:1149–1164, 1990.
135. U.B. Machado. *Statistical analysis of non-Gaussian environmental loads and responses*. PhD thesis, Mathematical statistiucs, Lund: Lund University, 2002.
136. H.O. Madsen, S. Krenk, and N.C. Lind. *Methods of structural safety*. New Jersey: Prentice-Hall, 1986.
137. S.N Majumdar, A. Pal, and G. Schehr. Extreme value statistics of correlated random variables: A pedagogical review. *Phys. Rep.*, 840(1), 2020.
138. C.S. Manohar and S. Gupta. Modeling and evaluation of structural reliability: Current status and future directions. In K.J. Jagadish and R.N. Iyengar, editors, *Research perspectives in civil engineering, Golden Jubilee Publications of Department of Civil Engineering*. Allied Publishers, 2004.
139. C.S. Manohar and R.A. Ibrahim. Progress in structural dynamics with stochastic parameter variations 1987–1998. *Appl. Mech. Rev.*, 52(5):177–197, 1999.
140. S.S. Manson and G.R. Halford. Practical implementation of a double linear damage rule and damage curve approach for testing cumulative fatigue damage. *Int. J. Fracture*, 17(2):169–192, 1981.
141. W. Mao, J.W. Ringsberg, I. Rychlik, and G. Storhaug. Development of a fatigue model useful in ship routing design. *J. Ship Res.*, 54:281–293, 2010.
142. W. Mao and I. Rychlik. Stochastic spatio-temporal model for wind speed variation in the arctic. *Ocean Eng.*, 165:237–251, 2018.
143. W. Mao, I. Rychlik, and G. Storhaug. Safety index of fatigue failure for ship structure details. *J. Ship Res.*, 54:197–208, 2010.
144. S.M. Marco and W.L. Starkey. A concept of fatigue damage. *ASME Transactions*, 76(627), 1954.
145. M. Matsuishi and T. Endo. Fatigue of metals subject to varying stress. Japanese Society of Mechanical Engineers Fukuoka, 1968.
146. R.E. Melchers. *Structural reliability analysis and prediction*. Chichester: John Wiley, 1999.
147. M. A. Miner. Cumulative damage in fatigue. *J. Appl. Mech.*, 12:A159–A164, 1945.
148. P. Mucka. Road waviness and the dynamic tyre force. *Int. J. Veh. Des.*, 36:216–232, 2004.
149. J.S. Nathwani, N.C. Lind, and M.D. Pandey. *Affordable safety by choice: The life qualty method.* Waterloo, Canada: University of Waterloo, 1997.
150. N.C. Nigam and S. Narayanan. *Applications of random vibrations*. Springer-Verlag, 1994.
151. M.K. Ochi and K. Ahn. Probability distribution applicable to non-Gaussian random processes. *Probab. Eng. Mech.*, 9:255–264, 1994.
152. K. Ortiz and A.S. Kiremidijian. Time series analysis of fatigue crack growth rate data. *Eng. Fract. Mech.*, 4:657–675, 1986.
153. T-L-. Paez. Random vibration – History and overview. In T. Proulx, editor, *Rotating machinery, structural health monitoring, shock and vibration, Volume 5*, pages 105–127, New York: Springer, 2011.
154. A. Palmgren. Die Lebensdauer von Kugellagern. *VDI Zeitschrift*, 68:339–341, 1924.
155. O.M. Phillips. The equilibrium range in the spectrum of wind generated waves. *J. Fluid Mech.*, 4:426–434, 1958.
156. J.W. Pierson and L. Moskowitz. A proposed spectral form for fully developed wind seas based on the similarity theory of S.A. Kitaigorodskii. *J. Geophys. Res.*, 69:5181–5190, 1964.

157. K. Podgórski and I. Rychlik. A model of significant wave height for reliability assessment of a ship. *J. Mar. Syst.*, 130:109–123, 2014.
158. K. Podgórski, I. Rychlik, and J. Wallin. Slepian noise approach for Gaussian and Laplace moving average processes. *Extremes*, 18:665–695, 2015.
159. K. Podgórski and J. Wegener. Estimation of stochastic models driven by Laplace motion. *Commun. Stat. – Theory Methods*, 40:3281–3302, 2011.
160. D.C. Polidori, J.L. Beck, and C. Papadimitriou. New approximations for reliability integrals. *J. Eng. Mech.*, 125(4):466–475, 1999.
161. D.C. Polidori, J.L. Beck, and C. Papadimitriou. Discussion on new approximations for reliability integrals. *J. Eng. Mech.*, 126(2):207–209, 2001.
162. R. Rackwitz and B. Fiessler. Structural reliability under combined random load sequences. *Comput. Struct.*, 9:489–494, 1978.
163. N. Raillard, M. Prevosto, and P. Ailliot. Modelling process asymmetries with Laplace moving average. *Comput. Stat. Data An.*, 81:24–37, 2015.
164. A.J. Rainal. Origin of Rice's formula. *IEEE Trans. Inf. Theory*, 34:1383–1387, 1988.
165. S.O Rice. Singing transmission lines. private note.
166. S.O Rice. Distribution of the maxima of a random curve. *Amer. J. Math.*, 61:409–416, 1939.
167. S.O. Rice. The mathematical analysis of random noise. *Bell Syst. Techn. J.*, 23:282–332, 1944.
168. S.O. Rice. The mathematical analysis of random noise II. *Bell Syst. Techn. J.*, 24:46–156, 1945.
169. I. Rychlik. *Statistical wave analysis with application to fatigue.* PhD thesis, Lund: Lund University, 1986.
170. I. Rychlik. A new definition of the rainflow cycle counting method. *Int. J. Fatigue*, 9:119–121, 1987.
171. I. Rychlik. Rain flow cycle distribution for ergodic load processes. *SIAM J. Appl. Math.*, 48:662–679, 1988.
172. I. Rychlik. Rainflow cycles in Gaussian loads. *Fatigue Fract. Eng. Mater. Struct.*, 15:57–72, 1992.
173. I. Rychlik. Note on cycle counts in irregular loads. *Fatigue Fract. Eng. Mater. Struct.*, 16:377–390, 1993.
174. I. Rychlik. On the 'narrow-band' approximation for expected fatigue damage. *Probab. Eng. Mech.*, 8:1–4, 1993.
175. I. Rychlik. Extremes, rainflow cycles and damage functionals in continuous random processes. *Stochastic Processes Appl.*, 63:97–116, 1996.
176. I. Rychlik. Fatigue and stochastic loads. *Scand. J. Statist.*, 23:387–404, 1996.
177. I. Rychlik. Simulation of load sequences from rainflow matrices: Markov method. *Int. J. Fatigue*, 18:429–438, 1996.
178. I. Rychlik. Rainflow cycles, Markov chains, and electrical circuits. In E. A. Johnson, editor, *Proceeding of 4th International Conference on Stochastic Structural Dynamics*, pages 229–234. Taylor and Francis, 1999.
179. I. Rychlik. Five lectures on reliability applications of Rice's formula for the intensity of level crossings. In S. Jendo and K. Doliński, editors, *Reliability-based design and optimisation, AMAS Course – RBO'03*. Polish Academy of Scienes, Warsaw, 2003.
180. I. Rychlik. Note on modeling of fatigue damage rates for non-Gaussian stresses. Technical report, Chalmers University of Technology, Mathematical Sciences, 2012.

181. I. Rychlik and S. Gupta. Rain-flow fatigue damage for transformed Gaussian loads. *Int. J. Fatigue*, 29:406–420, 2007.
182. I. Rychlik, P. Johannesson, and M.R. Leadbetter. Modelling and statistical analysis of ocean-wave data using transformed Gaussian processes. *Mar. Struct.*, 10:13–47, 1997.
183. I. Rychlik, G. Lindgren, and Y.K. Lin. Markov based correlations of damage cycles in Gaussian and non-Gaussian loads. *Probab. Eng. Mech.*, 10:103–115, 1993.
184. I. Rychlik and W. Mao. Probabilistic model for wind speed variability encountered by a vessel. *Natural Resources*, 4:837–855, 2014.
185. S. Sakai and H. Okamura. On the distribution of rainflow range for Gaussian random processes with bimodal PSD. *JSME Int. J. Ser. A*, 38:440–445, 1995.
186. M. Scheutzow. A law of large numbers for upcrossing measures. *Stoch. Proc. Appl.*, 53:285–305, 1994.
187. F.R. Shanley. A theory of fatigue based on unbonding during reversed slip. Technical report, The Rand Corporation, 1952.
188. M. Shinozuka. Basic analysis of structural safety. *J. Struct. Eng.*, 109(3):721–740, 1983.
189. S. Sjöström. On random load analysis. *Trans. Roy. Inst. Technol.*, 181:1–41, 1961.
190. R. Skjong and H.O. Madsen. Practical stochastic fatigue analysis of offshore platforms. *Ocean Eng.*, 14:313–324, 1987.
191. D. Slepian. On the zeros of Gaussian noise. In M. Rosenblatt, editor, *Time series analysis*, pages 104–115, New York, 1963. Wiley.
192. K. Sobczyk and B.F. Spencer Jr. *Random fatigue: From data to theory*. San Diego: Academic Press, Inc., 1992.
193. D.F. Socier and P. Kurath. Cycle counting for variable-amplitude crack growth. In *Symposia paper STP791B1983002*, pages 19–32, 2020.
194. G.P. Solomos. First passage solutions in fatigue crack propagation. *Probab. Eng. Mech.*, 4(1):32–39, 1989.
195. M. St. Denis and W. J. Pierson. On the motions of ships in confused seas. *Soc. of Naval Architects and Marine Engineers, Trans.*, 61:280–332, 1953.
196. R.I. Stephens, G.W. McBurney, and L.J. Oliphant. Fatigue crack growth with negative R-ratio following tensile overloads. *Int. J. Fract. Mech.*, 10(4):587–592, 1974.
197. H. Sun, Y. Qiu, J. Li, J. Bai, and M. Peng. Novel models for fatigue life prediction under wideband random loads based on machine learning. *Fatigue Fract. Eng. Mater. Struct.*, 47:3342–3360, 2024.
198. L. Sun. Computer simulation and field measurement of dynamic pavement loading. *Math. Comput. Simulat.*, 56:297–317, 2001.
199. S. Suresh. *Fatigue of materials*. Cambridge: University Press, 2015.
200. N.N. Taleb. *The black swan: The impact of the highly improbable*. Penguin Books Ltd, 2008.
201. S.P. Timoshenko. *History of strength of materials*. New York: Dover Publications, [1953] 1983.
202. K. Torsethaugen. A two peak wave spectral model. In *Proceedings OMAE*, Glasgow, 1993.
203. K. Torsethaugen. Model for a doubly peaked wave spectrum. Technical Report STF22 A96204, SINTEF, 1996.
204. R. Tovo. Cycle distribution and fatigue damage under broad-band random loading. *Int. J. Fatigue*, 24:1137–1147, 2002.

205. R. Tovo. Cycle distribution and fatigue damage under broad-band random loading. *Int. J. Fatigue*, 24(11):1137–1147, 2002.
206. A. Tsurui and H. Ishikawa. Applications of the fokker-plank equation to a stochastic fatigue crack growth model. *Struct. Saf.*, 4:15–29, 1986.
207. WAFO-group. WAFO – A Matlab toolbox for analysis of random waves and loads. Tutorial for WAFO 2017. http://www.maths.lth.se/matstat/wafo, 2017.
208. WAFO-group. Wafo-project. https://github.com/wafo-project, 2017.
209. S. Wang and J.Q. Sun. Effect of skewness on fatigue life with mean stress correction. *J. Sound Vib.*, 282:1231–1237, 2005.
210. W. Weibull. A statistical distribution function of wide applicability. *J. Appl. Mech.*, 18:293–296, 1951.
211. W. Weibull. *Fatigue testing and analysis of results*. Oxford: Pergamon Press, 1961.
212. S.R. Winterstein. Nonlinear vibration model for extremes and fatigue. *J. Eng. Mech.*, 114:1772–1790, 1988.
213. S.R. Winterstein, T.C. Ude, and G. Kleiven. Springing and slow drift responses: Predicted extremes and fatigue vs. simulation. In *Proceedings, Behaviour of Offshore Structures at Sea (BOSS-94)*, volume 3, pages 1–15, 1994.
214. P.H. Wirsching and M.C. Light. Fatigue under wide band random stresses. *J. Struct. Div.–ASCE*, 106:1593–1607, 1980.
215. P.H. Wirsching, T.L. Paez, and H. Ortiz. *Random vibration: Theory and practice*. Wiley, 1995.
216. A. Wöhler. Uber die Festigheits-Versuche mit Eisen und Stahl. *Zeitschrift für Bauwesen*, 20:73–106, 1870.
217. Y.-T. Wu and P.H. Wirsching. New algorithm for structural reliability estimation. *J. Eng. Mech.*, 113(9):1319–1336, 1987.
218. L. Yu, P.K. Das, and N.P.D. Barltrop. A new look at the effect of bandwidth and nonnormality on fatigue damage. *Fatigue Fract. Eng. Mater. Stuct.*, 27:51–58, 2004.
219. W. Zhao and M. Baker. On the probability density function of rainflow stress range for stationary Gaussian processes. *Int. J. Fatigue*, 14:121–135, 1992.
220. Y.-G. Zhao and T. Ono. A general procedure for first/second order reliability method (FORM/SORM). *Struct. Saf.*, 95–112, 1999.
221. Y.-G. Zhao and T. Ono. New approximations for SORM: Part I. *J. Eng. Mech.*, 125(1):79–85, 1999.
222. Y.-G. Zhao and T. Ono. New approximations for SORM: Part II. *J. Eng. Mech.*, 125(1):86–93, 1999.
223. W.Q. Zhu and Y.K. Lin. On fatigue crack growth under random loading. *Eng. Fract. Mech.*, 43:1–12, 1992.
224. W. Zucchini, I.L. MacDonald, and R. Langrock. *Hidden Markov models for time series – An introduction using R*. New York: Chapman and Hall/CRC, second edition, 2016.

# Index